Models in Hardware Testing

FRONTIERS IN ELECTRONIC TESTING

Consulting Editor
Vishwani D. Agrawal

Volume 43

For further volumes
http://www.springer.com/series/5994

Hans-Joachim Wunderlich
Editor

Models in Hardware Testing

Lecture Notes of the Forum in Honor of Christian Landrault

Springer

Prof. Dr. Hans-Joachim Wunderlich
Universität Stuttgart
Institut für Technische Informatik
Pfaffenwaldring 47
70569 Stuttgart
Germany
wu@informatik.uni-stuttgart.de

ISSN 0929-1296
ISBN 978-90-481-3281-2 e-ISBN 978-90-481-3282-9
DOI 10.1007/978-90-481-3282-9
Springer Dordrecht Heidelberg London New York

Library of Congress Control Number: 2009939835

Cover design: eStudio Calamar S.L.

Printed on acid-free paper

Springer is part of Springer Science+Business Media (www.springer.com)

Contents

Contributors

Jean Arlat LAAS-CNRS; Université de Toulouse; 7, avenue du Colonel Roche, F-31077 Toulouse, France

Daniel Arumí Universitat Politècnica de Catalunya (UPC), Electronic Engineering Dpt. ETSEIB, Diagonal 647, 08028 Barcelona, Spain

Florence Azais LIRMM-CNRS, 161 rue ada, 34392 Montpellier, France

Bernd Becker Albert-Ludwigs-University of Freiburg, Germany

Stefano Di Carlo Politecnico di Torino, Control and Computer Engineering Department, Corso duca degli Abruzzi 24, 10129, Torino, Italy

Yves Crouzet LAAS-CNRS; Université de Toulouse; 7, avenue du Colonel Roche, F-31077 Toulouse, France

Joan Figueras Universitat Politècnica de Catalunya (UPC), Electronic Engineering Dpt. ETSEIB, Diagonal 647, 08028 Barcelona, Spain

Patrick Girard LIRMM/CNRS, 161rue Ada, 34392 Montpellier, France

Stefan Holst Institut für Technische Informatik, Universität Stuttgart, Pfaffenwaldring 47, D-70569 Stuttgart, Germany

Ilia Polian Albert-Ludwigs-University of Freiburg, Germany

Paolo Prinetto Politecnico di Torino, Control and Computer Engineering Department, Corso duca degli Abruzzi 24, 10129, Torino, Italy

Sudhakar M. Reddy Department of Electrical and Computer Engineering, University of Iowa, Iowa City, Iowa, USA

Michel Renovell LIRMM-CNRS, 161 rue ada, 34392 Montpellier, France

Rosa Rodríguez-Montañés Universitat Politècnica de Catalunya (UPC), Electronic Engineering Dpt. ETSEIB, Diagonal 647, 08028 Barcelona, Spain

Hans-Joachim Wunderlich Institut für Technische Informatik, Universität Stuttgart, Pfaffenwaldring 47, D-70569 Stuttgart, Germany

Preface

Model based testing is one of the most powerful techniques for testing hardware and software systems. While moving forward to nanoscaled CMOS circuits, we observe a plethora of new defect mechanisms, which require increasing efforts in systematic fault modeling and appropriate algorithms for test generation, fault simulation and diagnosis. The text presented here treats models and especially fault models in hardware testing in a comprehensive way, considers the most recent state of the art and puts them into their historical context.

The first chapter by Joan Figueras et al. considers the fact that open defects are becoming the predominant failure mechanism as technologies are scaled down. It analyzes these defects according to their locations and resistive nature, and deduces the faulty behavior. This chapter lays foundations for the subsequently described algorithms and proposes test strategies to improve the detectability and diagnosability of open defects.

The second large class of defects is formed by bridges and treated in chapter 2 by M. Renovell et al. Bridging defects are also responsible for a large percentage of failure in CMOS technologies, and their impact in nanometer technologies with dense interconnect structures will increase. The chapter explores the logic detectability of bridging defects by taking into account different ranges of resistances. The concept of an Analog Detectability Interval (ADI) and its use for increasing the quality of test vectors and the fault coverage are introduced.

Both resistive bridges and resistive opens may result in timing faults. Chapter 3 on delay faults by S. Reddy describes methods to generate appropriate tests and design for test methods to improve delay fault coverage. So-called small delay faults are only observable at a subset of paths in the circuit, and they are increasingly relevant in nanoscaled technologies. This chapter treats them as a part of ongoing research.

Two chapters deal with the algorithmic aspects introduced by the complex fault models described so far. Chapter 4 on fault modeling for simulation and test pattern generation by B. Becker and I. Polian presents algorithms which can handle the resistive fault models described above. It covers in detail the abstraction mechanisms required, the algorithms and their optimizations.

Chapter 5 on generalized fault modeling for logic diagnosis by H.-J. Wunderlich and S. Holst deals with the problem that in contrast to ATPG and fault simulation, diagnosis algorithms should not make pre-assumptions on the appropriate fault model but have to identify the faulty behavior instead. A generalized fault modeling technique and notation are introduced, and diagnosis techniques are proposed which can handle this fault modeling at a higher level of abstraction.

Larger and larger portions of the IC area are occupied by memory, and semiconductor memories have always been used to push silicon technology at its limits. This makes these devices extremely sensitive to physical defects and environmental influences that may severely compromise their correct behavior. Chapter 6 on models in memory testing by S. Di Carlo and P. Prinetto provides an overview of models and notations currently used and highlights challenging problems awaiting solutions.

Chapter 7 by P. Girard and H.-J. Wunderlich introduces power consumption during test as an additional aspect. In test mode, power consumption is even more critical than in system mode, and has severe impact on reliability, yield and test costs. This chapter describes models of different types and sources of test power. Power-aware techniques for test pattern generation, design for test and test data compression are presented which require minimized hardware cost and test application time.

The last chapter by J. Arlat and Y. Crouzet discusses physical fault models and fault tolerance. Dependability, online test and fault tolerance techniques receive more and more attention for nanoscaled devices. This chapter focuses on the representativeness of fault models with respect to physical faults for deriving relevant test procedures and experimental assessment techniques. The chapter links physical fault models to fault injection based dependability assessment techniques.

The authors of this book provided this comprehensive treatment of models in hardware testing in appreciation of the achievements of Christian Landrault who laid the foundations of many of the concepts presented here during his research life, and had a leading role in the European test and research community. The authors of this book are close colleagues and friends of Christian Landrault, and dedicating this book to him is their way to say thank you for many years of friendship and fruitful collaborations.

Sevilla
May 28, 2009

Hans-Joachim Wunderlich

To Christian: a Real Test and Taste Expert

Dear Christian,

Writing and setting up this book has been our way to express our deep and sincere THANKS!! In fact, we all owe you many THANKS for so many things and at so many "levels". Let's try to focus on some of them, starting with the scientific ones.

Your research interests and activities spanned several topics and areas, in each getting significant results and providing original contributions. As evidence of this, one should simply look at all the references to your papers at the end of each chapter of this book. In addition to these very significant "written" contributions, we have to thank you for the "oral" ones: your discussions during the conferences you attended have always been characterized by a constructive approach, always aimed at understanding, helping, and providing hints.

Thanks to all your efforts and to your capability of selecting high quality researchers and co-workers, your team at LIRMM has grown to become one of the highly recognized key players not only at the European level but also in the international test research community!!

The list of scientific events you served as General Chair, Program Chair, Steering Committee member, and Program Committee member is too long to list here and if we try to list we would definitely forget a lot of them.

The scientific community in general and the overall test community in particular owe you a gigantic thank you for the unbelievable amount of time and efforts you spent to serve them.

You have been a father (if not the father) of the European Test Community. Your strength, your dedication, your patience, your leadership, and your efforts allowed the community to grow; from the first presence at the CAVE Workshops to the Design for Testability Workshops, from the European Test Conferences to DATE, from the European Test Workshops to the European Test Symposiums (ETS). Under your leadership, the European Group of the IEEE Test Technology Technical Council grew significantly, becoming one of the most active regional groups of the council.

Your vision led to the creation of the European Test Symposium Steering Committee. Under your chairpersonship, the Committee started playing a key role in assuring to maintain those high quality levels that are unanimously recognized as the hallmark of ETS not only in Europe, but worldwide as well.

Dear Christian, last but definitely not least, we have to thank you at the personal and human level. The so many hours spent together discussing, eating, tasting wine, talking of culture, sharing everyday problems of our private lives, telling us your experiences in fishing and hunting, have been invaluable. It will be very hard for all of us attending next scientific and technical events without your friendliness. We will look for you until we realize that, instead of attending yet another boring panel session, you will be most likely hunting, or fishing, or enjoying Titou, your sons, and your granddaughters... lucky you!!

Amicalement
Your friends of the test community

From LAAS to LIRMM and Beyond

For the contributors to this book, as well as for many researchers in the field of testing and testability of integrated digital circuits and systems, Christian Landrault is one of the key figures in the research, development and teaching of this very important field.

Christian Landrault began his scientific life at LAAS-CNRS in Toulouse where he stayed during 10 years (1970–1980), just after his graduation as an Engineer from the prestigious *Ecole Nationale d'Ingénieurs de Constructions Aéronautiques*.

During this period, he was a member of the "Digital Automatisms" research team that I headed and which was subsequently led by Jean-Claude Laprie, to become the research group on "Dependable Computing and Fault Tolerance" as it is known today. Christian Landrault obtained his Ph.D. (1973) and *Doctorat d'Etat* (1977) at LAAS, both from the National Polytechnic Institute of Toulouse (INPT), respectively on the design of control systems, and on the modeling and evaluation of fault-tolerant computer architectures. Then, by the end of the 1970s, he initiated his pioneering work in the domain of hardware digital technology, in particular on fault modeling and testability of MOS integrated circuits, as well as on the design of self-checking microprocessor chips. This seminal work has resulted in a couple of papers that are among the most referenced papers at LAAS and that form the main basis for a large part of the material reported in Chapter 8 of this book.

In 1980, Christian joined LIRMM in Montpellier, in the Microelectronics Department. The research activities he developed and the related results attained, span mainly the area of testing and testability of digital integrated circuits: fault simulation, ATPG, DFT, BIST and fault tolerance. The results he obtained on these topics were published in more than one hundred papers in leading journals and conferences worldwide. Christian Landrault has always contributed very actively to the discussions and reflections in line with the state-of-the-art, the challenges, the evolution and prospects of the field with his academic and industrial colleagues.

Christian Landrault has been a member of numerous Program Committees of major conferences and workshops in the area of testing, among which, ITC, VTS, ATS, ETS, DATE, etc., which confirms the leading role he has played in the emergence and blooming of the scientific community on testing and testability. In particular, he has been the founder of the European Test Workshop in 1996 for which he was the first Chairman and has subsequently chaired the PC in 1998 and 1999. This event has since become a Symposium and its 14th edition has taken place this year. Christian was until 2008 the Elected Chair of the Steering Committee of the Symposium. He was also the European representative at the ITC Program Committee for several years.

To conclude, I would like to emphasize that, beyond his well-recognized skills and professional competencies, Christian possesses a rather unique sense for dialogue and friendship, and this is as a friend that I would like to tell him that we are all proud and pleased for the outstanding scientific career he has conducted with his colleagues, both at LAAS and at LIRMM, but also with researchers from the entire world.

Toulouse,
March 16, 2009

Professor Alain Costes
Director of LAAS-CNRS (1984–1996)
Chairman of INPT (1996–2000)
Director of Technology
with the French Ministry of Research (2000–2003)

Chapter 1
Open Defects in Nanometer Technologies
Models, Test and Diagnosis

Joan Figueras, Rosa Rodríguez-Montañés, and Daniel Arumí

Abstract Open defects are responsible for a significant number of failures affecting present CMOS technologies. Furthermore, they are becoming more common as technologies are scaled down due to changes in materials and fabrication steps of ICs manufacturing processes. In this chapter, open defects are classified according to their location and resistive nature. The behavior of such defects affecting interconnect lines and logic gates is reviewed. Test strategies to improve the detectability of open defects and diagnosis methodologies are also presented.

Keywords Open defect · Full open · Resistive open · CMOS · VLSI · Test · Diagnosis · Nanometer technologies

1.1 Introduction

An open defect consists of the partial or total breaking of the electrical connection between two points in a circuit which should be electrically connected by design. Failures associated with open defects are common in CMOS technologies. This class of defects is becoming more frequent with technology shrinking due to the increasing number of vias/contacts (Thompson 1996) and the replacement of aluminum with copper in metal interconnections (Stamper et al. 1998). Figure 1.1 illustrates the photographs of two real opens in a copper interconnect technology.

During the last decades, an intensive research effort has been dedicated to CMOS Integrated Circuits (ICs) in the presence of open defects. Scaling trends of CMOS in the nanometer range require new models and analysis methods. In this context, the presence of an open defect, coupled with increasing leakage currents, leads to new behaviors not visible in older technologies.

J. Figueras (✉), R. Rodríguez-Montañés, and D. Arumí
Universitat Politècnica de Catalunya (UPC), Electronic Engineering Dpt. ETSEIB,
Diagonal 647, 08028 Barcelona, Spain
e-mail: figueras@eel.upc.edu

H.-J. Wunderlich (ed.), *Models in Hardware Testing: Lecture Notes of the Forum in Honor of Christian Landrault*, Frontiers in Electronic Testing 43,
DOI 10.1007/978-90-481-3282-9_1, © Springer Science+Business Media B.V. 2010

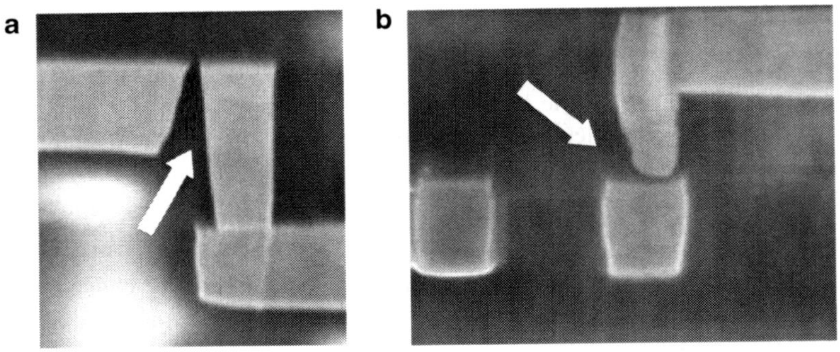

Fig. 1.1 Interconnect open defect photographs for a copper interconnect technology (Courtesy of NXP Semiconductors). (**a**) Defect in metal and (**b**) defect in via

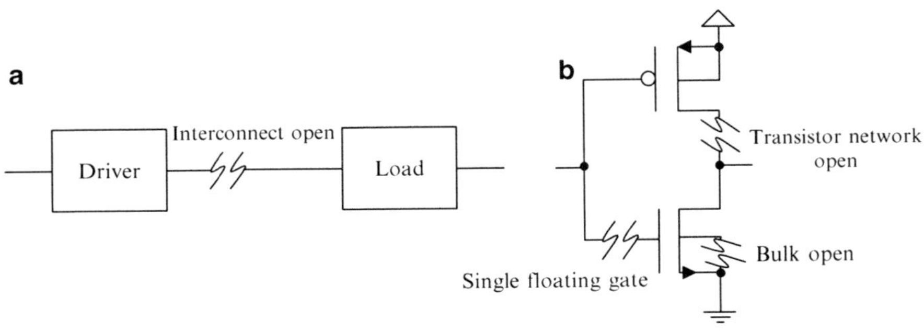

Fig. 1.2 Open defect classification based on location. (**a**) Interconnect and (**b**) intra-gate

An open defect can be classed according to its location (see Fig. 1.2), as interconnect and intra-gate opens, with the following subtypes:

- **Interconnect opens:**
- *Metal/Polysilicon open:* This break is located on metal or in polysilicon tracks.
- *Via open:* It is located in via that connects two metal tracks of different metal layers.
- *Contact open:* It is located in a contact between silicon and a metal track, or polysilicon and a metal track.
- **Intra-gate opens:**
- *Transistor network open:* It appears inside a logic gate and affects the connection between the drain/source of one or more transistors.
- *Bulk open:* In bulk CMOS technologies, the defect breaks or weakens the connection between the bulk of an nMOS transistor and GND, or the bulk of a pMOS transistor and V_{DD}.
- *Single/Multiple floating gate(s):* It disconnects a single or multiple transistor gate(s) from its (their) driver.

Depending on its resistance an open can also be classified into two different groups based on its electrical model:

- *Full (or strong) open:* The lack of conductive material causes a discontinuity, thus eliminating the electrical connection between the two end points of the defect site.
- *Resistive (or weak) open:* The discontinuity does not result in a complete electrical disconnection adding a finite resistance.

Other classifications based on the physical cause of the defect have also been used in the literature. What is considered in these categorizations are the basic operations in IC fabrication where open defects are more likely to appear: photolithography, mechanical planarization processes and chemical problems in contacts and vias.

1.2 Open Defect Models

Extensive work has been conducted to model opens and characterize the behavior of CMOS circuits with open defects. The first works on intra-gate opens appeared in the late 1970s. Stuck-open faults and the "two vector detection" of the defect were published (Wadsack 1978). Pioneering work on modeling and electrical analysis of gates with a single floating transistor gate were performed in the late 1980s (Renovell and Cambon 1986, 1992). Models and CMOS circuits with interconnect opens were electrically characterized later during the 1990s when the interconnect architecture of VLSI circuits started to become more prone to interconnect opens than intra-gate opens. The number of publications on interconnect opens has increased significantly since then.

In this section, the evolution of modeling and electrical characterization of circuits with opens is reviewed, presenting some key developments in the field. The section has been divided into two subsections based on open location, i.e., interconnect and intra-gate opens.

1.2.1 Interconnect Open Defects

The physical explanation of interconnect opens can be either a metal or polysilicon crack/void or a defective contact/via. These open defects result in gate input pairs being partially or totally disconnected from their drivers. Although opens may appear inside a logic module in CMOS technologies, the most likely place to appear is in an interconnect line (Xue et al. 1994). For this reason, special attention is paid to interconnect opens.

A review of interconnect open defects is provided next, following the classification according to defect resistance, i.e. *full* and *resistive* opens.

1.2.1.1 *Full* Open Defects in Interconnect Lines

In this subsection, we first review the classical model for *full* opens in interconnect lines capacitively coupled with neighboring lines. As traditionally considered, tunneling currents are assumed negligible. Next, thin open defects are described, and finally interconnect *full* open defects with gate leakage are modeled.

Full Open Defect Modeling in the Interconnect Paths

An interconnect line with a *full* open is disconnected from its driver and becomes electrically floating. This line may, in turn, drive one (or more) transistor pair(s). An illustrative example is shown in Fig. 1.3, where the interconnect line is driving an inverter. The floating line voltage (V_{FL}) is determined by (a) the surrounding circuitry, (b) the transistor capacitances of the driven gates, and (c) the initial trapped charge (Konuk 1997; Champac and Zenteno 2000; Arumí et al. 2005), as reviewed next.

a. Neighboring interconnect lines routed close to the floating line add parasitic coupling capacitances (C_{N1}, C_{N2}, C_{N3}, ... C_{Nm} in Fig. 1.3). There are also parasitic capacitances to the ground (C_{SUBS}) and to the power plane (C_{WELL}). Without loss of generality, an n-well CMOS process is considered in Fig. 1.3. The value of these capacitances depends on the dielectric filling the space, the distance between lines and their physical dimensions.
b. Another set of parasitic capacitances influencing the interconnect line is made up of the parasitic capacitances of the transistors driven by the floating line. These capacitances consist of gate drain (C_{gd}), gate_source (C_{gs}) and gate_bulk (C_{gb}) capacitances from both the pMOS and nMOS transistors of the downstream gate(s). The exact value of these transistor capacitances varies with the conduction state of the transistors.

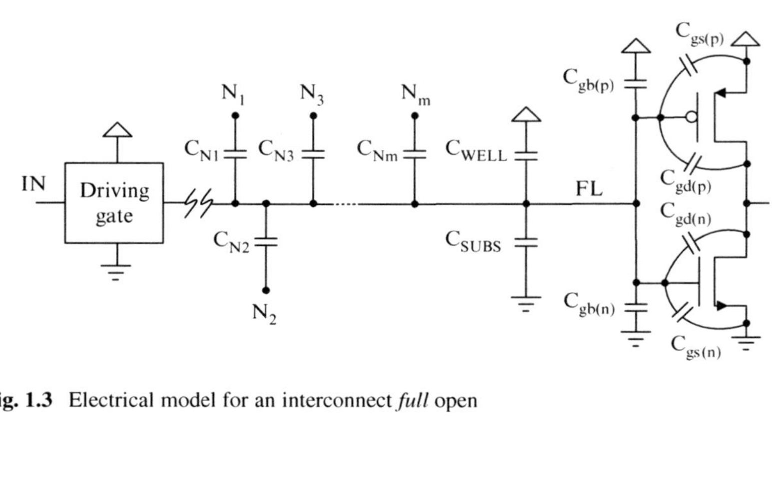

Fig. 1.3 Electrical model for an interconnect *full* open

c. The third factor influencing the floating line voltage is the trapped charge accumulated in the floating structure during the fabrication process. The trapped charge is an unknown, difficult-to predict parameter. In the work by Johnson (1994), measurements of the trapped charge were made on test structures consisting of floating-gate transistors with different polysilicon length extensions. These measurements always showed a positive charge on the floating polysilicon, generating voltages ranging from 0.1 to 2.3 V.

According to the charge conservation law, once the initial charge is trapped in the circuit, the total charge does not change and is redistributed among the connected capacitors. Therefore, for the example in Fig. 1.3, Eq. 1.1 must be satisfied:

$$\sum_{i=1}^{i=m} Q_{Ni} + Q_{VDD} + Q_{GND} + Q_M = Q_o \tag{1.1}$$

The sum of Q_{Ni} represents all the charges from the coupled neighbors, Q_{VDD} is the charge from capacitances tied to the power rail $\left(C_{WELL} + C_{gb(p)} + C_{gs(p)}\right)$, Q_{GND} is the charge from capacitances tied to the ground rail $\left(C_{SUBS} + C_{gb(n)} + C_{gs(n)}\right)$, Q_M is the charge related to the Miller capacitances $\left(C_{gd(n)} + C_{gd(p)}\right)$ and Q_o is the trapped charged accumulated during the fabrication process. Using the well-known expression relating the charge and the voltage across the capacitor terminals (Eq. 1.2) and Eq. 1.1, the expression in terms of V_{FL} and V_{OUT} reported in Eq. 1.3 is obtained:

$$Q = C \cdot \Delta V \tag{1.2}$$

$$\begin{aligned}(V_{FL} - V_{DD})\,(C_{NL1} + C_{VDD}) + V_{FL}\,(C_{NL0} + C_{GND})\\ + (V_{FL} - V_{OUT})\,C_M = Q_o\end{aligned} \tag{1.3}$$

C_{NL1} is the capacitance from all the neighbors set to logic 1 and C_{NL0} the capacitance from all neighbors set to logic 0. C_{NL1} and C_{NL0} are logic pattern dependent, since for every test pattern, a different state is set in the neighboring lines. In general, the drivers managing the neighboring lines are strong, hence these capacitances can be considered to be tied to V_{DD} or GND in steady state. In this way, Eq. 1.3 can be rearranged as follows:

$$(C_{UP} + C_{DOWN} + C_M)\,V_{FL} - C_{UP}V_{DD} - C_M V_{OUT} = Q_o \tag{1.4}$$

where C_{UP} is the sum of all the parasitic capacitances tied to V_{DD} $(C_{NL1} + C_{VDD})$ and C_{DOWN} is the sum of all the parasitic capacitances tied to GND $(C_{N0} + C_{GND})$.

For a wide range of input voltages (V_{FL}), the output voltage (V_{OUT}) is set to digital values (GND and V_{DD}). In these situations, C_M becomes part of C_{UP} or C_{DOWN}. Hence, V_{FL} can be isolated in Eq. 1.4, resulting in the simplified expression in Eq. 1.5:

$$V_{FL} = \frac{C_{UP}}{C_{UP} + C_{DOWN}} V_{DD} + \frac{Q_o}{C_{UP} + C_{DOWN}} \quad (1.5)$$

From Eq. 1.5 it is derived that the voltage of the floating line is determined by the ratio between the parasitic capacitances tied to the power supply (C_{UP}), and the sum of all the parasitic capacitances tied to the power supply and to ground ($C_{UP} + C_{DOWN}$), plus the influence of the trapped charge.

However, in some cases, both V_{FL} and V_{OUT} may be set to intermediate voltages not belonging to the digital domain. In such situations, V_{OUT} depends on the exact value of V_{FL} and the logic interpretation of the defective line is more difficult to be predicted.

Feedback capacitive paths may cause sequential behavior in some defective circuits. Konuk and Ferguson (1998) reported that Miller and wire-to-wire capacitances are the two types of capacitances responsible for these sequential behaviors.

Thin Open Defects

The behavior of interconnect *full* opens may vary depending on whether they have a small (thin) or a large (thick) lack of conducting material (Henderson et al. 1991; Hawkins et al. 1994). A large open decouples completely the two end points of the cavity created by the defect and its behavior is as reported in previous paragraphs. Nevertheless, if the open is small, the distance between the two electrically disconnected points causes the non-conductive material in between to be very thin. In this situation, electrons and holes are able to tunnel through, generating a slow charge transfer, which increases the rise and the fall times of the signal to be propagated through the line.

Open Defects with Gate Tunneling Leakage

Aggressive technology scaling trends have led to a significant increase in CMOS transistor gate leakage due to the reduction in gate oxide thickness. In nanometer technologies, high leakage current through the gate oxide is common in those devices due to direct tunneling mechanisms. Gate tunneling leakage affects the behavior of defective floating lines. The floating line cannot then be considered electrically isolated as it is subjected to transient evolutions until reaching the steady state, which occurs when the sum of all the gate leakage currents flowing into and out of the floating node is zero. This condition is determined by technology parameters and the topology of downstream gate(s) (Rodríguez-Montañés et al. 2007b). Arumí et al. (2008b) presented some simulation results where this behavior was observed. Figure 1.4 illustrates the SPICE simulation results corresponding to a floating line driving an inverter for a 90 nm technology. The dynamic evolution due to the impact of the gate leakage currents on the floating line (V_{FN}) and the response of the inverter (V_{OUT}) for two initial voltages at the input node (V_{FN0} equals 0 and

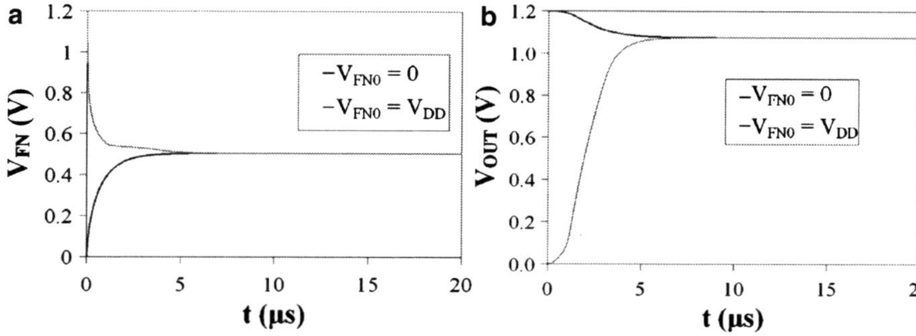

Fig. 1.4 Transient response of the inverter with its input floating for the 90 nm PTM technology (Arumí et al. 2008b). (**a**) Inverter input and (**b**) inverter output. V_{FN0} is the initial input voltage

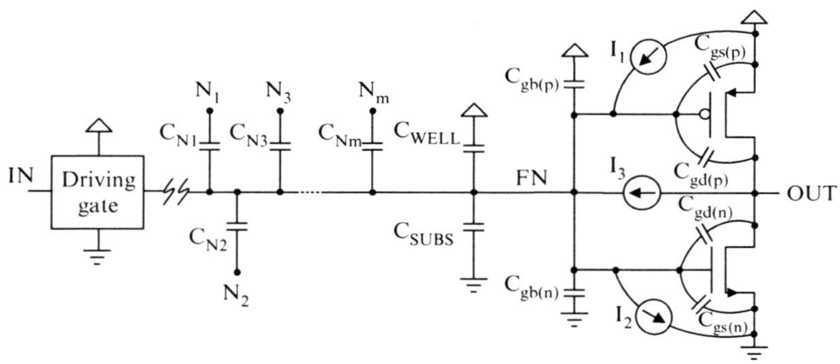

Fig. 1.5 Interconnect *full* open with the inclusion of the gate leakage currents (Rodríguez et al. 2008)

V_{DD}) are shown. A parasitic capacitance of 2 fF was assumed at the floating net. A transient evolution until reaching the final steady state, which does not depend on the initial voltage, is observed.

The time required for the defective inverter to reach the final steady state depends on the technology, the initial voltage value, the total capacitance of the floating node and the downstream transistors. Experimental results presented in the above work showed that, for a 0.18 μm technology, the transient evolutions were in the order of seconds. However, simulation results demonstrated that these evolutions were accelerated by several orders of magnitude for a 90 nm technology, being in the order of a few μs for a short net, as illustrated in Fig. 1.4. It is expected that these transient evolution times decrease even more as transistor dimensions are scaled down.

For nanometer technologies, the electrical model traditionally reported (see Fig. 1.3) is not accurate since the impact of gate leakage currents is ignored. These currents can be modeled by voltage controlled current sources. Without loss of generality, consider the example in Fig. 1.5, where the floating line is driving an inverter.

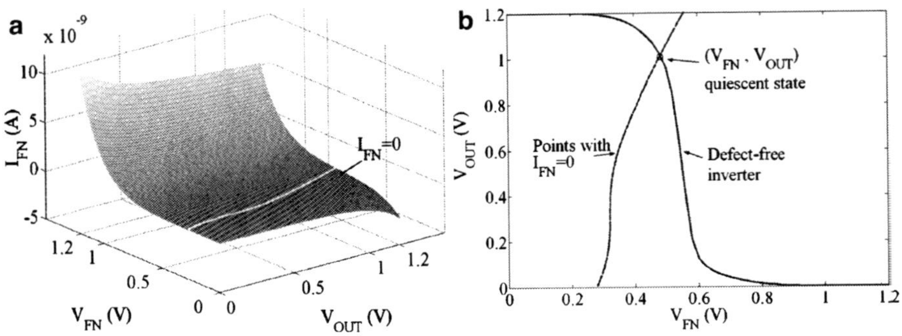

Fig. 1.6 (**a**) Sum of all the gate leakage currents (I_{FN}) at the floating input of a defective inverter, (**b**) prediction of the steady state voltage (Arumí et al. 2008b)

I_1 represents the total gate leakage currents flowing from the V_{DD} rail to the floating node. I_2 is equivalent to the total gate leakage currents flowing from the floating node to the GND rail and I_3 stands for the total gate leakage currents flowing between the floating node and the output node of the inverter.

With the knowledge of the gate leakage currents influencing the downstream gate, the steady state voltage of the floating line can be predicted. Assuming a floating line driving an inverter for a 90 nm technology, Fig. 1.6a illustrates the total gate current (I_{FN}) at the input floating node related to the input and output voltages of the downstream gate (inverter). The pairs (V_{FN}, V_{OUT}) where the resulting current is zero are shown in Fig. 1.6a as a level curve. Thus, as the transfer characteristic of the downstream gate is not modified, the steady state is determined by the intersection point between the (V_{FN}, V_{OUT}) pairs resulting in $I_{FN} = 0$ and the transfer characteristic of the gate, as shown in Fig. 1.6b. In this case, a logic low (high) level is generated at the input (output) of the defective inverter.

1.2.1.2 Resistive Open Defects in Interconnect Lines

When open defects cause a finite increment of line resistance, they are called *resistive* (or *weak*) opens. A *resistive* open weakens the affected signal, which has delay consequences on the transient behavior of the defective circuit (Moore et al. 2000). Two real *weak* open defects are exemplified in Fig. 1.7.

The electrical behavior of *resistive* opens relies on the value of the unknown resistance. Experimental measurements were carried out by Rodríguez-Montañés et al. (2002) on a set of test structures of a 0.18 μm technology in order to determine the open resistance values. Results showed that a high percentage of open defects were of full nature, with resistances higher than 1 GΩ, as illustrated in Figs. 1.8 and 1.9. Nevertheless, a non-negligible amount belonged to the class of *weak* or *resistive* opens, with resistances lower than 10 MΩ.

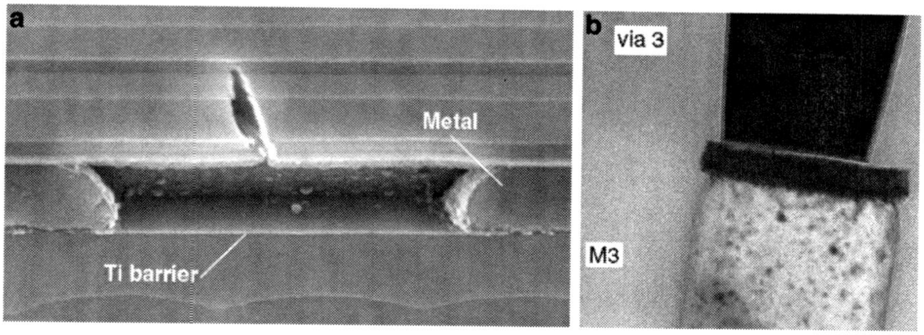

Fig. 1.7 *Weak* open defects (Rodríguez-Montañés et al. 2002). (**a**) Metal cavity and formation of a *weak* open defect due to the Ti barrier and (**b**) resistive via

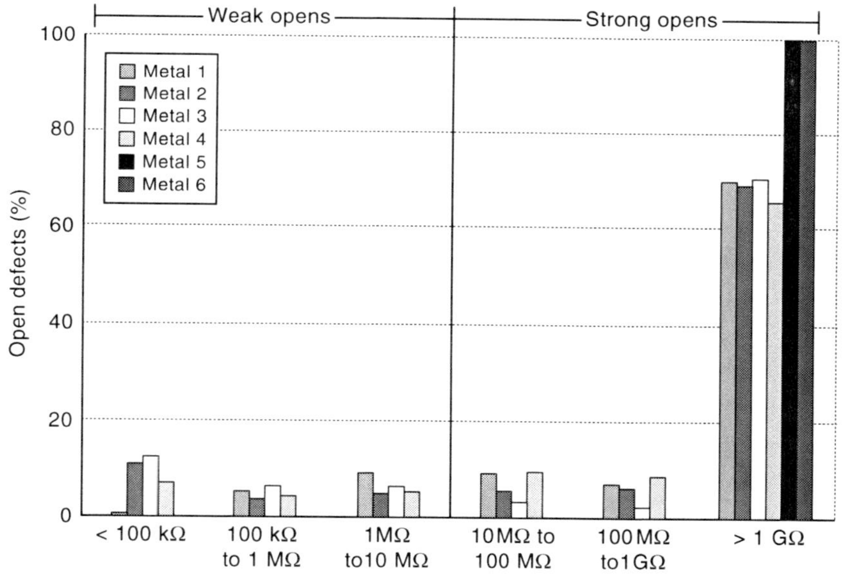

Fig. 1.8 Distribution of resistances for an open metal line (Rodríguez-Montañés et al. 2002)

Delay Model of *Resistive* Opens

Special attention has been paid to interconnect *resistive* opens. They can be modeled like interconnect *full* opens, but replacing the complete disconnection by an open resistance, as described in Fig. 1.10. In the presence of an interconnect *resistive* open, apart from the defect-free delay caused by the equivalent on resistance (R_{ON}) of the driving network and the total capacitance of the line (C), there is an extra delay caused by the open. This delay depends on the open resistance (R_o) and its exact location along the defective line (α), which determines the capacitance located after the open ($(1 - \alpha)\,C$).

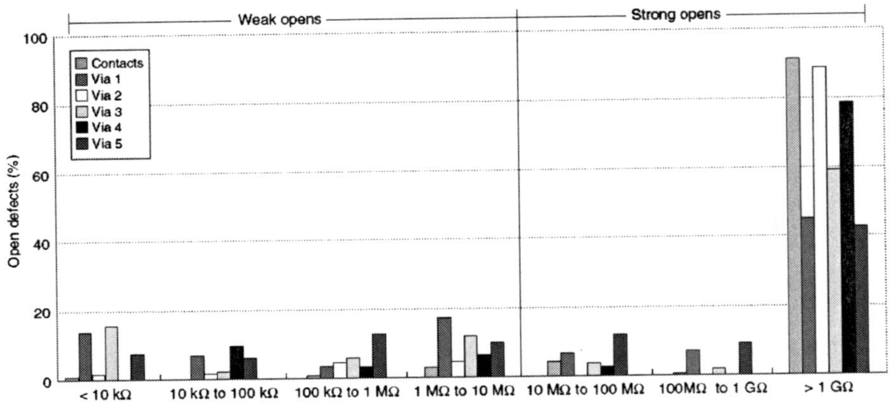

Fig. 1.9 Distribution of resistances for contact and via opens (Rodríguez-Montañés et al. 2002)

Fig. 1.10 Interconnect *resistive* open

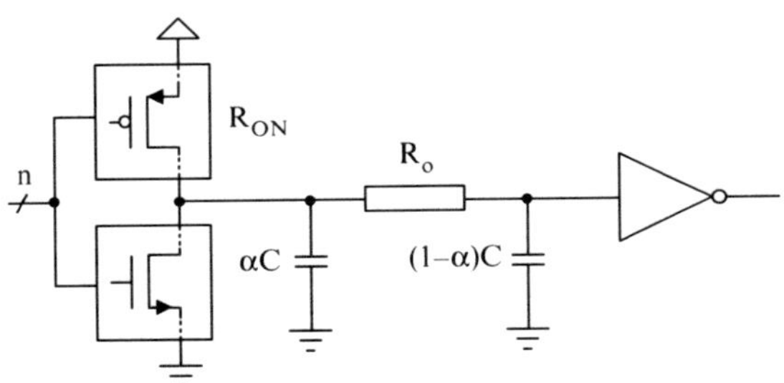

Using the Elmore model, the total delay for a transition propagation is approximated by Eq. 1.6:

$$\delta = R_{ON}C + R_o\,(1-\alpha)\,C \tag{1.6}$$

The factors influencing the delay added by an interconnect *resistive* open were experimentally analyzed by Arumí et al. (2008a). A set of *resistive* opens was injected into a test chip at different locations. Furthermore, the resistance was controllable because the opens were emulated by means of transmission gates. The delay measured on the tester for different resistances when transmitting a rising transition through the defective line can be seen in Fig. 1.11. The defective line was routed in metal 4 surrounded by two neighbors as close as allowed by the technology. Different open locations were considered (RN4–RN7), where RN4 has the minimum and RN7 the maximum coupling length, ranging from a few μm until a few mm of coupling length. The open resistance was controlled by the voltage of the transmission gate terminals (x-Axis). In this way, as we move from right to left on the x-axis, the equivalent resistance of the transmission gate increases. As expected, the delay increases for longer coupling capacitances and also for higher open resistances.

An interconnect *resistive* open defect weakens the signal propagated through the defective line. Thereby, the line is more vulnerable to crosstalk. Some of the

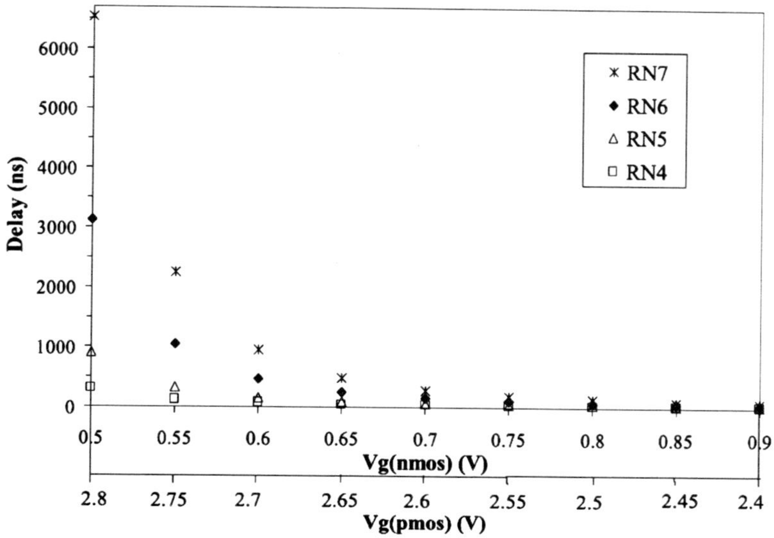

Fig. 1.11 Experimental delay results for *resistive* opens with quiescent neighbors (Arumí et al. 2008a)

parasitic capacitances affecting the defective line are related to neighboring lines, which may change their state when a new test pattern is applied. Thus, the effective coupled capacitances depend on the state of the neighboring lines. The experimental measurements in Fig. 1.12 show this phenomenon. A rising transition was transmitted through the defective line for every configuration of the two neighbors (N1 and N2) coupled to the defective line. As in the results of Fig. 1.11, the gate voltages of the transmission gates are controlled to obtain different resistance values. As expected, the delay is higher if the neighboring lines undergo the opposite transition related to the defective line. However, if both neighbors undergo the same transition as the defective one, the delay variability in the defect resistance is noticeably lower, since they help the defective line to reach the final (expected) state. When the neighboring lines have transitions of different sign, an intermediate behavior is observed.

The open resistance value has an important influence on the timing behavior of the defective circuit. Thus, when the resistance of the open is significantly higher than the on-resistance of the driving gate, i.e. $R_o \gg R_{ON}$, the delay can be simplified as follows:

$$\delta \approx R_o \,(1 - \alpha)\, C \tag{1.7}$$

The delay increases as the open is located close to the beginning of the line (low values of α). However, this simplification is not accurate for low *resistive* open defects. In these situations, a second order model must be considered. The maximum delay is not always found at the beginning of the net, but at an intermediate location, which is determined by the relationship between the open resistance, the on-resistance

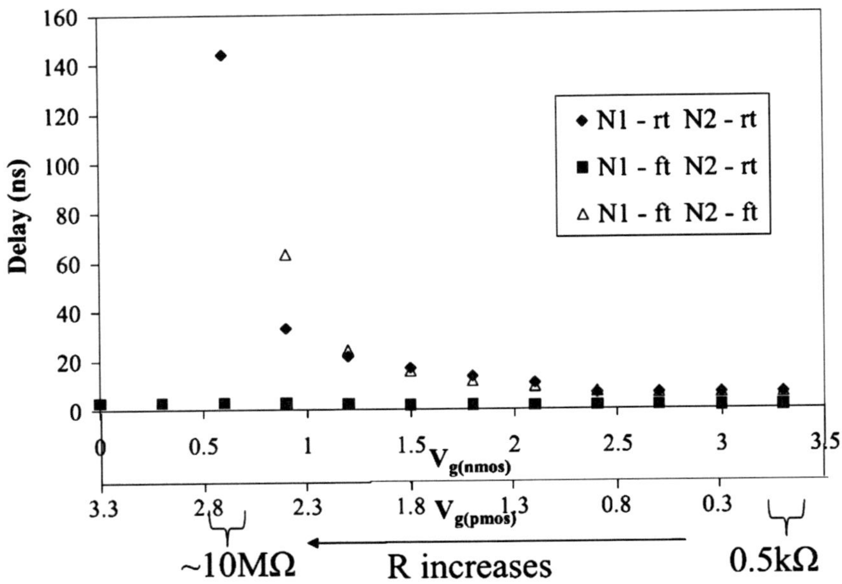

Fig. 1.12 Experimental delay results for *resistive* opens with neighbors changing their state, rt: rising transition, ft: falling transition (Arumí et al. 2008a)

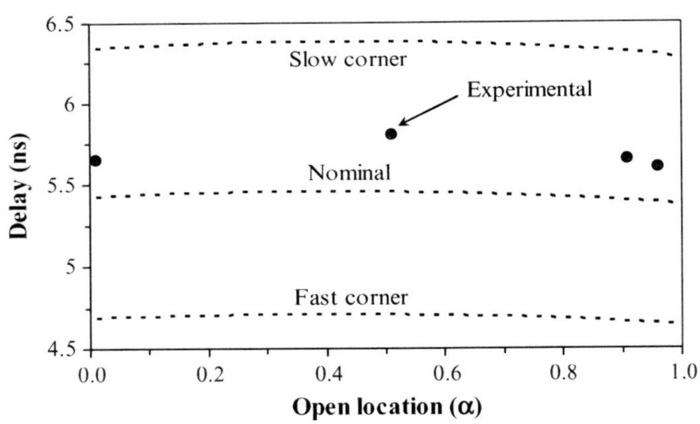

Fig. 1.13 Experimental and simulation delay results for low *resistive* open defects (Arumí et al. 2008c)

of the transistor network driving the defective line, the parasitic capacitances and the threshold voltage of the transistors driven by the defective net. Experimental evidence of this is presented by Arumí et al. (2008c). Figure 1.13 summarizes experimental and simulation results obtained with low *resistive* opens (a few kΩ). The delay is higher when the open is located in the middle of the interconnect line related to the rest of locations, i.e., the beginning and the end of the line.

1.2.2 Intra-gate Open Defects

The first research on open defects addressed the effect of stuck-open faults in CMOS circuits. The stuck-open fault model (Wadsack 1978) is a failure mechanism modeled as a loss of charge transfer in one transistor of the defective cell or gate so that the output is set to a high impedance state for at least one logic state. Thus, a sequential behavior is observed at the output node depending on its previous logic state. Figure 1.14 illustrates a 2-input NAND gate where the source terminal of one of the pMOS transistors is disconnected from the output. For the state (A B) = (1 0), the output (Z) is in a high impedance state. Hence, if the previous pattern applied is (A B) = (1 1), Z is interpreted as logic 0; otherwise it is interpreted as logic 1, as reported in Table 1.1.

Initial work on *full* opens disconnecting one single gate transistor terminal (Floating Gate Open) was carried out by Renovell and Cambon (1986, 1992). Champac et al. (1993, 1994) modeled this defect and validated the proposed model by experiments on test chips designed with intentional open defects. The floating gate voltage depends on the location of the poly break, modeled by the poly-bulk and metal-poly capacitances. The affected transistor may operate in the sub-threshold region, behaving as a stuck-open transistor or operate in the saturation and ohmic regions.

In the work by Soden et al. (1989), experiments were conducted to evaluate the transient response and current behavior of stuck-open faults. The results corroborated the predicted sequential behavior. Furthermore, high current consumption

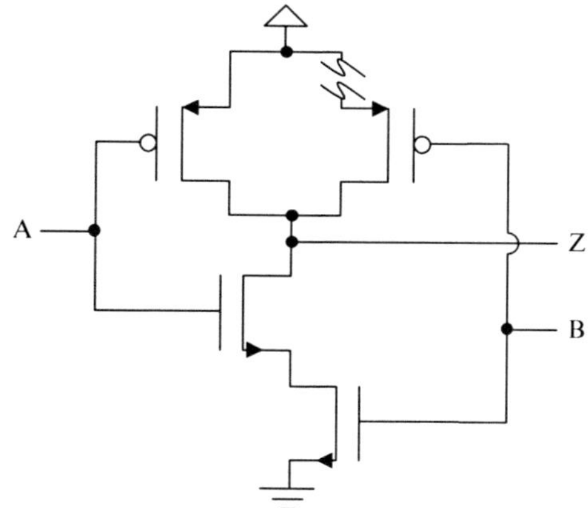

Fig. 1.14 Stuck-open fault in a NAND gate

Table 1.1 Stuck open behavior in a NAND gate

Input port (A)	Input port (B)	Output port (Z)
0	0	1
0	1	1
1	0	High impedance
1	1	0

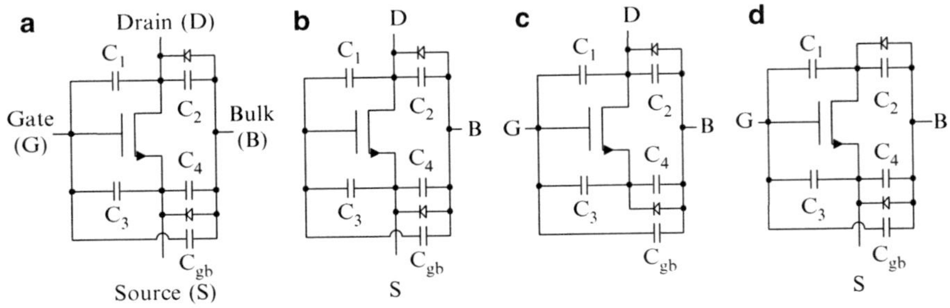

Fig. 1.15 nMOS transistor model (Maly et al. 1991). (**a**) Fault free, (**b**) open gate, (**c**) open source, and (**d**) open drain

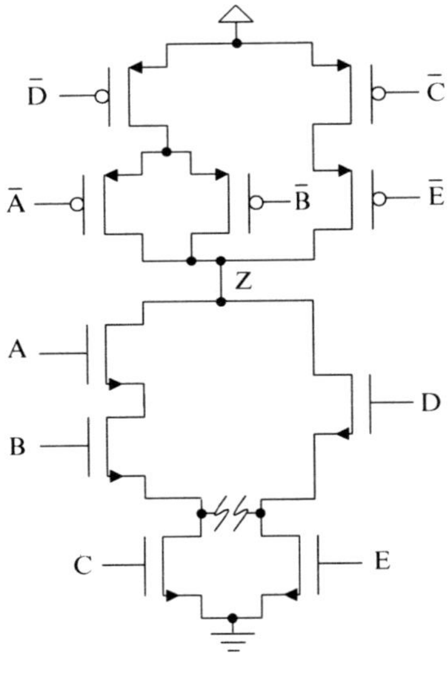

Fig. 1.16 Example of an intra-gate open fault

might be generated in the high impedance state. Although widely used, stuck-open faults only cover a small fraction of faults caused by actual opens. In this sense, in the work by Maly et al. (1991), the analysis of *full* opens was extended to faults located in any of the transistor terminals, as described in Fig. 1.15 for an nMOS transistor.

These previous models are not robust because they ignore both hazards and charge-sharing effects. In order to evaluate the impact of these effects, let us consider the example in Fig. 1.16 and the sequence of patterns in Table 1.2 (Di and Jess 1993). If Test 1 is applied to the circuit, the output (Z) is charged to V_{DD} and subsequently set to a high impedance state. With this sequence, the open could be detected. However, if some delay forces input D to change earlier than input E, a temporary leakage path from Z to GND is generated. Output Z can then be discharged, invalidating the test. This problem can be solved by applying a test sequence like Test 2, where only

Table 1.2 Test sequences for the gate in Fig. 1.14 (Di and Jess 1993)

	Test 1						
Pattern	A	B	C	D	E	Z (fault-free)	Z (faulty)
1	1	0	1	0	1	1	1
2	1	0	1	1	0	0	1
	Test 2						
3	1	1	1	0	0	0	0
4	1	1	0	0	1	1	1
5	1	1	0	0	0	0	1

input E changes. To overcome such problems, Di and Jess (1993) proposed a fault model considered at logic level. Detectability conditions were derived from Reduced Ordered Binary Decision Diagrams (ROBDD) data structures used during fault simulation. To avoid hazard effects, the two test sequences were chosen such that during the transition from the initialization phase to the test phase there was no temporary leakage path draining the charged output. In case of charge-sharing effects, a local circuit analysis was performed to estimate the voltage level after charge-sharing to check the validity of the test sequence. During the initialization phase, the charge-sharing path was set to conduct so that all nodes on the charge-sharing path were charged to a voltage level equal to that of the output node.

Favalli et al. (1996) presented the node break fault model, where broken connections were taken into account without any knowledge about the circuit layout. The conditions of a node break fault were derived from electrical considerations. The minimum number of patterns needed to test the fault was determined based on graph theory. Testing a node-break fault is an implicit test for stuck-open faults of every transistor whose drain or source is connected to that node.

1.3 Detectability of Open Defects

This section briefly presents the different methodologies to improve the detectability of open defects, following the classification based on defect location.

1.3.1 Detectability of Interconnect Open Defects

Logic-based methodologies are the most commonly used techniques for the detection of interconnect open defects. However, they are not always effective. Thereby, other alternatives have been developed to improve or complement the success of logic-based techniques. They are all summarized in the following subsections.

1.3.1.1 Logic Detectability of Interconnect Open Defects

In the presence of an interconnect *full* open, the floating line voltage is basically determined by the ratio between the parasitic capacitances related to the floating line tied to V_{DD} (C_{UP}) and the sum of all the parasitic capacitances ($C_{UP} + C_{DOWN}$) plus the influence of the trapped charge (V_{Qo}), as previously reported in Eq. 1.5. On the one hand, the trapped charge is an unknown but constant parameter. On the other hand, the ratio between the parasitic capacitances tied to V_{DD} and the total parasitic capacitance depends on a number of factors. One is the relationship between transistor and neighboring capacitances. When the floating line length (L_{FL}) is short, transistor capacitances generally dominate and set the floating line to an intermediate value. Nevertheless, for long L_{FL}, neighboring capacitances dominate and the floating line may achieve a wider range of values. The exact location of the open is also important since only parasitic capacitances located after the open influence the floating line. The last factor is the test pattern applied because it sets a certain state on the neighboring lines. Champac and Zenteno (2000) presented simulation results showing the influence of these factors. Furthermore, experimental evidence was provided in the work by Arumí et al. (2008a), where a set of open defects was intentionally injected into a test circuit. Every floating line was routed between two neighboring lines with different coupling lengths. Experimental results showed that when both neighbors had the same logic value, they determined the logic interpretation of the floating line, even for floating lines of a few tens of μm in length.

Therefore, in the presence of an interconnect *full* open defect, its detectability when carrying out a logic test can be improved in the following manner:

- Testing for an SA1 at the target node: Maximize the $C_{UP}/(C_{UP} + C_{DOWN})$ ratio.
- Testing for an SA0 at the target node: Minimize the $C_{UP}/(C_{UP} + C_{DOWN})$ ratio.

1.3.1.2 Delay Detectability of Interconnect Open Defects

In general, resistive defects have an impact on the time response of the circuit. Hence, delay testing is a widely used technique to detect such defects, including *resistive* opens. In the presence of an interconnect *resistive* open, signals propagated through the defective line undergo an extra delay. If the sum of the defect-free delay plus the one added by the defect exceeds the maximum delay permitted (test period), a malfunction can be caused and the defect is detected (Li et al. 2001; Kruseman and Heiligers 2006). However, performing a delay test at nominal conditions (speed) may lead to missing *resistive* opens. If the defect is sensitized along a short path, then the total delay may not exceed the maximum permitted. Nevertheless, if the defect were sensitized along a longer path, it could be detected. Therefore, the sensitization path is a key factor to detect *resistive* opens.

The other important factor is the open resistance. The higher the resistance, the larger the delay. Thus, given a fault site (location) and a test pattern (sensitization path), it is possible to predict the critical resistance, i.e., the minimum open resistance which generates sufficient delay so that a faulty behavior is observed. Baker et al. (1999) carried out simulations with a 0.25 μm technology to find the critical resistance. Results showed that, for interconnect opens, most critical resistances were about a few MΩ.

Another issue to be addressed is that the resistive line acts as a low pass filter. Assuming that $R_o \gg R_{ON}$, the time constant depends mainly on the open resistance (R_o) and location (parasitic capacitances located after the open). If the time constant due to the defect is lower than the signal period, the defective node reaches its final state before the next clock period is generated. However, if the time constant is higher than the signal period, the defective node has not reached the final state yet when the next transition has already been initiated. Hence, for every clock cycle, the defective node does not start from the expected logic 0 or 1 value but from some intermediate state. This effect is the so-called history (or memory) effect (Renovell et al. 2006; Arumí et al. 2008a). In Fig. 1.17, the evolution of the defective line (V_{def}) driven by an inverter is shown. An input sequence where 70% of values are logic 1s is considered. When starting from 0 V, the value of V_{def} increases until it evolves close to the region of 70% of V_{DD}. Therefore, for the next cycle, the initial voltage is at an intermediate value instead of the expected 0 or V_{DD}. Experimental results revealed the impact of this phenomenon. The experiment consisted of applying a rising transition at the defective node and measuring the propagation delay between the input and the output of the inverters for different initialization states and resistances (see Fig. 1.18). This initialization involved sequences of 0s and 1s being applied to the defective node from 0% to 100% of 1s prior to triggering the transition. The results are listed in Table 1.3, where 'd' denotes the defect detection and '•' denotes a test escape. The results show that the detectability interval of open resistances decreases as the initial state is closer to the final state (logic 1 for a rising transition).

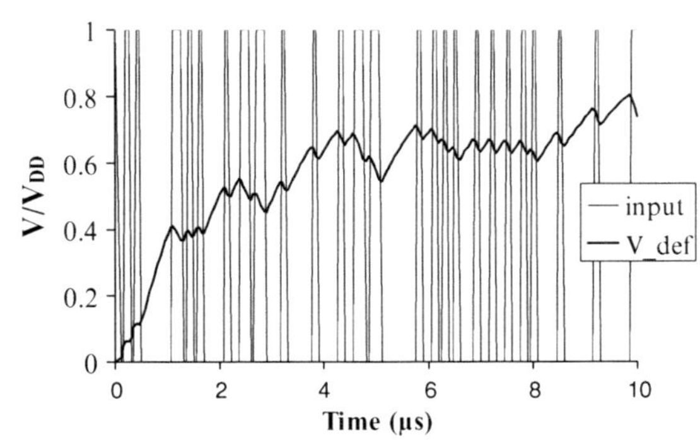

Fig. 1.17 Dynamic behavior of the defective line (Arumí et al. 2008a)

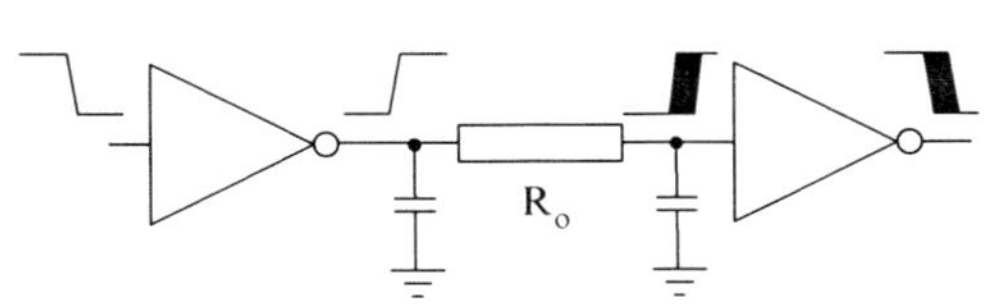

Fig. 1.18 Experiment performed to show the history effect

Table 1.3 Experimental results showing the history effect (Arumí et al. 2008a)

Sequence of 0s and 1s	R_O (100 kΩ < R1 < R2 < R3 < R4 < 100 MΩ)			
%1s	R1	R2	R3	R4
100	•	•	•	•
90	•	•	•	d
80	•	•	•	d
70	•	•	•	d
60	•	•	d	d
50	•	•	d	d
40	•	•	d	d
30	•	•	d	d
20	•	•	d	d
10	•	•	d	d
0	•	d	d	d

The history effect must be minimized when performing a delay test. Otherwise, *resistive* open defects may escape the test. For this reason, when a test is applied to a specific target net in order to test for a rising (falling) transition, the net must remain at a low (high) logic value for a sufficient number of cycles before the initialization pattern is applied. In this way, it is assured that the target node covers the maximum voltage excursion to reach its final logic state.

Finally, another factor is known to influence the detectability of *resistive* open defects, i.e. the dynamic behavior of neighboring lines coupled to the defective line. Figure 1.12 shows how the largest delay was obtained when the neighboring lines underwent the opposite transition related to the defective line. In fact, the effective capacitance between two nets depends on their state as well as on the skew between the transitions generated on every line. Let us assume that C_{Ni} is the capacitance between the neighboring line and the defective line when both lines are in a quiescent state. In the case of a null skew, when a transition is generated in the defective line, the effective capacitance $\left(C_{eff(Ni)}\right)$ between the defective line and its neighboring line N_i can be approximated as follows (Sakurai 1993):

$$C_{eff(Ni)} \begin{cases} 0 \text{ for the same transition in } N_i \\ C_{Ni} \text{ for } N_i \text{ in a quiescent state} \\ 2C_{Ni} \text{ for the opposite transition in } N_i \end{cases} \tag{1.8}$$

According to Eq. 1.8, obtaining the largest delay caused by a *resistive* open defect requires maximizing the total effective capacitances between the defective line and its neighboring lines.

Although usually applied to *resistive* opens, delay considerations can also be useful for interconnect *full* open defects. In nanometer technologies, it has been shown how, in the presence of an interconnect *full* open defect due to the impact of gate leakage currents, a transient evolution is induced in the floating line until it reaches the steady state, which is determined by the technology and the topology of the downstream gate(s). Experimental measurements show that these transient

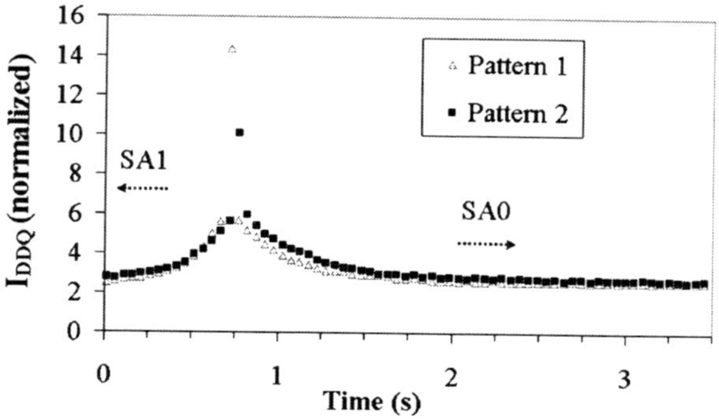

Fig. 1.19 I_{DDQ} time-dependent behavior of a 0.18 μm technology defective device (Arumí et al. 2008b)

evolutions are in the order of seconds for a 0.18 μm technology, as depicted in Fig. 1.19. The evolution of the floating line was observed by monitoring the current consumption of the circuit over time. This evolution influences the logic behavior of the floating line since its interpretation changes from logic 1 to logic 0 after few seconds. In this technology, these evolutions are too slow for testing purposes. However, analytical and simulation results report that these transient evolutions might be reduced by several orders of magnitude for future technologies, opening a new field of study on the detectability of such defects.

1.3.1.3 Alternative Techniques for the Detectability of Interconnect Open Defects

The modification of power supply voltage (V_{DD}) especially by High Voltage (HV) testing (Li et al. 2001; Kruseman and Heiligers 2006) has been successfully applied to detect interconnect open defects. The key in using high voltages stems from the idea that the delay added by a *resistive* open located in the interconnection is almost insensitive to power supply voltage. However, circuit delay depends on power supply voltage, increasing as V_{DD} decreases. Therefore, for high voltages, although the delay added by the defect is approximately the same, the circuit delay is smaller and consequently the defect delay becomes more observable. Figure 1.20 shows the shmoo plot for a defect free device in comparison with two defective devices with an interconnect resistive open of 1 and 3 MΩ, respectively.

In the presence of an open, the exact voltage-delay relationship depends strongly on the open location. In the work by Yan and Singh (2005), the difference between transistor-related defects and resistive interconnect defects was reported by sweeping the power supply value. Simulations were conducted for defective circuits at different V_{DD} values. The results showed that the delay added by transistor-related defects increased non-linearly when decreasing the power supply value whereas this had little impact on the delay added by resistive interconnect defects.

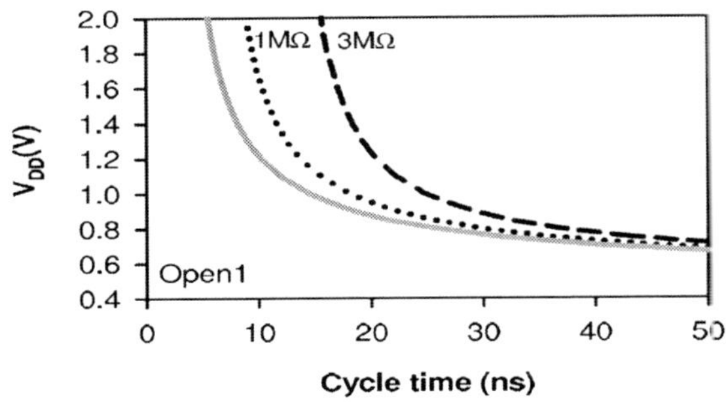

Fig. 1.20 Pass/fail boundary (Shmoo Plot) for defect-free silicon and with an interconnect open resistance of 1 and 3 MΩ, (Kruseman and Heiligers 2006)

In some cases, high voltages are also used as voltage stress testing for reliability screening (Kawahara et al. 1996; Chang and McCluskey 1997; Aitken 2002). Stressing the device with high voltages may improve the detection of some defects. This technique is particularly useful for detecting oxide thinnings and via defects, which shorten device lifetime. The goal of stressing devices is to make these flaws evident, causing via defects to become opens and oxide thinnings to become oxide breaks. However, two parameters must be thoroughly controlled, i.e., power supply voltage and stressing time. If any of these two parameters exceeds the allowed limit, defect-free devices could be damaged.

Observation of quiescent current consumption of the circuit (I_{DDQ}) may also be effective in technologies with reduced background leakage currents (i.e., low non-defective I_{DDQ}). In these circumstances, the detection of interconnect open defects may sometimes be possible although this technique is not as useful as for other types of defects, such as bridges. The detection of open defects by I_{DDQ} is strongly dependent on cell design and circuit topology. Assuming an interconnect *full* open defect, if an intermediate voltage is induced on the floating line, the two transistors driven by the floating line may be in a conduction state, generating a current path from V_{DD} to GND, and thus resulting in extra current consumption (Singh et al. 1995; Champac and Zenteno 2000).

Temperature can also help to detect *resistive* opens. Assume, as a first approximation, that the open resistance is not modified with temperature. As temperature decreases, the dominant effect is usually the increasing mobility, which decreases the on resistance of transistors. In such situation, the relative importance of the delay added by the defect increases. Hence, cold testing improves the observability of *resistive* opens. However, the open resistance does vary with temperature as well. Therefore, the delay induced by the open changes. The temperature coefficient of the resistance depends on the *resistive* open material. Hence, the delay added by the open may increase or decrease with temperature. In fact, *resistive* opens may pass the test at nominal conditions, but can be detected at a temperature different from the nominal one. For instance, the work of Needham et al. (1998) reported a *resistive* open between an interconnect and a via causing a functional failure at −20°C, which was not detected at room temperature.

1.3.2 *Detectability of Intra-gate Open Defects*

Early research to detect intra-gate open defects was founded on logic-based techniques. Nevertheless, these cannot always ensure the detectability of such opens. Logic based techniques and alternatives are presented in this section.

1.3.2.1 Logic Detectability of Intra-gate Open Defects

As already seen in Section 1.2.2, the detectability of stuck open faults depends on the pattern order. The output of the defective gate is in a high impedance state for at least one input combination. In this situation, the output voltage depends on the state induced by previous patterns. Therefore, with the appropriate pattern order, logic testing is suitable for the detection of such defects (Wadsack 1978; Soden et al. 1989).

If an open causes a single floating gate, its detectability depends on several factors (Champac et al. 1993, 1994; Ivanov et al. 2001), namely topological parameters, trapped charge and unpredictable poly-to-bulk capacitance (C_{pb}). The detectability of the fault can be ensured depending on the C_{pb} value. The final value of the output voltage of the affected gate increases with C_{pb}. Therefore, a critical value of the unpredictable parameter C_{pb} can be defined to detect a single floating gate. The detectability interval is defined as the range of C_{pd} values where the open fault can be detected.

1.3.2.2 Delay Detectability of Intra-gate Open Defects

Like interconnect *resistive* opens, intra-gate *resistive* opens influence the transient behavior of defective devices. In general, the higher the resistance, the larger the delay. Furthermore, the exact location of the intra-gate *resistive* open also has a significant impact on the transient behavior of the affected circuit, as analyzed by Baker et al. (1999). This work considered a 0.25 μm standard cell library. Transistor level netlists and interconnect parasitics were extracted from layout to find the critical resistances. For resistive drain/source faults, simulation results showed that most critical resistances were about 50 kΩ. However, for resistive single transistor gate faults, critical resistances ranged between MΩ and a few tens of MΩ depending on the duty cycle of the input waveform.

In some cases, time considerations can also be useful in the detectability of intra-gate *full* open defects. For single floating nMOS (pMOS) transistors, a rising (falling) transition applied to the defective input may detect the presence of such faults provided that the delay is large enough to generate a fault (Ivanov et al. 2001). This delay depends on topological parameters and C_{pb}. In general, the higher C_{pb}, the larger the delay.

1.3.2.3 Alternative Techniques for the Detectability of Intra-gate Open Defects

The modification of power supply voltage can also be useful for detecting intra-gate opens. Li et al. (2001) provided simulations where a *resistive* open was injected into the gate delay path of an inverter chain. The results showed that this class of fault was more easily detected at low power supply voltages. Furthermore, as previously reported, the delay added by transistor-related defects increased non-linearly when the power supply value was decreased. This behavior occurs when these transistor-related defects are due to intra-gate opens (Yan and Singh 2005).

I_{DDQ} testing is another alternative for detecting intra-gate opens for technologies with low background leakage currents. However, even in these technologies, the efficiency of I_{DDQ} testing is strongly dependent on cell design, circuit topology and open location. For example, the work by Champac et al. (1994) presented the I_{DDQ} detectability of a single floating transistor. It was reported that the location of the poly-break, modeled by the poly-bulk and metal-poly capacitances, determined the degree of conduction of the floating gate transistor and its detectability by current testing. For sufficiently high values of the poly-bulk capacitance, the defective transistor may work in the subthreshold region, where it can be modeled as a stuck-open transistor. It is therefore not detectable by an I_{DDQ} test. However, for sufficiently low values of the poly-bulk capacitance and sufficient metal track influence, the floating gate transistor operated above threshold, generating non-negligible I_{DDQ} values.

Singh et al. (1995) reported the results of an experimental test chip for analyzing the I_{DDQ} detectability of open defects. Open faults were divided into five different groups, see Fig. 1.21, namely open disconnecting a transistor pair (O_1), a single floating net belonging to a transistor being the only conduction path to the power rails (O_2), an open source/drain on the only conduction path to the power rails (O_3), a floating gate in a transistor on one of multiple conduction paths to V_{DD} or GND (O_4), and finally an open source/drain on one of multiple conduction paths to V_{DD} or GND (O_5). Based on the experimental results, the authors reported that opens O_1 and O_2 were the most likely to be detected by a I_{DDQ} test although their detectability could not be ensured for all configurations. For opens O_4 and O_5, if the affected

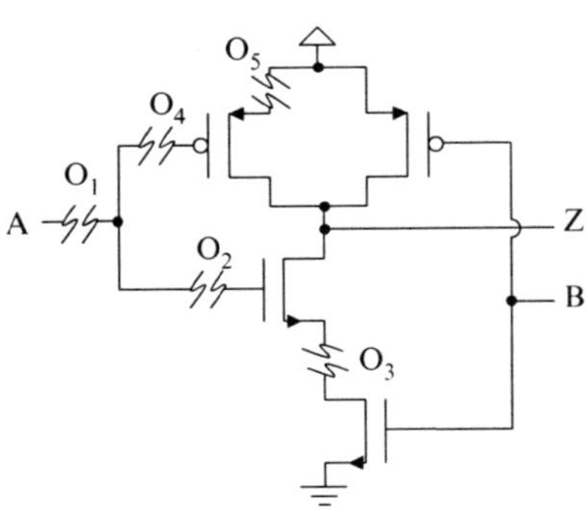

Fig. 1.21 I_{DDQ} detectability of open defects (Singh et al. 1995)

transistors were in the off state, it was possible to detect the defect by capturing an intermediate voltage at the floating node due to hazards that may affect the CMOS network. Finally, open O_3 was the most difficult to detect by current testing because this class of faults usually had a stuck-at behavior.

Finally, Nigh and Gattiker (2004) reported that I_{DDQ} versus time may give additional information about open defects. Some defective devices showed time-dependent I_{DDQ} behavior with evolution in the order of seconds. The authors conjectured that this dynamic behavior could be associated with an open defect and the subthreshold, gate and reverse bias pn junction leakage currents flowing into and out of the affected node.

1.4 Diagnosis of Open Defects

Accurate diagnosis of failure sites is important for solving process problems, analyzing failures and improving yields. The current diagnosis effort related to open defects has focused mostly on interconnect opens. Accordingly, in this section we will first analyze the strategies to diagnose interconnect opens followed by an overview of the techniques used to diagnose intra-gate opens.

1.4.1 Diagnosis of Interconnect Open Defects

One of the first works on diagnosis of interconnect open defects was conducted by Venkataraman and Drummonds (2000). The proposed methodology was based on logic information using the net diagnostic model. This model takes the different branches of the defective line into account. Let us now look at the example in Fig. 1.22. The line is composed of stem A and branches B and C. The logic errors caused by a 0/1 error at locations A, B and C are saved in the erroneous observation (EO) sets EO_1, EO_3 and EO_5, respectively, as described in Table 1.4. Similarly, the errors caused by a 1/0 error are saved in the sets EO_2, EO_4 and EO_6, respectively. The diagnostic signature EO for stem A is then computed as the union of sets EO_1, EO_2, EO_3, EO_4, EO_5 and EO_6. In the presence of an open on net ABC, only a subset of set EO is faulty. A path-tracing procedure can be used to identify the logic nets potentially associated with an interconnect open.

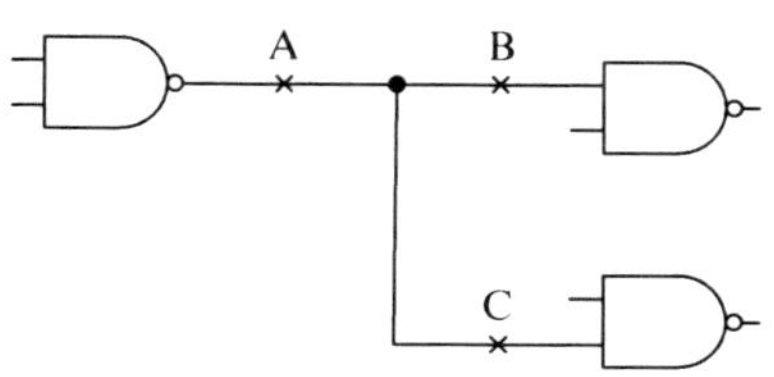

Fig. 1.22 Net diagnostic model (Venkataraman and Drummonds 2000)

Table 1.4 Net diagnostic model for Fig. 1.23 (Venkataraman and Drummonds 2000)

EO_1	EO_2	EO_3	EO_4	EO_5	EO_6
A (0/1)	B (0/1)	C (0/1)	A (1/0)	B (1/0)	C (1/0)

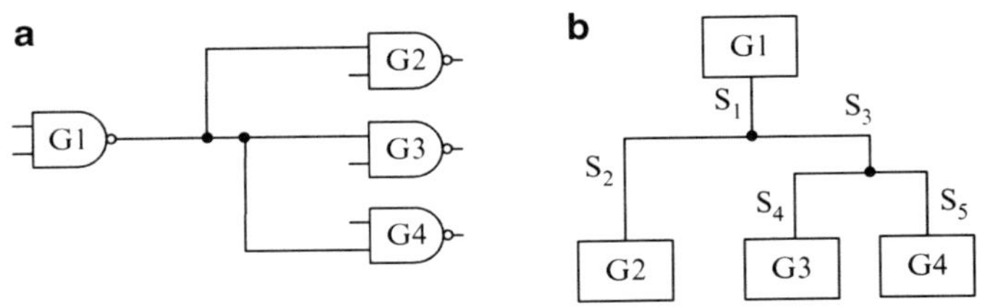

Fig. 1.23 Segment fault model (Huang 2002). (**a**) Target net driving three gates and (**b**) segment division according to layout information

In a subsequent work, Liu et al. (2002) presented a model-free diagnosis algorithm for multiple interconnect open faults. In the presence of an open fault, this procedure considered the worst case scenario. Each fan-out branch of the stem was assumed to behave randomly, that is, independent of the value on the stem. Hence, every branch could take an arbitrary logic 1 or 0 for each test pattern. An iterative algorithm using X values identified possible faulty locations. Subsequently, simulations were carried out to reduce the set of candidates.

Unlike these previous works, some recent studies have considered physical information to improve diagnosis resolution. Huang (2002) proposed a diagnosis procedure using the segment fault model. A segment (S_i) is a part of a net based on routing information. By knowing the layout, the target net can be divided into several segments, as shown in Fig. 1.23. Symbolic simulation is performed to find open segments on the target line. The main drawback of this methodology is that there are cases where segments are still too long and the open cannot be precisely located along the line.

In the work by Sato et al. (2002), a technique to find open vias by using physical information was proposed. The capacitances between the floating net and its neighboring lines were taken into account to predict changes in the floating node voltage for every test pattern (P), as described by Eq. 1.9:

$$E(P) = \frac{C_1(P)}{C_0(P) + C_1(P)} \tag{1.9}$$

C_1(P) is the sum of the capacitances between the floating net and coupled structures tied to logic 1 for a specific test pattern, and C_0(P) stands for the sum of the capacitances between the floating net and its coupled structures set to logic 0 for the same P pattern. The patterns exciting the fault are divided into two sets: Ω_0 and Ω_1, where Ω_0 (Ω_1) is composed of patterns which set the floating net voltage to a value lower (higher) than the threshold voltage of the downstream gate. Assuming that $E(\Omega_0) = [\min E(p), \max E(p)]$ for $\forall p \in \Omega_0$ and $E(\Omega_1) = [\min E(p), \max E(p)]$ for $\forall p \in \Omega_1$, to obtain consistent results in the presence of an open defect, Eq. 1.10 should be satisfied.

$$E(\Omega_0) < E(\Omega_1) \tag{1.10}$$

This methodology neglected capacitances between internal nodes. Its feasibility was also limited in situations where the floating net had fan-out and the threshold voltages of the inputs of the driven gates were different, since Eq. 1.10 may not be satisfied. Furthermore, this work focused on open vias only and discarded finding opens due to broken metal tracks.

The diagnosis technique presented by Zou et al. (2006) was founded on the segment fault model previously proposed by Huang (2002). In this methodology, the segment model was used as a first step to get the set of potential open segments responsible for the faulty behavior. Subsequently, SPICE simulations were carried out to calculate the input threshold voltages of the driven gates. With all this information and the charge conservation principle, a prediction of the initial trapped charge was made. According to the above principle, once the initial charge is trapped in the circuit during the fabrication process, the total amount of charge does not change and is redistributed among the capacitors when different test patterns are applied, as described by Eq. 1.11:

$$Q_{trap} = Q_{wire}\left(P, V_{fn}\right) + Q_{gate}\left(V_{fn}\right) \tag{1.11}$$

where $Q_{wire}(P, V_{fn})$ is the sum of charges stored in the capacitors between the floating node and its coupled neighbors. This factor depends on the test pattern applied (P) and the floating node voltage (V_{fn}). $Q_{gate}(V_{fn})$ is the charge stored in the capacitors of the downstream gates and it also depends on the floating node voltage (V_{fn}). For a set of test patterns, it was possible to determine an upper and lower bound for the Q_{trap} value. The consistency of these results was used to reduce the number of possible open vias within the segments explaining the faulty behavior. The application of such methodology requires the use of Q–V look-up tables for every gate.

In the proposal of Rodríguez-Montañés et al. (2007a), the target net was divided according to the FOS (Full Open Segment) model to diagnose interconnect full open defects in long floating lines where the impact of transistor capacitances are low. The FOS model considered any possible location of the open along the line. With this model, the floating line is partitioned into several segments (Seg_i). Segment breaks are caused by a change in the neighborhood layout. For the example in Fig. 1.24, the target line is divided into nine different segments. Hence, each segment consists of the target line and zero to two neighboring lines since only coupling neighbors between the same metal layer are considered. It is therefore possible to extract the parasitic capacitances for every segment easily. Given an open location (segment k) and a test pattern (P), the floating line voltage is determined by the parasitic capacitances of the segments located after the open, as reported in Eq. 1.12.

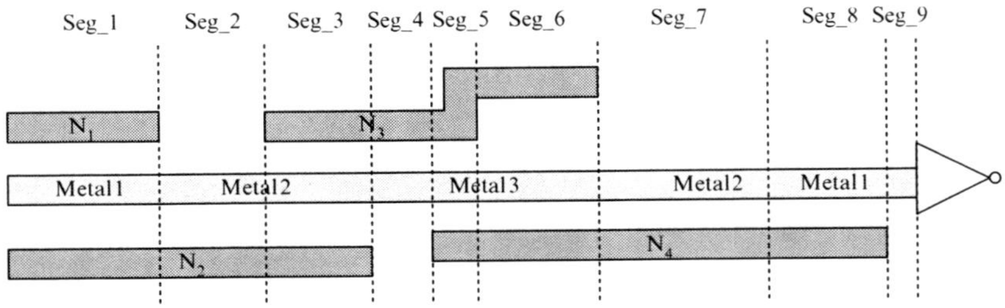

Fig. 1.24 Segment division according to the FOS model (Rodríguez-Montañés et al. 2007a)

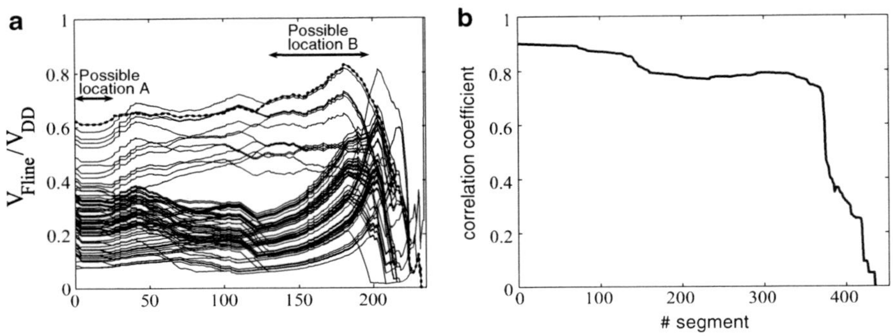

Fig. 1.25 Defective device of a 0.18 μm technology containing an open defect (Rodríguez-Montañés et al. 2007b). (**a**) Prediction of the floating line voltage and (**b**) correlation of the current-based results

$$V_{FL}(k,P) = \frac{\sum_{i=k+1}^{N} C_{up\_i}}{\sum_{i=k+1}^{N} C_{up\_i} + \sum_{i=k+1}^{N} C_{down\_i}} V_{DD} \tag{1.12}$$

The numerator stands for the sum of all neighboring parasitic capacitances tied to logic 1 (C_{up_i}) and located after segment k. The denominator is the sum of all neighboring parasitic capacitances tied to logic 1 (C_{up_i}) and logic 0 (C_{down_i}) and also located after segment k.

This methodology predicts the floating line voltage at the far end of every segment for every test pattern exciting the open fault (the voltage at intermediate locations within any segment is found by interpolating the voltage results at their end points). These predictions were associated with the experimental results obtained on the tester. The voltage predictions for a real defective device of a 0.18 μm technology can be seen in Fig. 1.25a. Patterns generating a floating line voltage interpreted as logic 1 on the tester are plotted in dotted lines, whereas patterns generating a logic 0 in the floating line are plotted in plain lines. To find a location where the predicted results are consistent with the experimental results obtained on the tester, the predicted voltage of the floating line for the dotted patterns must be above those for the plain patterns. Note that the methodology is based on relative predictions of the floating line. Thus, uncertainty due to the trapped charge and the threshold voltage of the downstream gate is eliminated. The predictions in Fig. 1.25a are consistent for two ranges of locations (A and B). The rest of locations can be discarded.

Based on the same methodology, the authors also proposed, when feasible, the use of I_{DDQ} measurements to improve the accuracy of diagnosis results. The predictions of the floating line voltage allowed, in turn, the extra current consumed by the downstream gate to be predicted by SPICE simulations. The predicted currents were compared with the results obtained from the I_{DDQ} test, and the correlation coefficient between the predicted and measured currents was calculated. Results for the same defective device are shown in Fig. 1.25b. By combining both logic and current results, the authors determined that the most likely location for the open is region A (at the beginning of the defective net, close to the driver).

Liu et al. (2007) presented a diagnosis methodology minimizing the layout information to locate open vias. Depending on the interpretation of the floating line voltage, one of the following equations must be satisfied:

$$\begin{aligned} V_{FL}(P) &= \frac{C_1(P)}{C_{TOT}} V_{DD} + V_{Qo} > V_{th}(P) \\ V_{FL}(P) &= \frac{C_1(P)}{C_{TOT}} V_{DD} + V_{Qo} < V_{th}(P) \end{aligned} \tag{1.13}$$

where C_1 (P) is defined as in Eq. 1.9. Considering that C_1 (P) is pattern dependent, it is possible to rearrange the previous inequalities in the following way:

$$\begin{aligned} C_{a1}(P)V_{DD} + k - V_{th}(P)C_{tot} > 0 \\ C_{a1}(P)V_{DD} + k - V_{th}(P)C_{tot} < 0 \end{aligned} \tag{1.14}$$

C_{a1} (P) is the part of C_1 (P) referring to the neighboring coupling capacitances tied to logic 1 for pattern P, and k is a pattern independent variable depending on Q_o and other known variables. These inequalities are linear. Hence, for every applied test pattern, an inequality like those in Eq. 1.14 is obtained. Then, given n test patterns, n inequalities are reported. A solver can be used to determine if these inequalities have a solution. If not, the suspected via is removed from the list.

Little research has addressed the diagnosis of *resistive* open defects since these are intrinsically included in methodologies for delay fault diagnosis. However, James and McCluskey (2005) proposed a methodology focused on the diagnosis of *resistive* opens, in particular based on the transition fault model. Transition faults are timing failures large enough to make the path delay through which the fault is propagated exceed the clock interval. Figure 1.26 shows the fourteen possible *resistive* open locations in a NAND gate. The eleven intra-gate *resistive* open defects (R_1–R_{11}) can be modeled as single-transition faults. The inter-gate *resistive* open defects (R_{12}–R_{14}) cannot be modeled as any single-transition fault. They must be

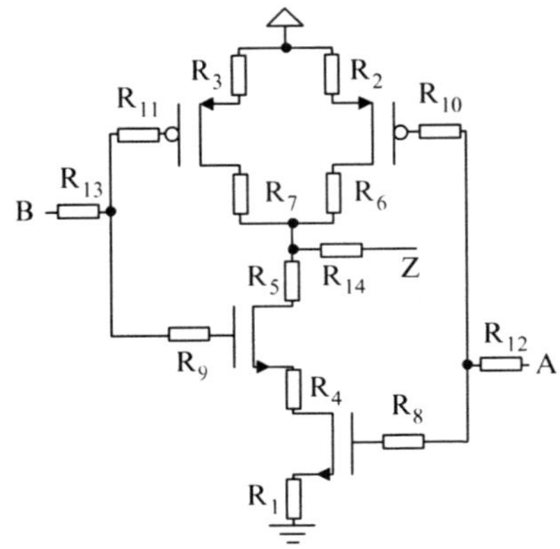

Fig. 1.26 *Resistive* open faults in a NAND gate

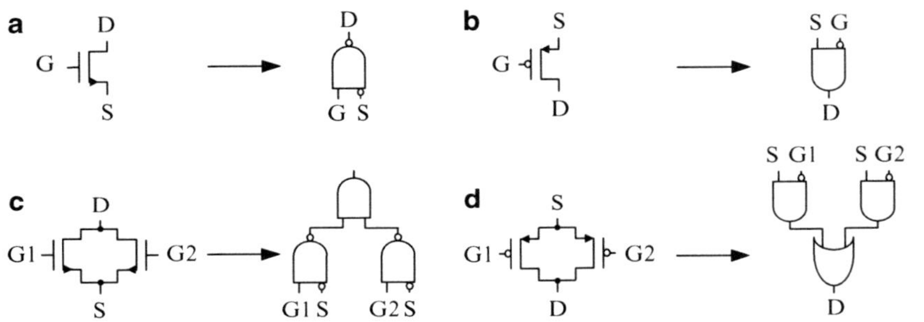

Fig. 1.27 Gate level equivalences (Fan et al. 2005). (**a**) n-transistor, (**b**) p-transistor, (**c**) parallel n-transistors, and (**d**) parallel p-transistors

modeled as the combination of two transition faults. The proposed technique uses a stuck-at fault diagnosis as a first step. In the second step, excitation condition tables are built for every gate and *resistive* open to find the fault gate input sequence.

1.4.2 Diagnosis of Intra-gate Open Defects

Stuck-open fault diagnosis has also been investigated. In the work by Li and McCluskey (2002), two tables were built for every type of gate. The first contained the gate input pair to excite stuck-open faults. The second listed input values of the gates for every test pattern applied on the tester. With this information, the possible sequence behavior of stuck-open defects was considered during diagnosis.

In the proposal by Fan et al. (2005), a transformation method was developed where transistors were replaced by a gate-level equivalent so that a stuck-open fault was represented with a stuck-at fault, as described in Fig. 1.27. For n-transistors (p-transistors), the idea was to propagate the 0 (1) voltage from the source to the drain when the transistor was on. This transformation method was the basis for a stuck-at fault diagnosis tool to be directly applied to the diagnosis of intra-gate stuck-open faults.

1.5 Summary

Open defects are very common and have been studied over a wide range of CMOS technologies. The first works were limited to stuck-open faults, which represent only a small part of the open defects that may affect CMOS devices. For this reason, extensive work was subsequently dedicated to the study of other classes of open defects, such as single-floating transistors and interconnect open defects. The latter are currently the most likely open defects to occur since interconnection structures occupy a significant part of the total area of VLSI chips.

Process variations and partial opens have an increasing impact on nanometer technologies, consequently *resistive* opens have dominated most research during the last years. The continuous CMOS scaling trend causes new failure mechanisms. Among these, the influence of leakage currents on the behavior of open defects has opened a new field of research, which is expected to contribute new techniques for test and diagnosis of these defects.

References

Aitken RC (2002) Test generation and fault modeling for stress testing. International symposium on quality electronic design, pp 95–99

Arumí D, Rodríguez-Montañés R, Figueras J (May 2005) Defective behaviours of resistive opens in interconnect lines. European test symposium, pp 28–33

Arumí D, Rodríguez-Montañés R, Figueras J (Jan 2008a) Experimental characterization of CMOS interconnect open defects. IEEE Trans Comput-Aided Des Integr Circuits Sys 27(1):123–136

Arumí D, Rodríguez-Montañés R, Figueras J, Eichenberger S, Hora C, Kruseman B (2008b) Full open defects in nanometric CMOS. VLSI test symposium, pp 119–124

Arumí D, Rodríguez-Montañés R, Figueras J (2008c) Delay caused by resistive opens in interconnecting lines, accepted for publication in Integration, the VLSI Journal, http://dx.doi.org/10.1016/j.vlsi.2008.11.001

Baker K, Gronthoud G, Lousberg M, Schanstra I, Hawkins C (1999) Defect-based delay testing of resistive vias contacts: a critical evaluation. International test conference, pp 467–476

Champac VH, Rubio A, Figueras J (1993) Analysis of the floating gate defect in CMOS. Defect and fault tolerance in VLSI systems, pp 101–108

Champac VH, Rubio A, Figueras J (Mar 1994) Electrical model of the floating gate defect in CMOS ICs: implications on I_{DDQ} testing. IEEE Trans Comput-Aided Des Integr Circuits Syst 13(3):359–369

Champac VH, Zenteno A (2000) Detectability conditions for interconnection open defects. VLSI test symposium, pp 305–311

Chang JTY, McCluskey EJ (1997) SHOrt Voltage Elevation (SHOVE) test for weak CMOS ICs. VLSI test symposium, pp 446–451

Di C, Jess JAG (1993) On accurate modelling and efficient simulation of CMOS opens. International test conference, pp 875–882

Fan X, Moore W, Hora C, Gronthoud G (2005) A novel Stuck-At Based method for transistor Stuck-Open Fault diagnosis. International Test Conference, paper 16.1

Favalli M, Dalpasso M, Olivo P (Jul 1996) Modeling and simulation of broken connections in CMOS ICs. IEEE Trans Comput-Aided Des Integr Circuits Sys 15(7):808–814

Hawkins CF, Soden JM, Righter AW, Ferguson FJ (1994) Defect classes-an overdue paradigm for CMOS IC testing. International test conference, pp 413–425

Henderson CL, Soden JM, Hawkins CF (1991) The behavior and testing implications of CMOS IC open circuits. International test conference, pp 302–310

Huang SY (2002) Diagnosis of byzantine open-segment faults. Asian test symposium, pp 248–253

Ivanov A, Rafiq S, Renovell M, Azaïs F, Bertrand Y (Jan 2001) On the detectability of CMOS floating gate transistor faults. IEEE Trans Comput-Aided Des Integr Circuits Sys 20(1)

James C-ML, McCluskey EJ (Nov 2005) Diagnosis of resistive and stuck-open defects in digital CMOS ICs. IEEE Trans Comput-Aided Des Integr Circuits Sys 24(11):1748–1759

Johnson S (1994) Residual charge on the faulty floating gate CMOS transistor. International test conference, pp 555–561

Kawahara R, Nakayama O, Kurasawa T (1996) The effectiveness of I_{DDQ} and high voltage stress for burn-in elimination. International workshop on I_{DDQ} testing, pp 9–13

Konuk H (1997) Fault simulation of interconnect opens in digital CMOS circuits. International conference on computer-aided design, pp 548–554

Konuk H, Ferguson FJ (Nov 1998) Oscillation and sequential behavior caused by opens in the routing in digital CMOS circuits. IEEE Trans Comput-Aided Des Integr Circuits Sys 17(11):1200–1210

Kruseman B, Heiligers M (2006) On test conditions for the detection of open defects. Design, automation and test in Europe, pp 896–901

Li JC-M, Tseng C-W, McCluskey EJ (2001) Testing for resistive opens and stuck opens. International test conference, pp 1049–1058

Li JC-M, McCluskey EJ (2002) Diagnosis of sequence-dependent chips. VLSI test symposium, pp 187–202

Liu JB, Veneris A, Takahashi H (2002) Incremental diagnosis of multiple open-interconnects. International test conference, pp 1085–1092

Liu C, Zou W, Reddy SM, Cheng W-T, Sharma M, Tang H (2007) Interconnect open defect diagnosis with minimal physical information. International test conference, pp 21–26

Maly W, Nag PK, Nigh P (1991) Testing oriented analysis of CMOS ICs with opens. International test conference, pp 302–310

Moore W, Gronthoud G, Baker K, Lousberg M (2000) Delay-fault testing and defects in deep sub-micron ICs – does critical resistance really mean anything? International test conference, pp 95–104

Needham W, Prunty C, Yeoh EH (1998) High volume microprocessor test escapes, an analysis of defects our tests are missing. International test conference, pp 25–34

Nigh P, Gattiker A (2004) Random and systematic defect analysis using I_{DDQ} signature analysis for understanding fails and guiding test decisions. International test conference, pp 309–318

Renovell M, Cambon G (Jan 1986) Topology dependence of floating gate faults in MOS circuits. Electron Lett 22(3):152–153

Renovell M, Cambon G (1992) Electrical analysis and modeling of floating-gate fault. IEEE Trans Comput-Aided Des Integr Circuits Sys 11(11):1450–1458

Renovell M, Comte M, Polian I, Engelke P, Becker B (2006) Analyzing the memory effect of resistive open in CMOS random logic. Design and test of integrated systems in nanoscale technology, pp 251–256

Rodríguez-Montañés R, Volf P, Pineda de Gyvez J (2002) Resistance characterization for weak open defects. IEEE Des Test Comput 19(5):18–26

Rodríguez-Montañés R, Arumí D, Figueras J, Eichenberger S, Hora C, Kruseman B, Lousberg M, Majhi AK (2007a) Diagnosis of full open defects in interconnecting lines. VLSI test symposium, pp 158–166

Rodríguez-Montañés R, Arumí D, Figueras J, Eichenberger S, Hora C, Kruseman B (Oct 2007b) Impact of gate tunneling leakage on CMOS circuits with full open defects. Electron Lett 43(21):1440–1441

Rodríguez-Montañés R, Arumí D, Figueras J, Eichenberger S, Hora C, Kruseman B (2008) Time-dependent behaviour of full open defects in interconnect lines. International test conference, pp 1–10

Sakurai T (Jan 1993) Closed-form expressions for interconnection delay, coupling, and crosstalk in VLSIs. IEEE transaction on electron devices, pp 118–124

Sato Y, Yamazaki L, Yamanaka H, Ikeda T, Takakura M (2002) A persistent diagnostic technique for unstable defects. International test conference, pp 242–249

Singh AD, Rasheed H, Weber WW (1995) I_{DDQ} testing of CMOS opens: an experimental study. International test conference, pp 479–489

Soden JM, Treece RK, Taylor MR, Hawkins CF (1989) CMOS IC stuck-open fault electrical effects and design considerations. International test conference, pp 423–430

Stamper A, McDevitt TL, Luce SL (1998) Sub-0.25-micron interconnect scaling: damascene copper versus subtractive aluminum. IEEE advanced semiconductor manufacturing conference, pp 337–346

Thompson KM (1996) Intel and the myths of test. IEEE Des Test Comput 13(1):79–81

Venkataraman S, Drummonds SB (2000) A technique for logic fault diagnosis of interconnect open defects. VLSI test symposium, pp 313–318

Wadsack RL (1978) Fault modelling and logic simulation of CMOS and MOS integrated circuits. Bell SysTech J 811(57):1449–1474

Xue H, Di C, Jess JAG (1994) Probability analysis for CMOS floating gate faults European design and test conference pp 443–448

Yan H, Singh AD (2005) A delay test to differentiate resistive interconnect faults from weak transistor defects. International conference on VLSI design, pp 47–52

Zou W, Cheng W-T, Reddy SM (2006) Interconnect open defect diagnosis with physical information. Asian test symposium, pp 203–209

Chapter 2
Models for Bridging Defects

Test and Diagnosis

Michel Renovell, Florence Azais, Joan Figueras, Rosa Rodríguez-Montañés, and Daniel Arumí

Abstract Bridging defects are responsible for a large percentage of failures in CMOS technologies and their impact in nanometer technologies with highly dense interconnect structures is expected to increase. In this chapter, a survey of the key developments in modeling bridging defects and their implications in test and diagnosis are presented. An overview of the historical developments of these models from the "wired AND/OR" and "voting" models to more realistic proposals taking into consideration the resistance values of the bridge are presented. The logic detectability of bridging defects considering the resistance of the bridge assuring its detectability is explored. The concept of Analogue Detectability Interval (ADI) as well as its applicability to increase the quality of the vectors detecting these defect classes is introduced. Quality of electronic circuits and systems requires the availability of effective diagnosis techniques. The basic concepts of logic as well as current-based (I_{DDQ}) diagnostic strategies are included in this chapter.

Keywords VLSI · Test · Diagnosis · Defect · Short · Bridging defect · CMOS · Realistic model · Analogue detectability interval

2.1 Introduction

Traditionally, test generation relies on fault models to produce tests that are expected to identify defects such as unintended bridges and opens. Test generation does not directly target defects for two main reasons. Firstly, many defects are not easy to analyze and no model exists to completely describe their behaviour. Secondly, there is a variety of possible defects in a circuit. Since available resources (like memory)

M. Renovell (✉) and F. Azais
LIRMM-CNRS, 161 rue ada, 34392 Montpellier, France
e-mail: renovell@lirmm.fr, azais@lirmm.fr

J. Figueras, R. Rodríguez-Montañés, and D. Arumí
Universitat Politècnica de Catalunya (UPC), Electronic Engineering Dpt. ETSEIB, Diagonal 647, 08028 Barcelona, Spain
e-mail: figueras@eel.upc.es, rosa@eel.upc.edu, arumi@eel.upc.edu

H.-J. Wunderlich (ed.), *Models in Hardware Testing: Lecture Notes of the Forum in Honor of Christian Landrault*, Frontiers in Electronic Testing 43,
DOI 10.1007/978-90-481-3282-9_2,

and CPU time limit test generation and application, generating tests for all defects is unfeasible. Consequently, a relatively small set of abstract defects, namely faults, is constructed and these faults are targeted to generate tests. With this approach, the test quality relies on detection of non-directly targeted defects.

In this chapter, we consider a very frequent defect encountered in today technologies: the undesired electrical connection between adjacent lines known as *bridging defect*. The advent of nanometer technologies with extensive interconnect structures contributes to the need of effective models for this defect.

Defects originated in the manufacturing process must be tested and detected before the chip is mounted in the application. In many cases as for example systematic defects, the defect has not only to be detected but also to be localized and identified. For this reason, test but also diagnosis are two major factors of the final product quality. Consequently, diagnosis techniques targeting bridging defects are also presented in this chapter.

2.2 Previous Work

The limitations of the Stuck-At fault model to adequately predict the behaviour of frequent realistic defects such as *bridges* and *opens* have triggered active research in the area since the 1970s. In this section, we analyze the evolution of the models for *bridges* highlighting some of the key contributions.

2.2.1 Wired-AND and Wired-OR Models

This popular model assumes a logic value at the defective bridged nodes generated by the AND or the OR function of the bridged nodes. A pioneering work on wired bridging fault models was reported by Mei (1974). These bridging fault models are known as the wired-AND and the wired-OR bridging fault models. In a bridging fault, each signal net tries to drive the bridged nets to a value equal to the logic value in the fault-free circuit. The wired-AND and the wired-OR fault models assume that the values on the bridged nets are both the same (zero bridge resistance) and are the result of an AND or an OR operation between the logic values of the nets, respectively.

Figure 2.1 shows an example of a bridge between the outputs of two NAND gates and their equivalent wired-AND and wired-OR fault models. On the one hand, the wired-AND fault model assumes that the nMOS transistor networks logically win and drive the bridged nets when they are excited. On the other hand, the wired-OR fault model considers that the pMOS transistor networks logically win when they are excited. These fault models, although widely used in the past, do not reflect the behaviour of bridging faults in CMOS technologies. The voltage on the bridged nets is not always logic 0 or logic 1, as the wired-AND and the wired-OR fault

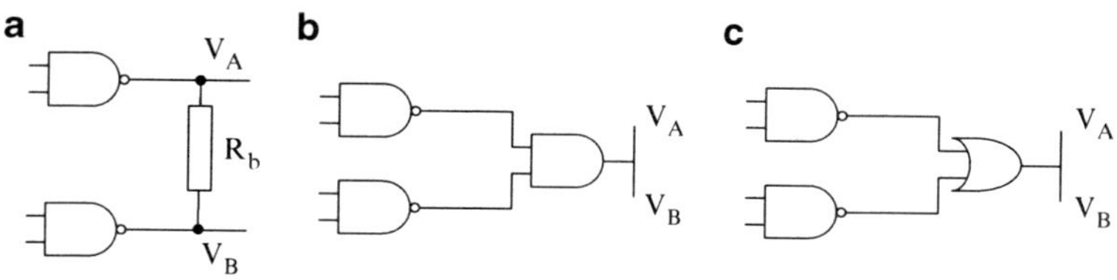

Fig. 2.1 Two NAND gates. (**a**) Bridging fault, (**b**) wired-AND, and (**c**) wired-OR

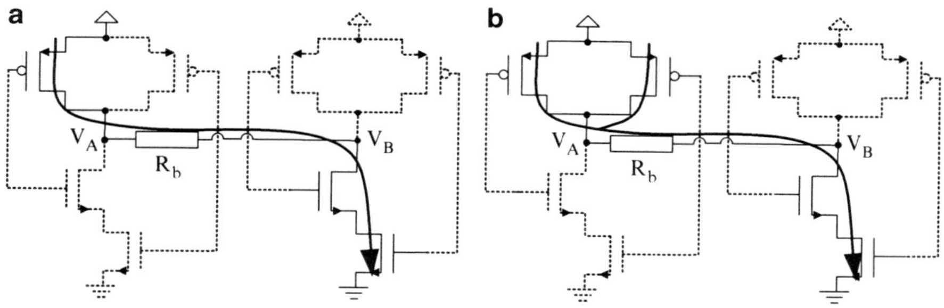

Fig. 2.2 Transistor description of a bridging fault between two NAND gates. (**a**) One pMOS transistor on and (**b**) both pMOS transistors on

models assume. These fault models are more suitable for technologies where one of the logic levels is clearly stronger than the other one. However, the wired-AND and wired-OR fault models are the easiest for simulation, pattern generation and diagnosis purposes.

2.2.2 *The Voting Model and Other Zero Resistance Defect Models*

A refinement of the wired-AND and the wired-OR fault models was subsequently presented by Acken and Millman (1991), *the voting model*. When the bridged nets are set to opposite logic values, the voting model considers the resultant circuit as a resistive divider between V_{DD} and RV_{GND}. In CMOS circuits, the electrical resistance to V_{DD} comes from a combination of conducting pMOS transistors, whereas the resistance to V_{GND} comes from a combination of conducting nMOS transistors. The voting model also assumes that the bridge resistance is negligible. The evaluation of the two networks strengths determines whether the net is considered as logic 1 or logic 0. Nevertheless, this fault model does not determine the actual values on the bridged nets.

Considering the transistor description of the bridge between the outputs of two 2-input NAND gates illustrated in Fig. 2.2, the voting model differentiates between the strengths of the pMOS networks depending on the number of conducting pMOS transistors to determine the logic interpretation of the bridged nets. The voting model evaluates the relative strengths of the different networks by means of SPICE simulations, which are stored in tables. During fault simulation, this information is accessible and no SPICE simulations are required. In a first approach, the model assumes that all the downstream gates have the same threshold. However, the same authors refined this fact in Acken and Millman (1992). The limitation of this fault model is that if any new logic element has a threshold voltage outside the range used to generate the tables, new simulations are required. Furthermore, results when the strengths of the pMOS and nMOS networks are similar are not accurate. To overcome the limitations of the voting model, an improved fault model was proposed by Maxwell and Aitken (1993) the biased voting model. In this case, the threshold voltage is not considered fixed. The biased voting model is able to calculate the voltage values of the bridged nets by means of an iterative procedure.

The biased voting model allows to perform fault simulation at the logic level using electrical information obtained through an electrical pre-characterization of the library. However, the fault simulation is considerably slowed down by the iterative procedure requested to precisely compute the effective voltage value Vx. The objective of the direct-voting model proposed later by Renovell et al. (1994a) is to keep the same accuracy as the biased-voting model while removing the iterative procedure. The direct-voting model allows to get the voltage values Vx directly without any computation or iteration.

The fundamental concept of the direct voting model is very simple and comes from a very simple observation:

1. Assuming a bridging defect between two logic nodes set to opposite logic values, the resulting intermediate voltage Vx could be computed as a function of the topological parameters of the p-transistors (W_p, L_p) and n-transistors (W_n, L_n), plus the technological parameters such as C_{ox}, μ_n, μ_p, V_{Tn}, V_{Tp}...
2. For the different possible bridging defects in a given circuit, all the technological parameters are the same.

As a consequence, for a given circuit with known technological parameters, the resulting intermediate voltage Vx of a bridging defect is only a function of the topological parameters of the bridged p- and n- ON transistors and, more precisely, of the so-called configuration ratio β:

$$Vx = f(\beta) \text{ with } \beta = \left(W_p/L_p\right)/(W_n/L_n)$$

The authors proposed to perform a pre-characterization of the library by drawing the Vx versus β characteristics using SPICE simulations. Note that these characteristics are constructed and memorized for different parallel and serial networks of transistors. During fault simulation, the fault simulator knowing the topological parameters and so knowing the configuration ratio of the bridged gates, can directly deduce from the above characteristics the voltage value Vx without any computation or iteration.

Other works tried to develop more accurate fault models. Rearick and Patel (1993) presented a fault model where the use of SPICE-derived data for every input of the bridged gates is taken into account. A different approach was presented by Di and Jess (1993), who proposed a method based on Faulty Boolean Expressions in order to calculate the voltage of the bridged nodes. For that purpose, a simplified transistor model is used.

2.2.3 Non-zero Resistance Defect Models

So far, none of the models had taken the value of the bridge resistance into consideration. All of them assumed it negligible. Although most of bridging defects have low resistance values, a non negligible percentage of bridges has high resistance (Rodríguez-Montañés et al. 1992).

Interesting models took the bridge resistance into account, e.g., the ones presented by Rodríguez-Montañés et al. (1990) and Renovell et al. (1994b). For this new model, the objective is no longer the computation of the intermediate voltages resulting from the bridging defect. Indeed, the resistance of the bridge is a parameter of the defect that cannot be predicted. Now, the voltage value of the bridged nodes depends on the random resistance of the bridge. The basic concept of this new model is the evaluation of the detectable resistance range as illustrated hereafter.

In presence of a zero bridge resistance, both nets have the same voltage value and the circuit exhibits faulty logic behaviour. However, as the bridge resistance increases, the voltage of the bridged nets gets closer to the defect-free value so that for high resistance values, the circuit operates properly. In this way, there is a critical resistance value (R_C) above which the circuit does not show faulty logic behaviour (Rodríguez-Montañés et al. 1990, 1991; Renovell et al. 1995). This behaviour is illustrated in Fig. 2.3. Suppose that the bridge in Fig. 2.3a is excited in such a way that V_A is set to logic 1 in the defect-free case, whereas V_B is set to logic 0. The plot in Fig. 2.3b represents the voltage of the bridged nets as a function of the bridge resistance. For a zero resistance bridge, both V_A and V_B have the same value. However, as R_b increases, V_A increases and V_B decreases, to the point that R_b is so high

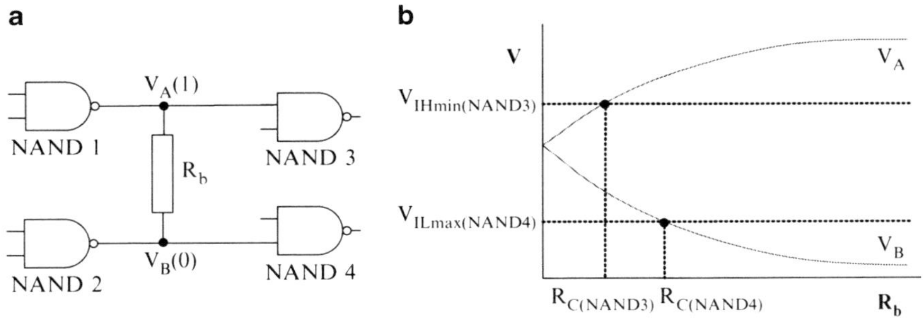

Fig. 2.3 Resistive bridge between two NAND gates. (**a**) Gate level and (**b**) V-R_b characteristics

that V_A is properly interpreted by *NAND3* ($R_{C(NAND3)}$), and for a higher resistance V_B is also properly interpreted by *NAND4* ($R_{C(NAND4)}$). Therefore:

- When $R_b < R_{C(NAND3)}$, logic errors are propagated through both *NAND3* and *NAND4*.
- When $R_{C(NAND3)} < R_b < R_{C(NAND4)}$, logic errors are propagated through *NAND4*.
- When $R_b > R_{C(NAND4)}$, the circuit does not show faulty logic behaviour.

For a given resistive bridging defect, the proposed model allows to easily compute during fault simulation, the different critical resistances ($R_{C(NAND3)}$, $R_{C(NAND4)}$) which, in fact, define the range of detectable resistance associated to the defect. This information is used during fault simulation to evaluate some quality metrics of the test vectors. The first model proposed in 1996 used simplified transistor equations. In a more recent work carried out by Polian et al. (2005), the critical resistance was calculated based on more accurate transistor models: the Fitted Model, which uses equations with free variables that are fitted in order to match actual SPICE data, and the Predictive Model, which is fully analytical and employs BSIM4 equations.

Finally, an analysis of the behaviour of bridging defects was presented by Sar-Dessai and Walker (1999). This work analysed five different bridging fault configurations, namely: a bridging fault between two primary inputs, between a primary input and a gate output, between two gate outputs, between two gates outputs driving the same gate and between two primary outputs. Based on the model for these five configurations, look-up tables can be constructed, where the information about the voltage on the bridged nets is stored for every vector. The detectable resistance interval and the propagating path are also taken into account. Furthermore, it is also determined whether the bridging fault is detectable at the driven gate outputs based on their logic thresholds.

2.2.4 Modeling Feedback Bridging Faults

A feedback bridging fault is a bridging fault such that both involved nets lie on the same path in the circuit (Mei 1974). The voltage value of one bridged net may depend on the value of the other bridged net. The bridged net with lower topological ordering is usually called the back net, while the other one is called the front net. The analysis of feedback bridging faults is complex. They can induce sequential behaviour in combinational circuits, depending whether the path is sensitized or not and depending also on the topological situation of the bridge. Thus, three different cases may appear (Rajsuman 1991; Koch and Muller-Glaser 1993; Chess and Larrabee 1998; Dahlgren 1988):

1. The logic path is not sensitized.
2. The logic path is sensitized and the feedback loop has an even number of inversions.
3. The logic path is sensitized and the feedback loop has an odd number of inversions.

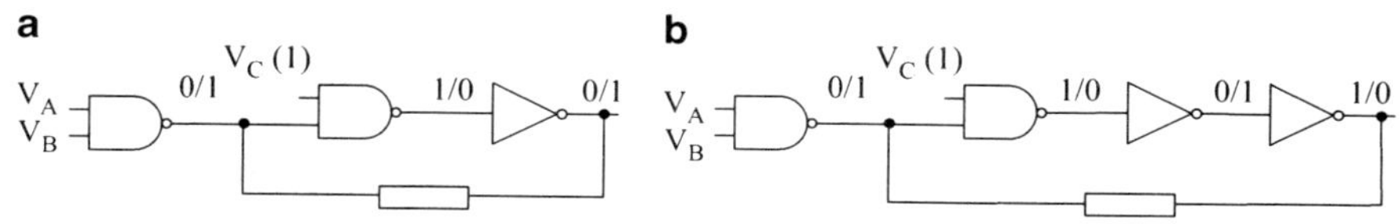

Fig. 2.4 Feedback bridge. (**a**) Even number of inversions and (**b**) odd number of inversions

When the logic path is not sensitized, it is equivalent to a non-feedback bridging fault. The logic value of the back net is independent of the logic value of the front net. Considering the examples shown in Fig. 2.4, this is accomplished as long as V_C is set to logic 0.

If the logic path is sensitized and the feedback loop has an even number of inversions, both nets have the same logic value. An example is illustrated in Fig. 2.4a provided that V_C is set to logic 1. This case is redundant as long as the back net is stronger than the front net, otherwise a circuit with asynchronous memory behaviour appears. It can be described as a latched state. The voltage on the bridged nets depends on the transistor strengths and the bridge resistance. The detectability of such fault cases relies on the sequence of test patterns applied.

Finally, if the logic path is sensitized with an odd number of inversions, the logic values of the bridged nets are opposite on a fault-free circuit (see Fig. 2.4b). Two different behaviours may appear depending on the gate strengths. If the back gate is stronger than the front gate, it behaves as a non-feedback bridging fault. However, if the front gate is stronger, the defect may cause oscillation in the circuit. The oscillation period is related to the delay of the logic connecting the bridged nodes and it is usually lower than the clock period.

The impact of the bridge resistance in feedback bridges is not a trivial issue, since it turns out to be computationally complex (Polian et al. 2003). However, bridge resistances with high values usually result in fewer situations of active feedback because the dominance conditions of the front net are less likely to be accomplished.

2.2.5 *Resistance Characterization of Bridging Defects*

For a better knowledge of the defect behaviour, early works have analyzed and characterized real bridges demonstrating that bridging defects have resistances with different values which can be modelled with a statistical distribution for each technology node.

Traditionally, conventional test monitors such as the comb-string-comb structure (Bruls et al. 1991) have been used to characterize the resistive nature of bridging defects and open defects, both of which are the main contributors to yield loss in wiring structures. This test monitor basically consists in a long string wire (meander-shaped) as shown in Fig. 2.5 (from pad S_1 to pad S_2) lying between two combs (C_1 and C_2). The string and the two combs are made up of the targeted layer of the

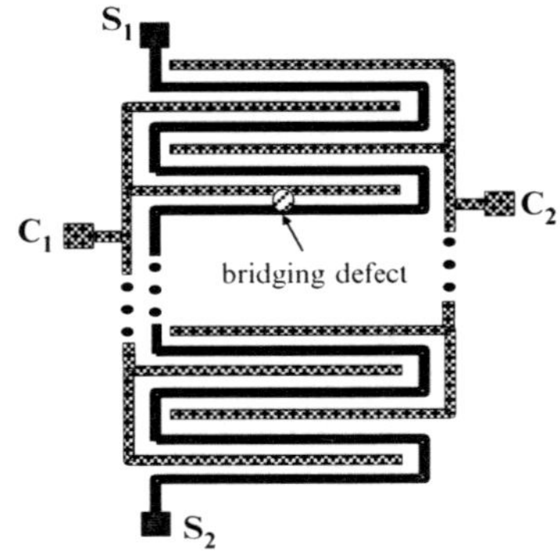

Fig. 2.5 Schematic representation of a basic test comb-string-comb structure. A bridging defect has been included between comb C_1 and the string

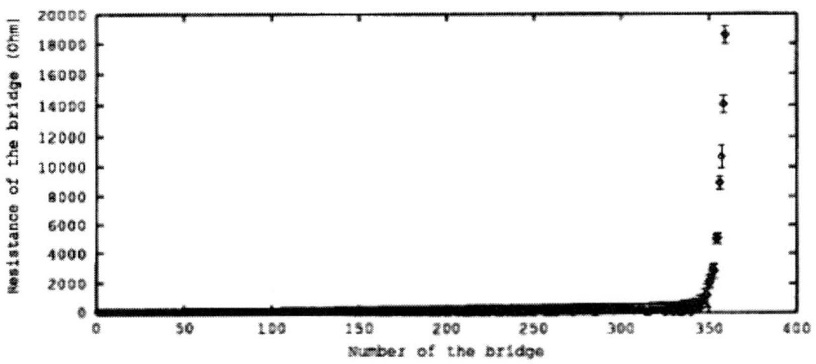

Fig. 2.6 Measurement of single bridges with the corresponding uncertainty interval (Rodríguez-Montañés et al. 1992)

manufacturing process. The length L of the wire follows from the line/space pitch and the test structure area, which is chosen such that the required defect density resolution is obtained at wafer or lot level.

The test monitor permits the identification of single bridges making a connection between the string and one of the combs (Fig. 2.5 illustrates the example of a bridging defect between the string and comb C_1). The resistive nature of the bridging defect is easily characterized (Rodríguez-Montañés et al. 1992, 1996) with the measurements of the resistance between each comb and the two end points of the string, provided the total resistance of the string is known.

Rodríguez-Montañés et al. (1996) analyzed 400 defective monitors made of metal 1 from different batches and production lines and found the resistance distribution shown in Fig. 2.6. More detailed information about the resistive values and their uncertainty intervals is shown in Tables 2.1 and 2.2.

From the above results, the majority of the bridging defects (64.5%) has a resistance below 500 Ω, even considering the worst case error analysis. On the other

Table 2.1 Bridging defects distribution (Rodríguez-Montañés et al. 1996)

Guaranteed range (KΩ)	Total number of bridges
Rb ≤ 0.5	258 (64.5%)
Rb ≤ 1	379 (94.8%)
Rb ≤ 5	394 (98.5%)
Rb ≤ 10	397 (99.3%)
Rb ≤ 20	400 (100%)

Table 2.2 High resistance bridging defects (Rodríguez-Montañés et al. 1996)

Guaranteed range (KΩ)	Total number of bridges
Rb ≥ 0.5	14 (3.5%)
Rb ≥ 1	12 (3.0%)
Rb ≥ 5	4 (1.0%)
Rb ≥ 10	2 (0.5%)
Rb ≥ 20	0 (0%)

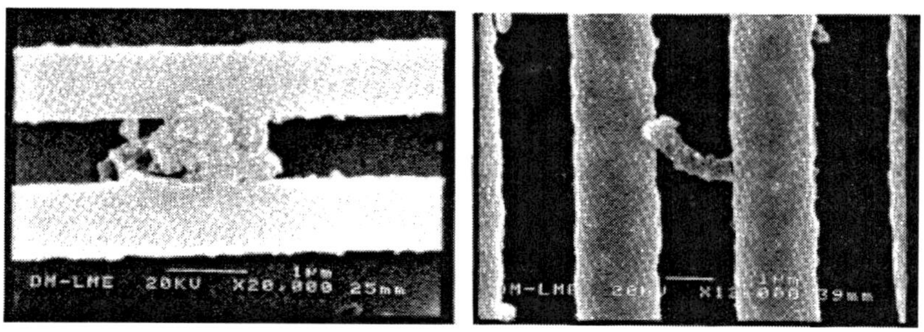

Fig. 2.7 (**a**) Low resistive and (**b**) high resistive bridging defect (Rodríguez-Montañés et al. 1996)

side, 3.5% of the bridges have a resistance above 500 Ω. The maximum resistance value was found to be around 20 kΩ. Two pictures of a low resistive and a high resistive defect are shown in Fig. 2.7.

2.3 Detectability of Bridging Defects

As the quality demands increase, the effectiveness of test generation without any defect consideration becomes questionable. High quality test generation requires a better knowledge of defect behaviour. As a matter of fact, the analysis of defect behaviour is a quite difficult task. One of the main difficulties comes from the presence in the defect of random value parameters such as the bridging resistance preventing any prediction of the defect behaviour. The mechanisms of defect appearance are obviously not controlled, resulting in electrical situations with unknown parameters. The question is how to predict the voltage created by a bridge when the value of the bridge resistance is not known a priori. The classical assumptions such as

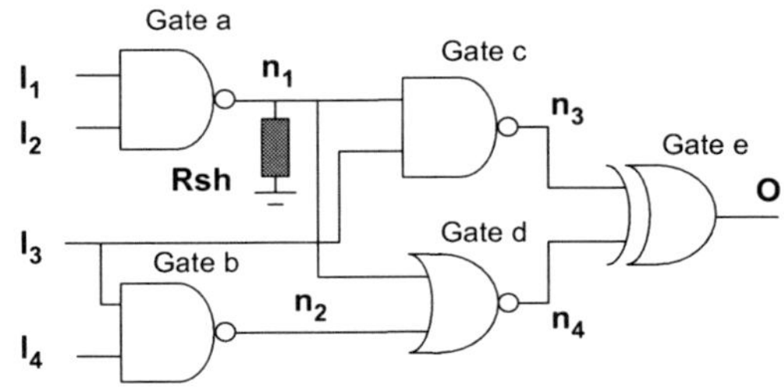

Fig. 2.8 Didactic defective circuit

zero-resistance bridge can no longer be used and a realistic analysis of the defect behaviour is required. Besides, a realistic model of defect behaviour must incorporate the unpredictable parameters.

2.3.1 Impact of the Resistance on the Defect Behaviour

To illustrate the impact of the bridging resistance on the defect behaviour and subsequently on its detecting conditions, let us consider as an illustrative example the small defective circuit given in Fig. 2.8. Note we prefer to use here the term defective (in relation with defect) than faulty (in relation with fault). The circuit is composed of five logic gates and comprises four primary inputs (I_1 to I_4) and one primary output (O). The bridging defect under consideration corresponds to the logical node n1 bridged to ground through an Rsh resistance. Of course, the value of the intrinsic bridge resistance Rsh is not known a priori.

Considering a classic Boolean test technique, the detection of this bridge requires both defect excitation and propagation of its effect to a primary output. Regarding excitation, a bridge-to-ground is excited by any input vector trying to set the bridged node to logic '1'. The bridged node n_1 depends on the primary inputs $I_1 I_2$ through the NAND gate 'a'; the defect excitation is therefore guaranteed by any of the following 12 vectors: $I_1 I_2 I_3 I_4 = 00XX, 01XX$ or 10XX. The defective value due to the bridge has then to be propagated through the succeeding logic gates. The bridged node n_1 is connected to the input of the NAND gate 'c' and to the input of the NOR gate 'd'; consequently, the effect can be propagated through the NAND gate 'c' setting its side input to a logic '1' ($I_1 I_2 I_3 I_4 = XX1X$) or through the NOR gate 'd' setting its side input to a logic '0' ($I_1 I_2 I_3 I_4 = XX11$). Note that propagation through gate 'd' necessarily implies propagation through gate 'c'. Table 2.3 summarizes the defect excitation and propagation characteristics associated with each one of the 2^4 possible input vectors. It appears that six vectors allow both defect excitation and effect propagation: vectors #2, #3, #6, #7, #10 and #11 (grey arrays in Table 2.3).

When considering classical faults such as stuck-at or zero-resistance bridging faults, the two conditions of fault excitation and fault propagation completely determine fault detection. This is not the case when realistic bridging defects with

Table 2.3 Defect excitation and propagation (Renovell et al. 1999)

#	$I_1I_2I_3I_4$	Excitation	Propagation
0	0000	Y	N
1	0001	Y	N
2	0010	Y	Y
3	0011	Y	Y
4	0100	Y	N
5	0101	Y	N
6	0110	Y	Y
7	0111	Y	Y
8	1000	Y	N
9	1001	Y	N
10	1010	Y	Y
11	1011	Y	Y
12	1100	N	N
13	1101	N	N
14	1110	N	Y
15	1111	N	Y

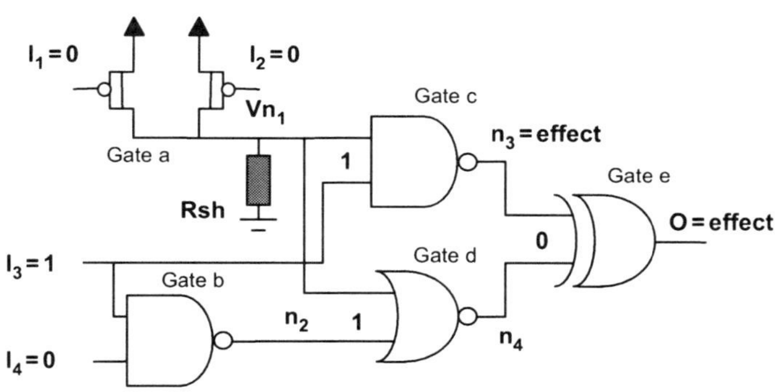

Fig. 2.9 Effect of the defect (Renovell et al. 1999)

a bridge resistance are under consideration. Indeed, an excited defect can produce either a defective effect or a defect-free effect depending on the bridge resistance value Rsh, which is evidently an unpredictable parameter. For this reason, we refer to "effect" propagation.

To illustrate this point, let us consider the circuit of Fig. 2.8 with vector #2 on its inputs. This vector guarantees both defect excitation and effect propagation. In Fig. 2.9, only the 'ON' transistors of the NAND gate 'a' are represented.

Vector #2 tries to set node n_1 to a logic '1'. Under such conditions, a conducting path is created from Vdd to Gnd through node n_1 including the resistance Rsh of the defect. An intermediate voltage Vn_1 appears on node n_1. The logic interpretation of this intermediate voltage depends on the logic threshold Vth_c of the driven NAND gate 'c':

- Vn_1 is interpreted as a logic '0' if it is smaller than the gate 'c' logic threshold Vth_c. A *defective effect* appears on the output of gate 'c' and is propagated to the primary output of the circuit.
- Vn_1 is interpreted as a logic '1' if it is higher than the gate 'c' logic threshold Vth_c. A *defect-free effect* appears on the output of gate 'c' and is propagated to the primary output of the circuit.

This small example highlights that the bridge defect may be detected or not by a given test vector depending on the value of the intermediate potential Vn_1. This potential depends on the technological and topological transistor parameters of Gate 'a' (C_{ox}, V_T, μ_P, W_P, L_P...) but also on the value of the bridge resistance Rsh. As the bridge resistance value is obviously an unpredictable parameter, it is not possible to compute the intermediate voltage Vn_1. Consequently it is not possible to predict the effect of the defect.

However, it is possible to analyze the impact of the bridge resistance value on the effect of the defect. Figure 2.10 shows the Vn_1 vs. Rsh characteristics obtained through electrical simulations. Two distinct regions can be distinguished according to the Rsh value. In the high resistance range, the bridging defect does not modify the logical behaviour and the circuit operates as a defect-free circuit. In the low resistance range, the bridging defect induces a Vn_1 value smaller than the logic threshold Vth_c of the driven gate 'c' and this value is logically interpreted as a defective '0'. The frontier between these two regions corresponds to a *critical resistance* value R_C. It is worth noting that although the bridge resistance value is an unpredictable

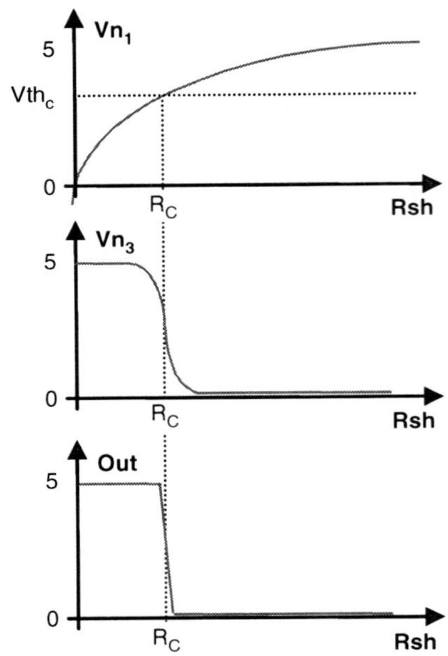

Fig. 2.10 Analogue detectability interval (Renovell et al. 1999)

parameter, the critical resistance value R_C associated with a given test vector is a deterministic parameter that depends on the technological and topological transistor parameters.

So in this very simple example, a faulty logic value appears on the output of the driven gate if the unpredictable parameter Rsh is smaller than the critical resistance R_C. Consequently even if it is not possible to guarantee the detection of the bridge, it is demonstrated that the bridge defect is detected by vector #2 if the unpredictable parameter Rsh falls into the interval $[0, R_C]$. This interval is called the '*Analogue Detectability Interval*' (ADI) associated to vector #2.

2.3.2 Logic Detectability Techniques

The analysis performed in the previous section has established that a bridging defect may result in a defective or defect-free effect depending on the value of the unpredictable bridge resistance. The concept of Analogue Detectability Interval has then been introduced to represent the value of the bridge resistance creating a defective effect. It is now interesting to study the dependence between ADIs and test vectors. Indeed in the previous example of Section 2.3.1, the ADI $[0, R_C]$ associated to vector #2 is defined by the value of the critical resistance R_C. This critical resistance corresponds to the intersection of the Vn_1 characteristics with the logic threshold Vth_c of the driven gate. Consequently, the value of the critical resistance depends on both the shape of the Vn_1 vs. Rsh characteristics (defect excitation) and the location of the logic threshold Vth_c (defect propagation).

2.3.2.1 Defect Excitation

The Vn_1 vs. Rsh characteristic depends on the electrical parameters of the transistor(s) driving the bridged node. These driving transistors are fully determined by the input vector. As an example, the vector #2 in Fig. 2.9 turns 'ON' the two p-transistors of the driving NAND gate 'a' (with $I_1 I_2 = 00$), while the vector #6 turns 'ON' only one p-transistor (with $I_1 I_2 = 01$). As a consequence, excitation with vectors #2 and #6 results in different values of the critical resistance, as illustrated in Fig. 2.11. Two different ADIs are therefore defined: $[0, R_C^1]$ corresponding to vector #2 and $[0, R_C^2]$ corresponding to vector #6. This simple example demonstrates that for a given defect, different ADIs exist depending on the excitation defined by the input vector.

2.3.2.2 Defect Propagation

The Analogue Detectability Interval also depends on the logic threshold Vth of the driven gate(s). Here again, the gates that propagate the effect of the defect are fully

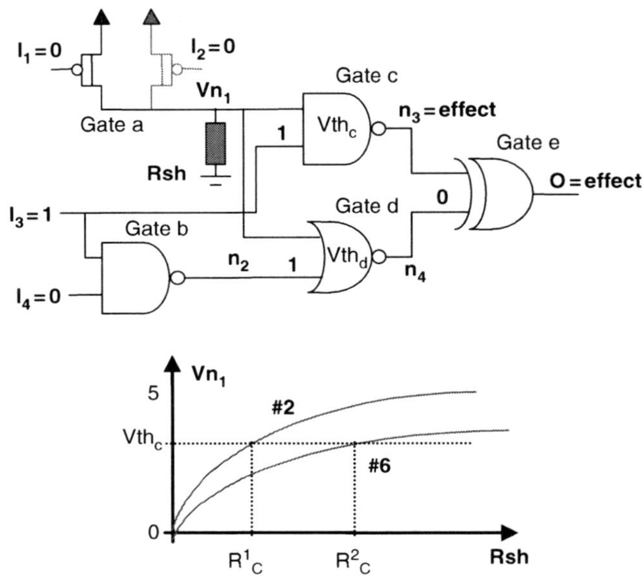

Fig. 2.11 Different excitations (Renovell et al. 1999)

determined by the input vector. As an example, the vector #2 in Fig. 2.9 propagates the effect through gate 'c' (with $I_3I_4 = 10$), while the vector #3 in Fig. 2.12 propagates the effect through both gates 'c' and 'd' (with $I_3I_4 = 11$). Different logic gates have different logic thresholds. The value of the critical resistance is therefore different through gate 'c' and through gate 'd'.

A very important point for propagation concerns the re-convergence of effects. Considering vector #2, the effect of the defect is propagated through a single gate (gate 'c') implying no re-convergence: the ADI is equal to $[0, R^1_C]$. Considering now vector #3 in Fig. 2.12, the effect of the defect is propagated through two different gates (gates 'c' and 'd'). The two effects re-converge on gate 'e'. According to the unpredictable resistance value, a defect-free or a defective effect may be propagated, as illustrated in Fig. 2.13:

- Between 0 and R^1_C, the two defective effects cancel each other producing a defect-free value on the primary output.
- Between R^1_C and R^3_C, a defective and defect-free effects re-converge producing a defective value on the primary output.
- Between R^3_C and infinity, the two defect-free effects obviously produce a defect-free effect on the primary output.

The Analogue Detectability Interval corresponding to vector #3 is consequently equal to $[R^1_C, R^3_C]$. This small example illustrates that the propagation of the effect has to be taken into account when defining the ADI.

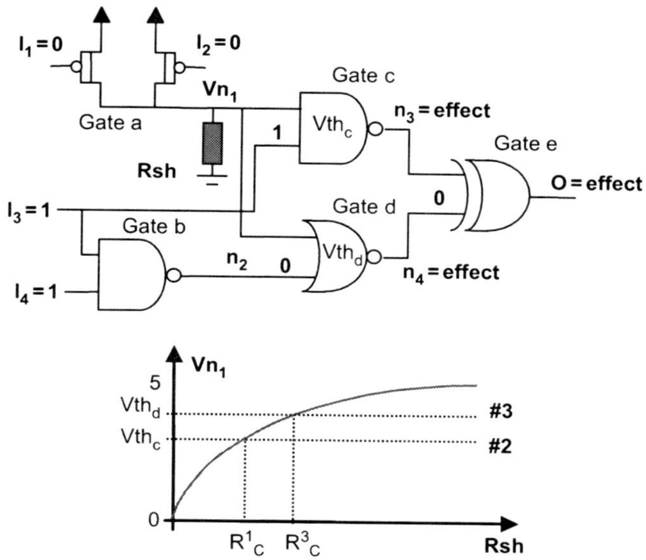

Fig. 2.12 Different propagations (Renovell et al. 1999)

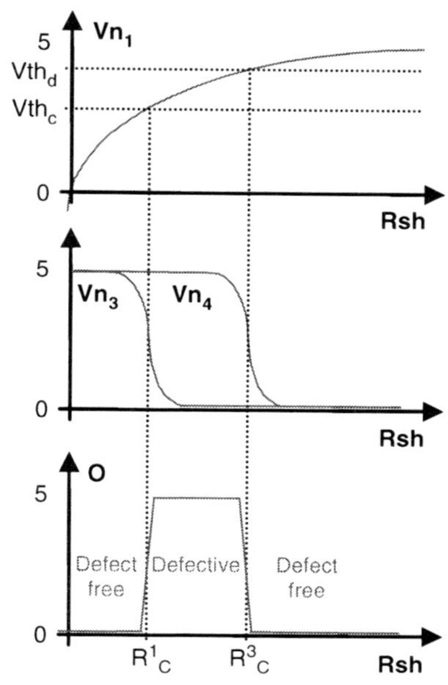

Fig. 2.13 Re-convergence of effects during defect propagation (Renovell et al. 1999)

This small example also points out that classical fault propagation procedures used in ATPG are inappropriate to handle non-zero resistance bridging defects. Indeed, a classical ATPG would allow fault propagation only through gate 'c' as double propagation through gates 'c' and 'd' will cause the faulty values to re-converge on the XOR gate 'e', thus cancelling each other. In other words, a classical ATPG would not generate the vector #3 to detect a stuck-at-0 fault on node n_1, whereas this vector allows the detection of a bridge-to-ground defect on node n_1 provided that the value of the bridge resistance falls into the interval $[R^1_C, R^3_C]$.

2.3.2.3 Generalization

From the analysis conducted on the small illustrative example of the preceding sections, general comments can be drawn. An essential concept when dealing with non-zero resistance bridging defects is the concept of critical resistance. Indeed although the bridge resistance value is an unpredictable parameter that is not known a priori, the critical resistance is a deterministic value that can be easily computed for a given defect and a given test vector. Analogue Detectability Intervals can then be calculated for different test vectors defining the range of detectable bridge resistance values associated to the defect.

An important point is that Analogue Detectability Intervals are pattern dependent: a specific ADI can be associated to each input vector taking into account both the defect excitation and the effect propagation. As an illustration, Table 2.4 gives the exhaustive list of Analogue Detectability Intervals associated to each input vector for the small defective circuit of Fig. 2.8. Note that an empty interval is associated to the vectors that do not excite or/and propagate the defect.

Despite its unpredictable parameter, defect detection can be optimized taking into account the concept of ADIs. From the ADIs given in Table 2.4, it appears that six vectors are potentially able to excite the defect and propagate its effect to

Table 2.4 Associated ADIs (Renovell et al. 1999)

#	$I_1I_2I_3I_4$	Associated ADIs
0	0000	Ø
1	0001	Ø
2	0010	$[0, R^1_C]$
3	0011	$[R^1_C, R^3_C]$
4	0100	Ø
5	0101	Ø
6	0110	$[0, R^2_C]$
7	0111	$[R^2_C, R^4_C]$
8	1000	Ø
9	1001	Ø
10	1010	$[0, R^2_C]$
11	1011	$[R^2_C, R^4_C]$
12	1100	Ø
13	1101	Ø
14	1110	Ø
15	1111	Ø

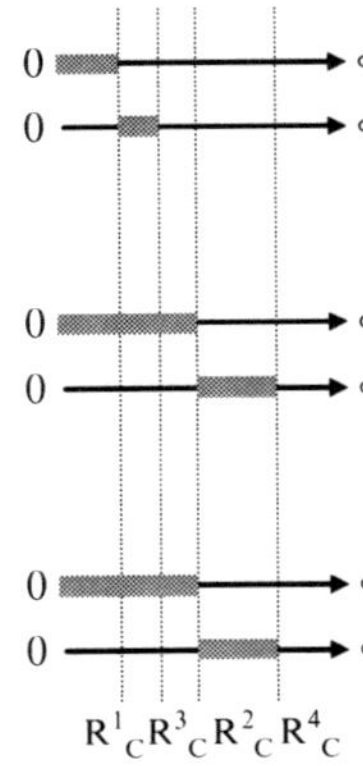

the primary output. However, all these vectors are not equivalent in terms of defect detection. Several points can be discussed.

First, it should be noted that some vectors have larger ADIs than others. For instance, the ADI associated to vector #6 is larger than the ADI associated to vector #7. This means that vector #6 covers a larger range of detectable bridge resistance value than vector #7, i.e. vector #6 is more efficient than vector #7 in terms of defect detection domain.

The second point that should be noted is that some vectors have ADIs contained in the ADI of another vector, whereas others have ADIs that cover different ranges. For instance, the ADI associated to vector #2 is contained in the ADI associated to vector #6, whereas the ADIs associated to vector #2 and #7 are fully disjoint. Consequently regarding defect detection, it is completely useless to use both vectors #2 and #6 while the use of both vectors #2 and #7 permits to cover a larger range of detectable bridge resistance value. In other words, using several vectors may permit to enlarge the defect detection domain but these vector have to be adequately selected.

Finally, the last point that should be highlighted in the example of Table 2.4 is that it exists a domain for the bridge resistance value that is not covered by any vector: $[R^4_C, \infty]$. Obviously, such a domain must not be considered from the point of view of the optimization process.

All these points can be generalized and formalized by introducing the concepts of '*Global-ADI*' and '*Covered-ADI*'.

Definition 2.1. Given a circuit under test and the list of Analogue Detectability Intervals ADI^V associated to each possible input vector V for a considered defect, the Global Analogue Detectability Interval G-ADI is given by the union of all ADIs:

$$G - ADI = \bigcup ADI^V$$

The Global ADI represents the complete domain of the unpredictable parameter for which the defect can be detected considering the given test technique. On the example, the G-ADI represents the complete domain of the bridge resistance that can be detected by the input vectors using the static voltage test technique. This Global ADI is equal to G-ADI $= [0, R^4_C]$. If the bridge resistance of the defect falls into the G-ADI, then it exists at least one input vector able to detect the defect.

In opposition, if the bridge resistance of the defect falls out of the G-ADI, there is no input vector able to detect this defect. In that case, the defect can be deemed as a *redundant defect* for the test technique under consideration. This concept of redundancy must be carefully considered as it differs from the usual concept of redundancy. The classical concept of a redundant fault refers to a fault that cannot be excited and/or propagated for a given test technique. In case of a bridging defect, two situations may arise. The first situation is similar to the classical concept used for redundant faults: the defect cannot be excited and/or propagated whatever the input vector, which means that the global-ADI is empty for this particular defect.

The defect is therefore redundant whatever the value of the bridge resistance. In the second situation, it exists some vectors able to excite and propagate the defect. The defect redundancy therefore depends on the value of its bridge resistance. If the bridge resistance falls within the Global-ADI, the defect is detectable since it exists at least one input vector to propagate the defective value to a primary output. In contrast, if the bridge resistance falls out the Global-ADI, the defect is redundant since a defect-free value is propagated to the primary outputs whatever the input vector. It is clear that in both situations, redundant defects cannot be detected and thus are not in the optimization focus.

The actual objective of the optimization process is detectable defects, i.e. defects with unpredictable parameters falling into the Global-ADI. In other words, the objective of the optimization is to cover the Global-ADI. A given test sequence may cover or not the Global-ADI. Therefore the concept of '*Covered-ADI*' related to a test sequence can be introduced in the following way:

Definition 2.2. Given a circuit under test and the list of Analogue Detectability Intervals ADI^V associated to each vector V for a considered defect, the Covered Analogue Detectability Interval C-ADI related to a test sequence is given by the union of the ADIs associated with all the vectors of the test sequence V^T :

$$C - ADI = \bigcup ADI^{V_T}$$

As an illustration, let us assume a test sequence including 3 vectors: #2, #3, and #13 of Table 2.4. The Covered-ADI for this sequence is:

$$C - ADI = \left[0, R^1{}_C\right] \cup \left[R^1{}_C, R^3{}_C\right] \cup \emptyset = \left[0, R^3{}_C\right]$$

It can be observed that this test sequence does not cover the Global-ADI $[0, R^4{}_C]$. Three possible situations exist according to the unpredictable value of the bridge defect resistance:

- The unpredictable value of the bridge defect resistance falls into C-ADI; the sequence will detect the defect.
- The unpredictable value of the bridge defect resistance falls into G-ADI but out of C-ADI; the sequence will not detect the defect.
- The unpredictable value of the bridge defect resistance falls out of G-ADI; the sequence will not detect the defect.

The third situation corresponds to the case of a redundant defect for which it exists no sequence able to detect the defect. Consequently, nothing can be done to optimize the detection of this defect for the test technique under consideration. In contrast, the second situation does not correspond to the case of a redundant defect, implying that it exists one or several input vectors able to detect the defect. Consequently, the considered test sequence is not the most favourable and better vectors could be used to ensure an optimal detection range of the bridge defect resistance value.

This example leads to the following key definition:

Definition 2.3. Given a circuit under test and the list of Analogue Detectability Intervals ADI^V associated to each possible input vector V for a considered defect, optimizing the defect detection process consists in finding a minimum number of input test vectors that cover the Global-ADI.

At this point, the problem is equivalent to any coverage problem and can be treated by classical algorithms. Considering the simple example of Table 2.4, it is clear here that several minimum solutions exist. For instance, a test sequence composed of the two vectors #6 and #7 covers the complete Global-ADI:

$$C - ADI = \left[0, R^2_C\right] \cup \left[R^2_C, R^4_C\right] = \left[0, R^4_C\right] = G - ADI$$

It is worth highlighting that vector #7 appears as an 'essential' vector to cover the Global-ADI whereas this vector would not be generated by a classical ATPG. In other words, the detection of a bridge defect with a resistance falling into the interval $[R^2_C, R^4_C]$ is not guaranteed using the stuck-at fault model, while it exists some input vectors (#7 and #11) able to detect the defect.

2.3.3 *Alternative Detectability Techniques*

Some of the alternative techniques to the logic-based detectability strategies are presented in this subsection. Among them, the widely used detectability techniques based on the surveillance of quiescent current consumption are reviewed in more detail.

2.3.3.1 Quiescent Current (I_{DDQ}) Testing

At the early 1980s, an alternative testing technique for bridging and other defects in CMOS technologies was proposed by Levi (1981). According to his proposal, the increased testability of CMOS technologies is based on their negligible static current consumption if no defect is present in the circuit. This characteristic is derived from the complementary nature of the n-network versus the p-network that avoids the simultaneous conduction of both networks provided a quiescent state has been reached.

The quiescent current (I_{DDQ}) testing technique (Malaya and Su 1982) has been widely used for the detection of bridging faults. It is based on the fact that the defect causes an increase in the quiescent current consumption of the circuit provided the appropriate excitation is applied. It consists in monitoring the power supply current (I_{DDQ}) once all the transient currents in the circuit have settled-down. The measured current is compared to a threshold value and if it is higher than this reference current, the device is considered faulty.

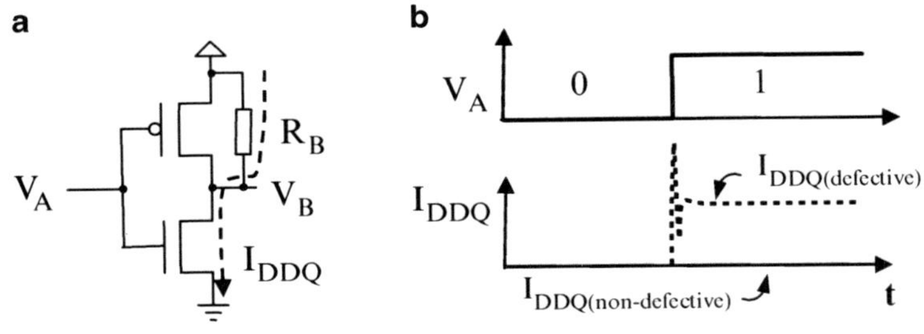

Fig. 2.14 Bridging defect affecting the output of an inverter at the (**a**) gate level and its (**b**) $\mathbf{I_{DDQ}}$ consumption versus the logic signal at the input of the defective gate

The effectiveness of I_{DDQ} testing has been reported in a wide range of works (Baschiera and Courtois 1984; Turner et al. 1985; Rodríguez-Montañés et al. 1991) to detect different bridging defect classes, such as interconnect bridges, gate oxide bridges and inter-gate bridges. Two bridged nodes set to the opposite logic value create a current path between the power and ground rails (Acken 1983). Consequently, an extra current above the defect-free case is generated flowing from power to ground nodes. An example of the I_{DDQ} testing technique applied to the detection of bridging defects is illustrated in Fig. 2.14, where an inverter contains a bridging defect (R_B) between its output and the power node (or equivalently, between the source and the drain of the pMOS transistor). When the inverter input (V_A) is in a low logic state, the nMOS transistor is off. The current consumption is only due to leakage current, as shown in Fig. 2.14b. However, if V_A transitions from logic 0 to logic 1, the nMOS transistor turns on and the pMOS transistor turns off. In the fault free case, once all the signals have settled-down, the current consumption is again the leakage current. Nevertheless, due to the bridge, during the high logic state of V_A there is a current flowing from the power rail to ground through the nMOS transistor, increasing the quiescent current value.

I_{DDQ} testing provides high defect observability. Indeed, it requires only fault sensitization, since the fault-effect is always observable through the power supply current measurement. Hence, the fault propagation requirement during test generation is not needed unlike logic based testing techniques. However, I_{DDQ} testing technique has some drawbacks (Soden and Hawkins 1996; Sachdev 1997; Ferré and Figueras 1997; Figueras and Ferré 1998). Among them, it must be pinpointed that it offers low test application times since it needs to wait for the level to settle and then perform the sensing and the comparison of the current level with the test threshold value. This drawback can be partially solved due to the lower required number of vectors compared to the required in voltage based techniques.

The most important drawback is due to the shrinking of minimum feature size for CMOS technology nodes. The theoretical basis of I_{DDQ} consists in the appropriate estimation of the leakage current for the defect-free circuit in order to determine the threshold value above which the circuit will be considered defective. Due to statistical variations of process parameters, the defect-free current consumption can not

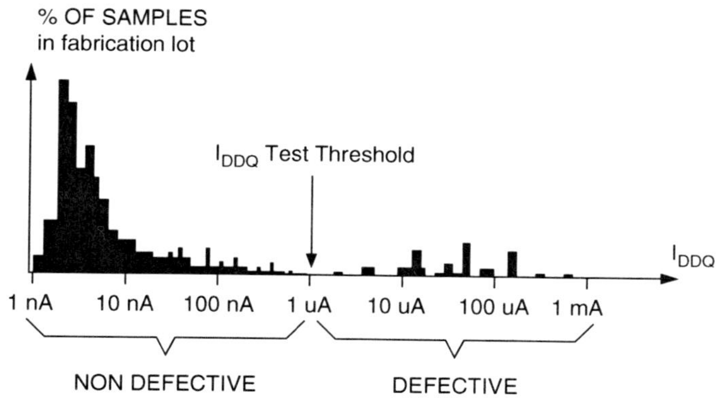

Fig. 2.15 I_{DDQ} histogram of a fabricated lot (Ferré and Figueras 1998)

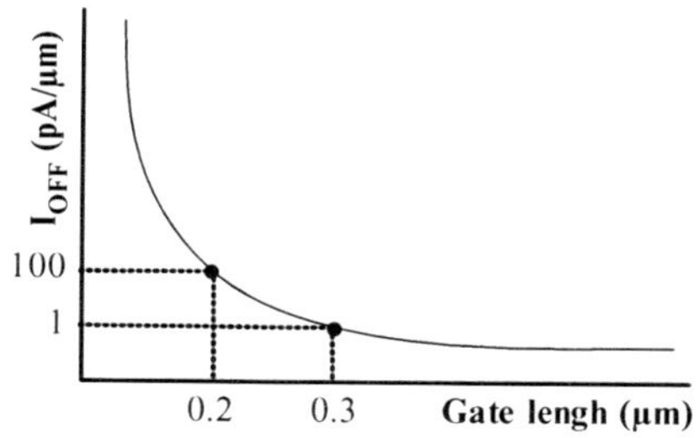

Fig. 2.16 Evolution of leakage current (Rajsuman 2000)

be considered nominal anymore but ranging within a statistical distribution (Ferré and Figueras 2002). Assuming also a statistical distribution for defective devices, only a clear distinction between defect-free and defective devices could be made if both distributions are far enough. However, this is far from reality as shown in Fig. 2.15 where the histogram of the current consumption for a real fabricated lot of circuits illustrates the fact that the defect-free circuits consumption overlap the defective circuits consumption (Figueras and Ferré 1998). Indeed, the leakage current increases in an exponential way for every new CMOS technology node, as described in Fig. 2.16. Therefore, it becomes comparable or higher than the defect current. The current mean value of the distribution of the fault-free devices increases and gets closer to the mean of the defective devices, overlapping both distributions. Thereby, it becomes more difficult to determine whether a variation in the I_{DDQ} value is due to the leakage current or due to a defect.

2.3.3.2 Other Quiescent Current Testing Techniques

Some techniques based on the post-processing of I_{DDQ} data have been developed to extend the effectiveness of I_{DDQ}. One of these techniques is the so-called 'Delta

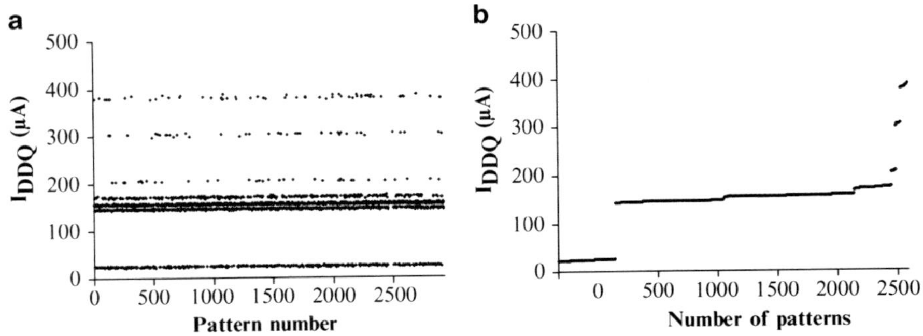

Fig. 2.17 I_{DDQ} test for a real 0.18 μm defective device. (**a**) Non-ordered and (**b**) current signature

I_{DDQ}' (Thibeault 1997; Miller 1999; Kruseman et al. 2001). Instead of observing the absolute value of the power supply current, 'Delta I_{DDQ}' considers the difference of the power supply current among successive test vectors. This difference is treated probabilistically to determine if the circuit is defective or not.

Another extension of the I_{DDQ} testing technique is based on the use of current signatures, which was proposed by Gattiker and Maly (1996). The measured I_{DDQ} data is not compared to a single threshold value, but the current for the whole test set is measured (Gattiker and Maly 1996; Nigh and Gattiker 2004). A current signature is generated by ordering all the obtained measures from the smallest to the highest value. This technique looks for sharp changes (or steps) in the current signature, which indicates some kind of defect in the device. In case of bridges, the number of steps may give information about the number of network excitations that have been activated. Figure 2.17 illustrates the I_{DDQ} data for a real CMOS 0.18 μm defective device. On one hand, Fig. 2.17a shows the I_{DDQ} values in the same order as in the test procedure. On the other hand, the values are ordered in Fig. 2.17b. Notice that different steps are observed for the current signature of the defective device. Current signatures avoid the problem of I_{DDQ} and Delta I_{DDQ} testing when deciding the current threshold limit.

The current ratios technique (Maxwell et al. 1999) is based on the same idea as current signatures, but tolerating parameter variations. The basic idea relies on the fact that the slopes of the rank-ordered current signatures for dies having differences in the absolute I_{DDQ} values are quite similar. Therefore, it is possible to set a test limit based on the ratio of the maximum to minimum I_{DDQ} value. This value is more or less constant and independent of the mean of the I_{DDQ} measurements for each die. This ratio is determined by means of an iterative process. Once obtained the ratio, the vector which typically gives the minimum current is identified. The current for that vector is measured. Subsequently, the maximum current is computed due to the ratio previously obtained. Outliers are then identified.

Other solutions have been proposed in order to overcome the leakage problem (Keshavarzi et al. 1997; Sachdev 1997; Figueras and Ferré 1998; Meijer et al. 2004), namely: lowering test temperature, controlling the back vias voltages, partitioning the device using multiple power sources, multiple transistors thresholds or Silicon On Insulator (SOI) technologies for a sharper sub-threshold swing.

2.3.3.3 Very Low Voltage Testing

In case of bridges, different voltage-based techniques have been developed to improve the observability of their effect on defective circuits. One of the most common techniques is lowering the power supply voltage below the nominal operation value. In fact, this technique has been demonstrated to detect defects which are not detected by means of other testing techniques (Hao and McCluskey 1993; Chang 1998; McCluskey and Tseng 2000).

Different works reported the effectiveness of lowering the power supply voltage in logic tests when detecting bridges. Very Low Voltage (VLV) logic testing is suitable in order to detect resistive bridges. Some works reported that lowering V_{DD} is appropriate to detect bridges (Engelke et al. 2004), since the critical resistance (the highest bridging resistance which can be detected by means of logic tests) increases as V_{DD} decreases (Kruseman et al. 2002). Chao-Wen et al. (2001) proposed a different concept when lowering the power supply value based on the minimal V_{DD}. At a given clock frequency, this technique consists in lowering the V_{DD} value until obtaining the minimum V_{DD} at which the device still functions. The authors showed that some defective devices had a higher minimum V_{DD} than the fault free ones.

In general, lowering the power supply value is a technical condition easy to implement, since it does not require any extra equipment or performance. However, it decreases the speed of the circuit-under-test. Thus, there is an increase in test time because the clock frequency is lower than the one at nominal conditions.

2.3.3.4 Shmoo Plots

Shmoo plotting analyzes the performance of a digital IC compared to the major analogue parameters that influence the characteristics of the electrical behaviour of the circuit (Baker and van Beers 1997). Shmoo plotting offers a way to visualize the relationship between the performance of an IC and changes in the external environment, such as temperature, V_{DD}, and timing (Huott et al. 2000; Chao-Wen et al. 2001; Kruseman et al. 2002).

Figure 2.18 illustrates two Shmoo plots (V_{DD} vs Period) for a combinational CMOS circuit. As the example shown in this figure, given a working frequency of the defective circuit, the majority of bridging defects allows the circuit to work properly provided a sufficiently high V_{DD} is applied. This is due to the fact that the resistance of the connected n-network and p-network decreases with increasing V_{DD} making the effect of the (fixed) resistance of the bridge less visible. However, not all the bridging defects cause the same type of shmoo plot. For two balanced n and

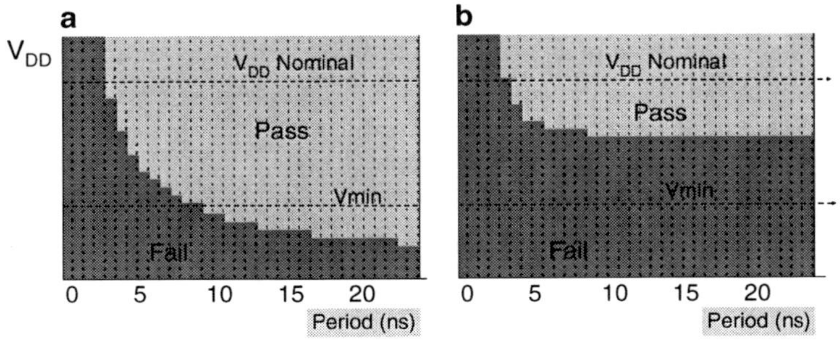

Fig. 2.18 Shmoo plots of a combinational CMOS circuit with a bridging defect: (**a**) Defect-free case and (**b**) defective case (Rodríguez-Montañés 2006)

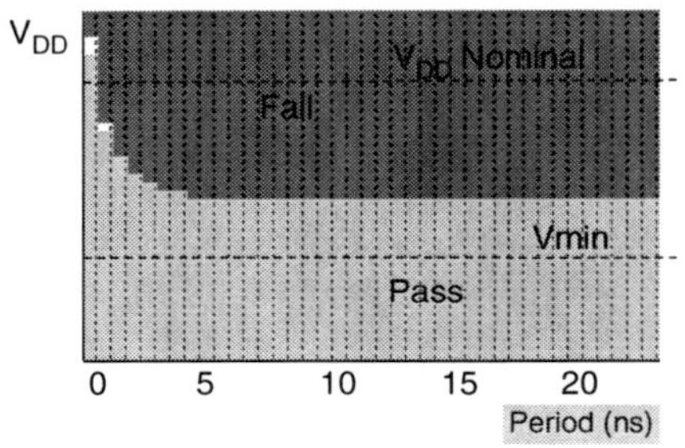

Fig. 2.19 Shmoo plots for two inverters with their outputs bridged together (Rb = 1 kΩ). Each inverter belongs to a different inverter chain (Rodríguez-Montañés 2006)

p-bridged networks a shmoo plot like the one in Fig. 2.19 is obtained (Rodríguez-Montañés et al. 2006). In this particular example, the bridge is connecting the output of the inverters.

The main drawback of this technique is the number of ATE measurements required. Each individual measurement performed on the ATE can result in a pass-fail (e.g. of a functional test) or in a numerical result (e.g. failure counts or bit error rate).

Some approaches have been proposed to reduce the time required to generate a Shmoo plot. In this direction, Patten (2004) proposes a robust and efficient fast Shmoo algorithm that extracts the device characteristics from a minimum number of coordinate points that actually have to be measured. This algorithm reduces the Shmoo execution time significantly.

2.3.3.5 Temperature Based Testing Techniques

Temperature may also give additional information to detect bridges which are not observable at nominal conditions. Resistive bridges are temperature sensitive because their electrical resistance varies with temperature (Semenov and Sachdev 2000). Bridging defect materials having a positive resistance temperature coefficient (RTC) such as metals and polysilicon increase their resistance with temperature. Thereby, at low temperatures the bridging resistance induces higher I_{DDQ} values. Furthermore, the probability of causing faulty logic behaviour is also higher. Some works (Kundu 1998; Schuermyer et al. 2004) give experimental evidence of this phenomenon and show how testing results at two different temperatures are useful to detect device outliers.

However, there are some drawbacks when introducing temperature in the production testing environment. Techniques based on temperature variation are time consuming and expensive, especially for low temperatures, which furthermore require specialised equipment.

2.4 Diagnosis of Bridging Defects

Diagnosis is the process which identifies the type of fault and locates the failure site of a faulty device. Subsequently, failure analysis can be performed to physically examine the defect. Precise diagnosis is important since it helps manufacturers to solve process problems, improving yield and saving time on physical failure analysis, which is time consuming and require significant investment in equipments, tools and qualified personnel.

Diagnosis techniques combine simulation results with the data obtained from the ATE (Automatic Test Equipment). Most of the techniques involve two main elements: a fault model and a comparison algorithm. Thus, using accurate fault models is a key factor. If models are not accurate, the result may be an imprecise or even an incorrect location of the failure site.

Fault diagnosis techniques can be broadly classified into two groups: cause-effect and effect-cause techniques (Abramovici et al. 1994). Cause-effect diagnosis techniques are based on fault simulations to determine the possible response of a circuit in the presence of faults. This information is compared with the response obtained from the tester in order to obtain the fault location. Some cause-effect techniques use a pre-computed fault dictionary, which is a database containing the faulty responses of each fault. The algorithm then determines which fault from the dictionary best matches the faulty behaviour observed on the tester. Techniques using a fault dictionary are also known as static diagnosis techniques. However, with the increasing complexity and number of transistors in today's ICs, sometimes it is not feasible to build a dictionary for every possible fault, since the size of the dictionary would be prohibitive. Thereby, a lot of effort is focused on reducing and compressing the size of fault dictionaries (Pomeranz and Reddy 1992; Boppana et al. 1996; Chess and Larrabee 1999). Another possibility is using dynamic diagnosis techniques, which analyse the response of the faulty circuit. The list of fault candidates is reduced based on the response of the circuit and only the most probable faults are considered.

The effect-cause approach (Abramovici 1980) backtracks logic errors from the primary outputs to the location of the fault deducing the internal values of the circuit. In principle, most of these diagnosis techniques do not require neither fault dictionary nor fault enumeration.

2.4.1 Logic Diagnosis Techniques

Diagnosis of bridging faults by using information from single SA faults was common in the past, since processing SA faults is computationally simpler than processing bridging faults, both in terms of fault list size and fault simulation complexity. Different logic diagnosis methodologies have been developed using fault dictionaries and fault simulation. In the fault dictionary method (Millman et al. 1990; Chakravarty and Gong 1993, 1995; Chess et al. 1995; Aitken and Maxwell 1995; Lavo et al. 1998), the faulty response of each considered bridging fault is stored for every test pattern. The diagnosis process is carried out comparing the output response of the failing device to the information contained in the fault dictionary of bridges. In the works by Chakravarty and Gong (1993, 1995) the initialization graphs are used for generating the initial set of bridging fault candidates. Subsequently, a set of pruning rules are considered to reduce the candidates set. The first work Chakravarty and Gong (1993) is based on the wired-AND and the wired-OR bridging fault model, whereas the second work (Chakravarty and Gong 1995) is based on the voting model. Other works (Millman et al. 1990; Chess et al. 1995; Aitken and Maxwell 1995) took benefit from composite signatures. A composite signature (Millman et al. 1990) is the bridge fault signature resulting from the union of the four stuck-at fault signatures associated with the bridged nodes. The main improvement in the work by Chess et al. (1995) related to previous work in Millman et al. (1990) is the restriction of the number of faults under consideration, which increases the efficiency of the methodology. This is achieved by eliminating from the composite signature entries that cannot be used to detect the bridging fault and also defining the set of vectors which should detect a particular bridge. In Aitken and Maxwell (1995), quality measurements were defined to create a ranking criterion for bridging faults diagnosis. These quality measurements were subsequently used for other works and even applied to other fault models. The criterion is based on the comparison between the results obtained on the tester and the prediction of the bridging fault model. The part of the tester results which is also included in the fault model prediction is called *Intersection* (see Fig. 2.20). Failing vectors predicted by the fault model which have not failed on the tester are called *Mispredictions*. Vectors

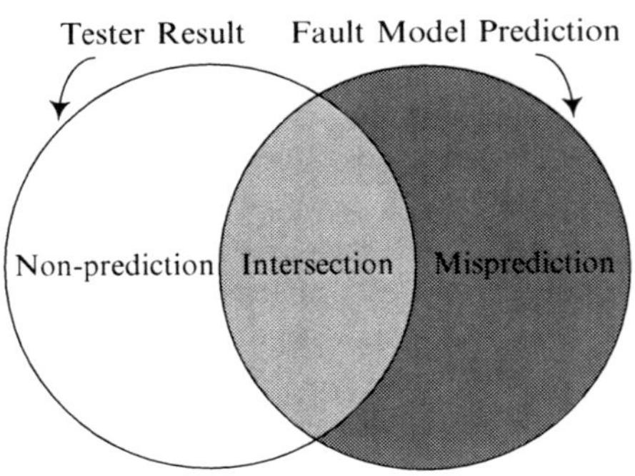

Fig. 2.20 Matching algorithm (Aitken and Maxwell 1995)

which have failed on the tester, but are not predicted by the fault model, are called *Nonpredictions*. The ranking criterion is based on the *Intersection* value; the higher the *Intersection*, the better the diagnosis.

Fault dictionaries are feasible when the diagnosis is performed repeatedly for a given design. However, their main drawback is the storage space. A circuit with n number of nets has $\binom{n}{2}$ possible bridging faults. Thus, considering every possible bridging fault is infeasible. Physical layout information is usually considered to eliminate bridges between nets that are extremely unlikely to be bridged together due to their physical location (Aitken and Maxwell 1995; Lavo et al. 1998). If the two nets are farther than some minimum distance or if there is another net between them (that would also be involved in the bridge), the corresponding bridging fault is discarded. However, there are also some techniques to reduce the number of candidates without using layout information, as the two techniques reported by Lavo et al. (1997). The first technique uses the SA fault diagnosis to identify one of the bridged nets. If this is accomplished, assuming a circuit with n nets, knowing the d net candidates to be one of the nets involved in the bridge, the number of bridged pairs is then reduced to $n{\cdot}d$. The second technique identifies the candidates that can have an intersection with the behaviour observed on the tester. Candidates with no intersection are then discarded.

Zou et al. (2005) proposed a diagnosis methodology based on dictionaries which take the bridge resistance into account. The methodology is divided into two steps. The first step consists in a logic diagnosis to find the potential candidates that can explain the faulty behaviour. In the second step, layout information as well as the resistive bridging fault model using the concept of critical resistance are used to prune the candidates list. The intersection between resistive intervals is utilized to discard bridging candidates. As an example, consider the bridged outputs (net A and B) of two gates ($G1$ and $G2$), as depicted in Fig. 2.21. They drive, in turn, gates $G3$ and $G4$, respectively. Assume that test patterns TP_1 and TP_2 cause $G3$ and $G4$ to fail, respectively, whereas TP_3 passes although it also activates the bridge. Gate $G3$ should have failed in this case. In the fault free case, consider that TP_1 and TP_3 set net A and net B to logic 1 and 0 respectively. On the contrary, TP_2 sets them to logic 0 and logic 1, respectively. As test patterns TP_1 and TP_2 make the circuit to fail, the bridge resistance should be lower than the minimum of the two critical resistances $R_c(TP_1, G3)$ and $R_c\,(TP_2, G4)$. Nevertheless, for passing pattern TP_3 the bridge resistance should be higher than the critical resistance $R_c\,(TP_3, G3)$. Thus, it must be accomplished that the bridge resistance is $R_c\,(TP_3, G3) < \mathrm{R_b} < min\,(R_c\,(TP_1, G3)\,, R_c\,(TP_2, G4))$. On the other hand, in case that $R_c\,(TP_3, G3) > min(\,R_c\,(TP_1, G3),\, R_c\,(TP_2, G4)$, there is no bridge

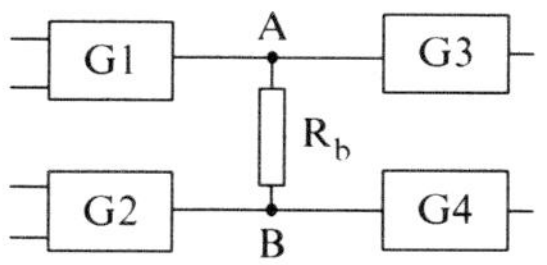

Fig. 2.21 Resistive bridging fault diagnosis

resistance explaining the faulty behaviour, and therefore, this candidate can be removed from the list. In a similar way, Khursheed et al. (2009) also presented a methodology where resistive intervals were used to diagnose resistive bridges. However, in this work resistive intervals at different power supply values are used to improve the accuracy of the diagnosis procedure.

Instead of using the pre-computed information stored in a table used by fault dictionaries, the fault simulation procedures Wu and Rudnick (1999, 2000) consist in comparing the actual output response of the failing device to the expected response for each possible bridge. A list of fault candidates is then generated. Faults whose effects most closely match the response of the failing device are identified as candidates. The advantage of this approach compared to fault dictionaries is that fault simulation is faster. In the work developed by Wu and Rudnick (2000), information from single SA faults is used. Single SA fault simulations are performed during fault diagnosis for a more accurate result.

All the methods discussed above are implemented at inter-gate level. However, bridging faults at intra-gate level are also possible. The work by Fan et al. (2006) addresses the logic diagnosis of intra-gate bridging faults by means of a transformation method.

2.4.2 *Current Diagnosis Techniques*

The quiescent current-based techniques for the diagnosis of bridging defects are reviewed in this subsection. The quiescent current flowing through the defect is analyzed in terms of diagnosis purposes. The impact of the consumption generated at the downstream gates is also analyzed.

2.4.2.1 I_{DDQ}-Based Diagnosis

I_{DDQ} has demonstrated to be also effective for the diagnosis of bridging faults although, at the beginning, it was believed that I_{DDQ} could not provide enough information for diagnosis purposes (Acken and Millman 1992). Subsequently, different works demonstrated the effectiveness of I_{DDQ} for bridging fault localisation. The main advantage of current methodologies is that fault signatures are easy to generate. The first works (Aitken 1991, 1992; Chakravarty and Suresh 1994; Nigh et al. 1997) were based on the simple I_{DDQ} bridging fault model, which assumes that abnormal high current is generated when the bridged nets are set to different logic values. Aitken (1991) demonstrated that combining logic and current information diagnosis resolution was improved. Subsequently, the same author presented diagnosis results without using logic information (Aitken 1992). Chakravarty and Suresh (1994) proposed an I_{DDQ} based diagnosis algorithm which considers also whether one of the nodes involved in the bridge is internal or not. Subsequently, Nigh et al. (1997) relied on a set of realistic bridges based on layout information with good results. As the number of possible bridges to be considered is huge, most of the works relied on a set of limited realistic bridges obtained by extraction tools. Nevertheless, although this idea has been demonstrated to be effective, most of the extractors only identify possible bridges between nets in the same metal layer. However, reality has shown that bridges between nets in different metal layers are also possible (Aitken 1992). To avoid this loss of accuracy, Heaberlin (2006) proposed a heuristic method for high-speed diagnosis feasible for large industrial designs, which considers a priori all possible bridges in the circuit.

The application of the simple I_{DDQ} bridging fault model has mainly two drawbacks. The first one is the increase of leakage current in present and future technologies. The second drawback is that a bridge may have many equivalent faults which cannot be distinguished.

Some works have been proposed to overcome the limitations caused by the leakage current. Gattiker and Maly (1996, 1998) presented the amount of diagnostic information present in current signatures and how the number of current levels may distinguish bridging faults, which are equivalent under the assumption of the simple I_{DDQ} bridging fault model. Furthermore, Thibeault and Boisvert (1998) and Thibeault (2000) proposed a method based on differential or 'Delta I_{DDQ}' probabilistic signatures for bridging faults. The method is performed into two steps. The first one is a pre-processing step, where the most probable faults are listed. This is the starting point of the second step, where the fault location is carried out by finding the location that causes the expected current values match experimental measures. In subsequent works, Hariri and Thibeault (2003, 2006) proposed a diagnostic method combining three data sources, namely: I_{DDQ} measures to identify the most probable bridging faults, parasitic capacitances extracted from layout to create a list of realistic bridges and finally, logical errors produced by logic fault simulation to perform fault isolation. The I_{DDQ} stage procedure is based on 'Delta I_{DDQ}' probabilistic signatures previously proposed in Thibeault and Boisvert (1998) and Thibeault (2000).

In relationship with the incapacity of distinguishing two equivalent (in terms of current consumption) bridging defects, let us consider the two examples illustrated in Fig. 2.22. The quiescent current consumption of both circuits is shown in Table 2.5. Since the I_{DDQ} behaviour is the same for both defective circuits, the two different bridging defects could not be distinguished (diagnosed) with the single

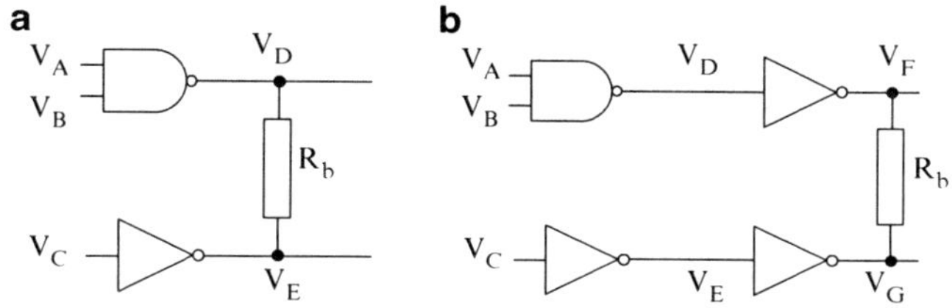

Fig. 2.22 Defective circuits equivalent in terms of I_{DDQ} consumption. (**a**) Bridge connecting an inverter and a NAND gate. (**b**) Bridge connecting two inverters

Table 2.5 Identical I_{DDQ} behaviour of the two examples in Fig. 2.22

V_A	V_B	V_C	V_F	V_G	I_{DDQ}
0	0	0	0	0	Leakage
0	0	1	0	1	High
0	1	0	0	0	Leakage
0	1	1	0	1	High
1	0	0	0	0	Leakage
1	0	1	0	1	High
1	1	0	1	0	High
1	1	1	1	1	Leakage

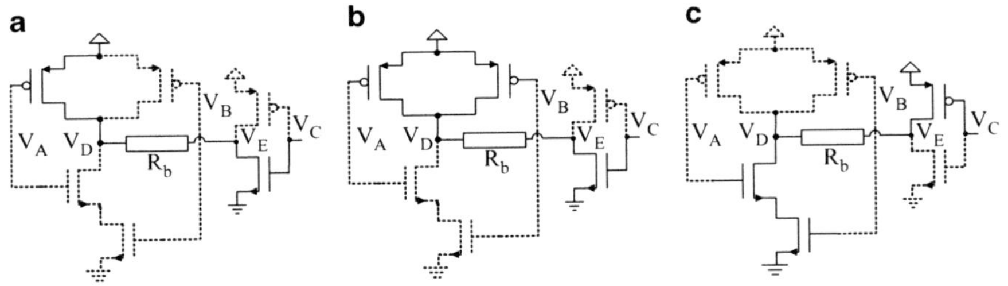

Fig. 2.23 Network excitations for example in Fig. 2.22a. (**a**) One pMOS transistor on (NAND gate), (**b**) both pMOS transistors on (**c**) nMOS network on (NAND gate)

I_{DDQ} threshold method. Indeed, the discrimination between these two faults is possible provided that the bridged network strengths are considered (Arumí et al. 2007) as shown below.

For the example in Fig. 2.22a, there are three different network excitations, as depicted in Fig. 2.23. Every excitation adds a different equivalent resistance between power and ground, generating thus different quiescent currents. If a set of patterns are applied so that all the possible combinations of the bridged networks are excited, the I_{DDQ} measurements would follow the behaviour of Fig. 2.24a, where four current levels are clearly observed. The lowest level corresponds to those patterns which do not excite the bridge. The three upper levels correspond to the patterns which activate the bridge. In these cases, apart from the leakage current, extra current is flowing through the bridged networks. According to Fig. 2.24a, letters a, b and c relate the current level with their corresponding excited network in Fig. 2.23. The highest current level corresponds to the case where both pMOS transistors of the NAND gate are in the on state (Fig. 2.23b), since the equivalent resistance composed by the parallel pMOS transistors of the NAND gate, the bridge resistance and the nMOS transistor of the inverter is lower than in the other two cases.

Regarding the bridge between the outputs of the inverters in Fig. 2.22b, there are only two equivalent network excitations. In fact, assuming identical inverters, there is only one different excitation. For that reason, the I_{DDQ} measurements would only show two current levels, the lowest one corresponding to the leakage current and the upper level when the bridge is activated, as depicted in Fig. 2.24b.

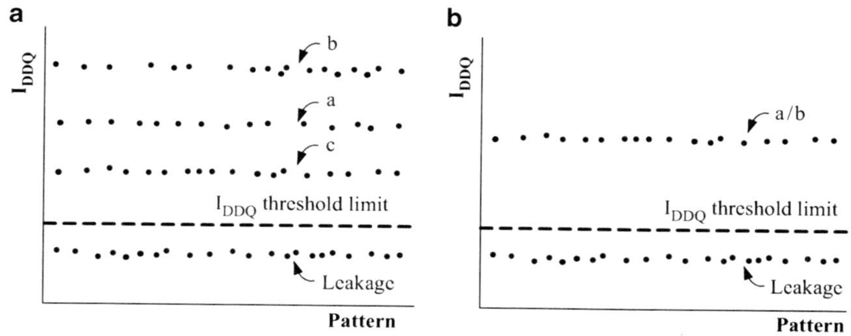

Fig. 2.24 Current measurements of circuit in (**a**) Fig. 2.22a, (**b**) Fig. 2.22b

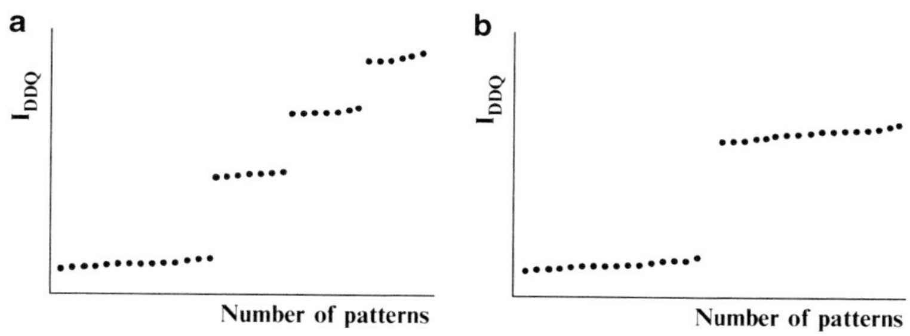

Fig. 2.25 Current signatures of the defective circuits in (**a**) Fig. 2.22a, (**b**) Fig. 2.22b

Reordering the I_{DDQ} measurements in increasing order, the current signatures (Gattiker and Maly 1996; Nigh and Gattiker 2004) corresponding to the two examples are presented in Fig. 2.25. On one hand, the current signature corresponding to the bridge between the output of the NAND gate and the inverter has three current steps. However, the one from the bridge between the inverters has only one current step. Notice that the two bridges, equivalent under the assumption of the simple I_{DDQ} bridging fault model, become now distinguishable (Arumí et al. 2007a)

2.4.2.2 Downstream Current Contribution

It has been illustrated how it is possible to discriminate between different bridging faults if the current information given by the network strengths is treated properly. However, it is well known that a bridge may cause voltage degradation on the bridged nodes (Rodríguez-Montañés et al. 1991). This voltage degradation causes, in turn, the gates driven by the bridged nodes (downstream gates) to consume more current than expected, as long as the proper conditions are given. Hence, the total current (I_t) caused by a bridging fault comprises two components: the bridge current (I_b) and the downstream current (I_d) (Rubio et al. 1991; Arumí et al. 2007b).

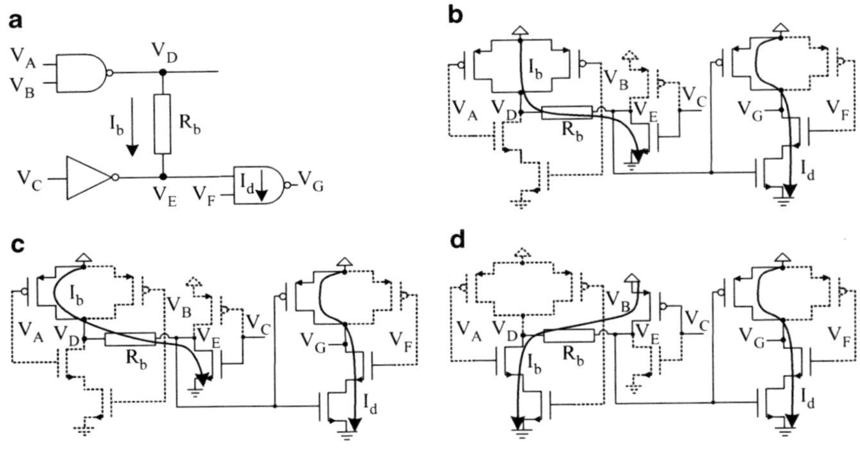

Fig. 2.26 Bridging fault with downstream current. (**a**) Gate level and (**b**) both pMOS transistors on (NAND gate), (**c**) one pMOS transistor on (**d**) nMOS network on (NAND gate)

In order to evaluate the impact of the downstream current, let us consider the example in Fig. 2.26a. It is similar to the one in Fig. 2.22a, but now the inverter is driving, in turn, a NAND gate. When the bridge is activated, the current flowing through V_D and V_E generates the bridge current. Due to voltage degradation on V_E, there may be also downstream current flowing through the NAND gate provided that V_F is set to logic 1. The three possible network excitations in the presence of downstream current ($V_F = 1$) are described in Fig. 2.26b, c, and d, respectively.

The magnitude of the downstream current depends on different factors, namely: the bridged networks, the topology of the downstream gate and the bridge resistance. The relationship between the downstream current and the voltage of the bridged node is similar to the current behaviour in the presence of a floating node caused by an interconnect full open. If an intermediate voltage value is induced between V_{Tn} and ($V_{DD} - V_{Tp}$), downstream current is generated. The amount of current depends on the exact voltage value, as observed in Fig. 2.27a.

Neglecting the downstream current, the current signatures for the examples in Figs. 2.22a and 2.26a would be identical. Nevertheless, regarding the example in Fig. 2.26a, for every network combination, there are two possibilities depending on the V_F value. If V_F is set to logic 0, there is not downstream current. In this case, the total current is the same as in Fig. 2.25a. However, for the patterns that set V_F to logic 1, there is downstream current, which increases the total current caused by the bridge. In this sense, every of three upper current levels may unfold into two sub-levels, as illustrated in Fig. 2.27b. Notice that the current signature is quite different since seven current levels are now reported.

Arumí et al. (2007, 2008) proposed a method to minimise the effect of the downstream current that consists in decreasing V_{DD}, to the point that the downstream current becomes negligible and the bridge current (I_b) is practically equal to the total current (I_t). Figures 2.28 and 2.29 illustrate the current signature for a real

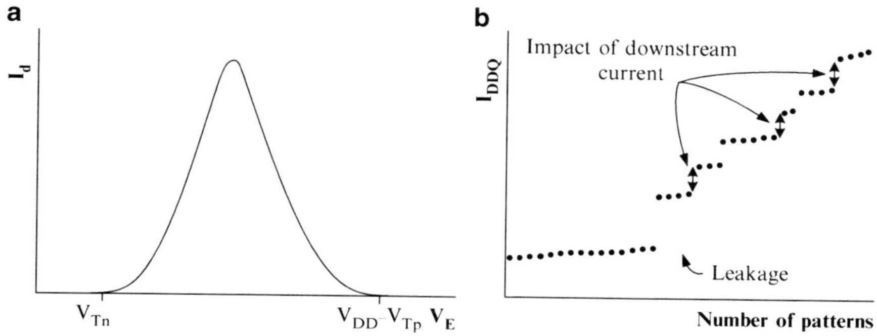

Fig. 2.27 Example in Fig. 2.26a. (**a**) Downstream current vs. node voltage and (**b**) current signature

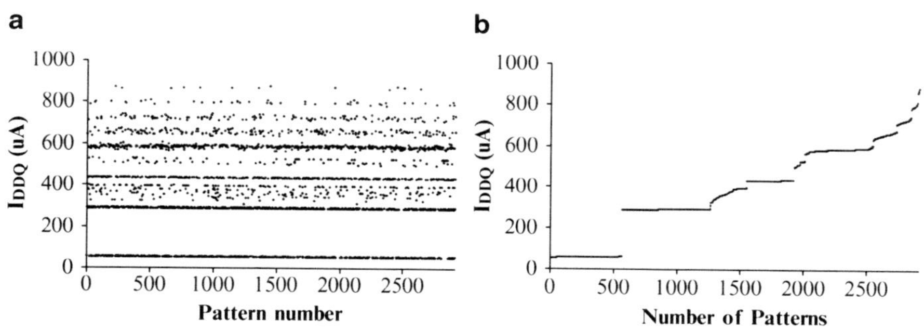

Fig. 2.28 Current signature at nominal V_{DD} for a real defective device. (**a**) Non-ordered and (**b**) ordered (Arumí 2008)

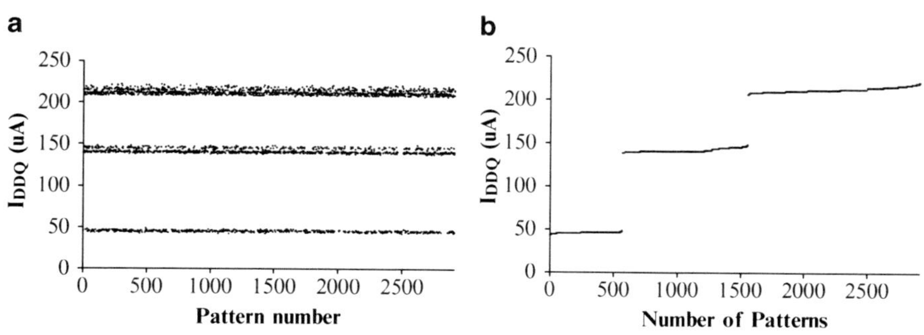

Fig. 2.29 Current signature at very low V_{DD} for a real defective device. (**a**) Non-ordered and (**b**) ordered (Arumí 2008)

defective circuit (0.18 μm NXP Semiconductors) obtained at the nominal V_{DD} and low V_{DD}, respectively. Figure 2.30 illustrates the bridging defect diagnosed with the proposed signature-based methodology.

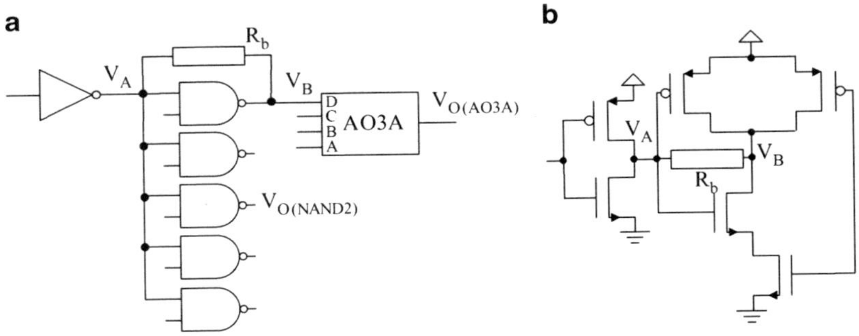

Fig. 2.30 Bridging defect diagnosed from the current signatures of Figs. 2.28 and 2.29. (**a**) Gate level and (**b**) transistor level

2.5 Summary

Traditional test techniques use the 'universal' stuck-at fault model to generate tests that are expected to detect various types of real defects. In this context, the test quality relies on detection of non-directly targeted defects. As the demand for high quality test increases, the limitations of the stuck-at fault model to adequately represent and predict the behaviour of frequent realistic defects such as bridges and opens, have demonstrated the need for more accurate and specific fault models. Under such conditions, extensive research works have been devoted to the modeling, detection and diagnosis of bridging defects.

The first models proposed for bridging defects were the so-called *wired models* which were inherited from previous non-CMOS technologies. Due to their intrinsic limitations, they were rapidly replaced by the family of voting models. All these models assume that the unpredictable resistance of the bridge is very small and can be neglected. However it has been experimentally proved that the resistance of a bridging defect is not small and has to be considered to understand its behaviour.

From these observations and considering the unpredictability of the resistance, new concepts were proposed which take into account the range of resistance that can be detected by a given vector: the Analogue Detectability Interval (ADI). Implementation of these new concepts into ATPG tools and Fault Simulators is described in a following chapter.

It is well-known that defects can be detected using logic-based techniques as well as current-based techniques. Indeed the quiescent current testing technique has been widely used for the detection of bridging defects. The effectiveness of I_{DDQ} has been reported in a wide range of works to detect various bridging defect classes. I_{DDQ} testing provides high defect observability but suffers from low test application frequency and the presence of high background leakage currents in present CMOS technologies. To improve the I_{DDQ} test technique efficiency, refined techniques have been developed such as the 'Delta I_{DDQ}', current signature, current ratio... In addition to the logic-based and current-based techniques, some works have investigated the possibility of varying the test conditions such as temperature or power supply voltage.

Fault models can be used for testing purposes but also for diagnosis purposes. In this way, the wired and voting models have initially been used for diagnosis. Some works take into consideration the resistance of the bridging defect and more precisely the value of the critical resistance.

As it is well-known, the main objective of diagnosis is to increase accuracy by reducing the list of initial candidates. Many logic-based techniques have been proposed to reduce the set of candidates as much as possible. However, it is clear that additional information is required to further decrease the list of candidates.

This chapter gives a broad view of the state-of-the-art of modeling bridging defects. From the initial and simplistic models to the most recent ones which are realistic, accurate and easy to handle, we observe a drastic improvement of the knowledge of the bridging defect behaviour which translates to better quality test sequences. The advances observed for the case of bridging defects have to be considered as an exemplary guideline for defect-oriented test strategies. Similar research and developments are today dedicated to other types of defect such as resistive opens.

References

Abramovici M, Breuer MA (Jun 1980) Multiple fault diagnosis in combinational circuits based on an effect-cause analysis. IEEE Trans Comput C-29(6):451–460

Abramovici M, Breuer MA, Friedman AD (1994) Digital system testing and testable design. IEEE Press

Acken JM (1983) Testing for bridging faults (shorts) in CMOS circuits. Design automation conference, pp 717–718

Acken JM, Millman SD (1991) Accurate modelling and simulation of bridging faults. Custom integrated circuits conference, pp 17.4.1–17.4.4

Acken JM, Millman SD (1992) Fault model evolution for diagnosis: accuracy vs precision. Custom integrated circuits conference, pp 13.4.1–13.4.4

Aitken RC (1991) Fault location with current monitoring. International test conference, pp 623–632

Aitken RC (1992) A comparison of defect models for fault location with I_{DDQ} measurements. International test conference, pp 778–787

Aitken RC, Maxwell PC (Feb 1995) Better models or better algorithms? Techniques to improve fault diagnosis. Hewlett-Packard J, 110–116

Arumí D, Rodríguez-Montañés R, Figueras J, Eichenberger S, Hora C, Kruseman B, Lousberg M, Majhi AK (2007a) Diagnosis of bridging defects based on current signatures at low power supply voltages. VLSI test symposium, pp 145–150

Arumí D, Rodríguez-Montañés R, Figueras J, Eichenberger S, Hora C, Kruseman B, Lousberg M (Mar 2007b) I_{DDQ} based diagnosis at Very Low Voltage (VLV) for bridging defects. IEE Electr Lett 43(5):25–26

Arumí D (2008) Enhancement of defect diagnosis based on the analysis of CMOS DUT behaviour. PhD Dissertation, http:/www.tdr.cesca.es/es/UPC.html, UPC

Baker K, van Beers J (Jul–Sep 1997) Shmoo plotting: the black art of IC testing. IEEE design and test of computers, pp 90–97

Baschiera D, Courtois B (1984) Testing CMOS: a challenge. VLSI design, pp 58–62

Boppana V, Hartanto I, Fuchs WK (1996) Full fault dictionary storage based on labeled tree encoding. VLSI test symposium, pp 174–179

Bruls EMJG, Camerik F, Kretschman HJ, Jess JAG (1991) A generic method to develop a defect monitoring system for IC processes. International test conference, pp 218–227

Chakravarty S, Gong Y (1993) An algorithm for diagnosing two-line bridging faults in combinational circuits. Design automation conference, pp 520–524

Chakravarty S, Gong Y (1995) Voting model based diagnosis of bridging faults in combinational circuits. International conference VLSI design, pp 338–342

Chakravarty S, Suresh S (1994) I_{DDQ} measurement based diagnosis of bridging faults in full scan circuits. International conference on VLSI design, pp 179–182

Chang JT-Y, Tseng C-W, Chu YC, Wattal S, Purtell M, McCluskey EJ (1998) Experimental results for I_{DDQ} and VLV testing. VLSI test symposium, pp 118–123

Chao-Wen T, Chen R, Nigh P, McCluskey EJ (2001) MINVDD testing for weak CMOS ICs. VLSI test symposium, pp 339–344

Chess B, Lavo DB, Ferguson FJ, Larrabee T (1995) Diagnosis of realistic bridging faults with single stuck-at information. International conference on computer-aided design, pp 185–192

Chess B, Larrabee T (1998) Logic testing of bridging faults in CMOS integrated circuits. IEEE transaction on computers 47 March:338–345

Chess B, Larrabee T (1999) Creating small fault dictionaries. IEEE Trans Comput-Aided Des Integr Circuits Sys 18(3):346–356

Dahlgren P (1988) Switch-level bridging fault simulation in the presence of feedback. International test conference, pp 363–371

Di C, Jess JAG (1993) On CMOS bridge fault modelling and test pattern evaluation. VLSI test symposium, pp 116–119

Engelke P, Polian I, Renovell M, Seshadri B, Becker B (2004) The pros and cons of very-low-voltage testing: an analysis based on resistive bridging faults. VLSI test symposium, pp 171–178

Fan X, Moore W, Hora C, Konijnenburg M, Gronthoud G (2006) A gate-level method for transistor-level bridging fault diagnosis. VLSI test symposium, pp 266–271

Heaberlin D (2006) The power of exhaustive bridge diagnosis using I_{DDQ} speed, confidence, and resolution. International test conference, pp 1–10

Ferré A, Figueras J (1997) I_{DDQ} characterization in submicron CMOS. International test conference, pp 136–145

Ferré A, Figueras J (Jun 2002) Leakage power bounds in CMOS digital technologies. IEEE Trans Comput-Aided Des Integr Circuits Sys 21(6):731–738

Figueras J, Ferré A (Nov 1998) Possibilities and limitations of I_{DDQ} testing in submicron CMOS. IEEE Trans Components Packaging Manufacturing Technol, Part B: Adv Packaging 21(4):352–359

Gattiker A, Maly W (1996) Current signatures. Proceedings VLSI test symposium, pp 112–117

Gattiker A, Maly W (1998) Current signatures: application. Proceedings international test conference, pp 1168–1177

Hao H, McCluskey EJ (1993) Very-low-voltage testing for weak CMOS logic ICs. International test conference, pp 275–284

Hariri Y, Thibeault C (2003) 3DSDM: a 3 data-source diagnostic method. International symposium on defect and fault tolerance in VLSI systems, pp 117–123

Hariri Y, Thibeault C (2006) Improving a 3 data-source diagnostic method. IEEE North–East workshop on circuits and systems, pp 149–152

Huott W, McManus M, Knebel D, Steen S, Manzer D, Sanda P, Wilson S, Chan Y, Pelella A, Polonsky S (2000) The attack of the Holy Shmoos: a case study of advanced DfD and Picosecond Imaging Circuit Analysis (PICA). International test conference, pp 883–891

Keshavarzi A, Roy K, Hawkins CF (1997) Intrinsic leakage in low power deep submicron CMOS ICs. International test conference, pp 146–155
Koch B, Muller-Glaser K (1993) An examination of feedback bridging faults in digital CMOS circuits. IEEE international symposium circuits systems, pp 1527–1530
Kruseman B, van Veen R, van Kaam K (2001) The future of delta I_{DDQ} testing. International test conference, pp 101–110
Kruseman B, van den Oetelaar S, Rius J (2002) Comparison of I_{DDQ} testing and very-low voltage testing. International test conference, pp 964–973
Kundu S (1998) I_{DDQ} defect detection in deep submicron CMOS IC's. Asian test symposium, pp 150–152
Khursheed S, Al-Hashimi BM, Reddy SM, Harrod P (2009) Diagnosis of multiple-voltage design with bridge defect. IEEE Trans Comput-Aided Des 28(3):406–416
Lavo DB, Chess B, Larrabee T, Ferguson FJ, Saxen J, Butler KM (1997) Bridging fault diagnosis in the absence of physical information. International test conference, pp 887–893
Lavo DB, Chess B, Larrabee T, Ferguson FJ (1998) Diagnosing realistic bridging faults with single stuck-at information. IEEE Trans Comput-Aided Des 17:255–268
Levi MW (1981) CMOS is most testable. International test conference, pp 217–220
Malaya YK, Su SYH (1982) A new fault model and testing technique for CMOS devices. International test conference, pp 25–34
Maxwell PC, Aitken RC (1993) Biased voting: a method for simulating CMOS bridging faults in the presence of variable gate logic thresholds. International test conference, pp 63–72
Maxwell P, O'Neill P, Aitken R, Dudley R, Jaarsma N, Quach M, Wiseman D (1999) Current ratios: a self-scaling technique for production I_{DDQ} testing. International test conference, pp 738–746
McCluskey EJ, Tseng C-W (2000) Stuck-fault tests vs. actual defects. International test conference, pp 336–342
Mei KY (1974) Bridging and stuck-at faults. IEEE Trans Comput C23(7):720–727
Meijer M, Pessolano F, Pineda de Gyvez J (2004) Technology exploration for adaptive power and frequency scaling in 90nm CMOS. International symposium on low power electronics and design, pp 14–19
Miller AC (1999) I_{DDQ} testing in deep submicron integrated circuits. International test conference, pp 724–729
Millman SD, McCluskey EJ, Acken JM (1990) Diagnosing CMOS bridging faults with stuck-at fault dictionaries. International test conference, pp 860–870
Nigh P, Forlenza D, Motika F (1997) Application and analysis of I_{DDQ} diagnostic software. International test conference, pp 319–327
Nigh P, Gattiker A (2004) Random and systematic defect analysis using I_{DDQ} signature analysis for understanding fails and guiding test decisions. International test conference, pp 309–318
Patten P (2004) Divide and conquer based fast Shmoo algorithms. International test conference, pp 197–202
Polian I, Engelke P, Renovell M, Becker B (2003) Modeling feedback bridging faults with non-zero resistance. European test workshop, pp 91–96
Polian I, Kundu S, Galliere J-M, Engelke P, Renovell M, Becker B (2005) Resistive bridge fault model evolution from conventional to ultra deep submicron technologies. VLSI test symposium, pp 343–348
Pomeranz I, Reddy SM (1992) On the generation of small dictionaries for fault location. Proceedings international conference on computer-aided design, pp 272–279
Rajsuman R (1991) An analysis of feedback bridging faults in MOS VLSI. VLSI test symposium, pp 53–58
Rajsuman R (Apr 2000) I_{DDQ} testing for CMOS VLSI. Proc IEEE 88(4):544–568
Rearick J, Patel J (1993) Fast and accurate CMOS bridging fault simulation. International test conference, pp 54–62
Renovell M, Huc P, Bertrand Y (1994a) A unified model for inter-gate and intra-gate CMOS bridging fault: the configuration ratio. Asian test symposium, pp 170–174

Renovell M, Huc P, Bertrand Y (1994b) CMOS bridging fault modelling. VLSI test symposium, pp 392–397

Renovell M, Huc P, Bertrand Y (1995) The concept of resistance interval: a new parametric model for realistic resistive bridging fault. VLSI test symposium, pp 184–189

Renovell M, Azaïs F, Bertrand Y (1999) Detection of defects using fault model oriented test sequences. J Electron Testing: Theory Appl 14:13–22

Rodríguez-Montañés R, Segura J, Champac V, Figueras J, Rubio A (1990) Bridging faults in CMOS: possibilities of current testing. European solid-state circuit conference, pp 117–120

Rodríguez-Montañés R, Segura JA, Champac VH, Figueras J, Rubio JA (1991) Current vs. logic testing of gate oxide shorts, floating gate and bridging failures in CMOS. International test conference, pp 510–519

Rodríguez-Montañés R, Bruls EMJG, Figueras J (1992) Bridging defects resistance measurements in CMOS process. International test conference, pp 892–899

Rodríguez-Montañés R, Bruls EMJG, Figueras J (1996) Bridging defects resistance in the metal layer of a CMOS process. J Electron Testing: Theory Appl 8:35–46

Rodríguez-Montañés R, Arumí D, Figueras J (2006) Effectiveness of very low voltage testing of bridging defects. IEE Electron Lett 42(19):1083–1084

Rubio A, Figueras J, Champac V, Rodríguez R, Segura J (1991) I_{DDQ} secondary components in CMOS logic circuits preceded by defective stages affected by analogue type faults. IEE Electron Lett 27(18):1656–1658

Sachdev M (1997) Deep sub-micron I_{DDQ} testing: issues and solutions. European design and test conference, pp 271–278

Sar-Dessai VR, Walker DMH (1999) Resistive bridge fault modeling, simulation and test generation. International test conference, pp 596–605

Schuermyer C, RuMer J, Daasch R (2004) Minimum testing requirements to screen temperature dependent defects. International test conference, pp 300–308

Semenov O, Sachdev M (2000) Impact of technology scaling on bridging fault detection in sequential and combinational CMOS circuits. International workshop on defect based testing, pp 36–42

Soden JM, Hawkins CF (1996) I_{DDQ} testing: issues present and future. IEEE Des Test Comput 13:61–65

Thibeault C (1997) A novel probabilistic approach for IC diagnosis based on differential quiescent current signatures. VLSI test symposium, pp 80–85

Thibeault C, Boisvert L (1998) Diagnosis method based on ΔIddq probabilistic signatures: experimental results. International test conference, pp 1019–1026

Thibeault C (Jun 2000) On the adaptation of Viterbi algorithm for diagnosis of multiple bridging faults. IEEE Trans Comput 49:575–587

Turner ME, Leet DG, Prilik RJ, McLean DJ (1985) Testing CMOS VLSI: tools, concepts and experimental results. International test conference, pp 322–328

Wu J, Rudnick EM (1999) A diagnostic fault simulator for fast diagnosis of bridge faults. International conference on VLSI design, pp 498–505

Wu J, Rudnick EM (Apr 2000) Bridge fault diagnosis using stuck-at fault simulation. IEEE Trans Comput-Aided Des Integr Circuits Sys 19(4):489–495

Zou W, Cheng W-T, Reddy SM (2005) Bridge defect diagnosis with physical information. Asian test symposium, pp 248–253

Chapter 3
Models for Delay Faults

Sudhakar M. Reddy

Abstract In this chapter fault models used to model the effects of defects causing excessive circuit delays are discussed. Methods to generate tests to detect modeled faults and design for test methods to improve fault coverage are reviewed. Current work in detecting what are called small delay defects is discussed.

Keywords Delay faults · Delay fault testing · Small delay defects · Design for test

3.1 Introduction

Delay faults model defects that affect performance of logic circuits in contrast to other fault models such as line stuck-at faults that model defects that affect the functionality of circuits. Delay faults were first considered by Breuer in 1974 (Breuer 1974) and have been extensively studied since early 1980s. Electronic gates and interconnects introduce finite delays in signal propagation in digital integrated circuits. Process variations and defects in manufacturing could cause delays in some gates and interconnects to be larger or smaller than the delays used for the design. Common defects in VLSI circuits are opens and unwanted shorts/bridges between circuit nodes. Resistive opens and bridges cause signal propagation delays to increase at the defect site. Opens and bridges are discussed in detail in Chapters 1 and 2, respectively. Larger delays cause data set up time violations at the inputs to the storage elements (latches and flip-flops) of the design causing the manufactured circuits to fail to operate correctly at the desired frequency of operation. These are the faults that are typically called delay faults in the literature and are the subject of this chapter. Smaller than modeled delays cause data hold time violations at storage element inputs and cause the circuit to malfunction at all frequencies of operation. Models for data hold time violations and tests to detect them have been investigated in (Reddy et al. 2000) and more extensively investigated in the context of diagnosing

S.M. Reddy (✉)
Department of Electrical and Computer Engineering, University of Iowa, Iowa City, Iowa, USA
e-mail: reddy@engineering.uiowa.edu

H.-J. Wunderlich (ed.), *Models in Hardware Testing: Lecture Notes of the Forum in Honor of Christian Landrault*, Frontiers in Electronic Testing 43,
DOI 10.1007/978-90-481-3282-9_3,

scan chain failures (Guo et al. 2006; Sinanoglu et al. 2007). This chapter contains description of basic fault models to represent delay defects together with methods to generate and apply tests to detect them. Some of the recent works on improving delay defect coverage are reviewed.

In this section, fault models proposed to model the effects of delay defects and test methods to detect the modeled faults are briefly described. In the next two sections recently developed methods to improve delay defect coverage are discussed. Included are methods, discussed in Section 3.2, to detect small delay defects and coverage metrics. Another topic discussed is related to the following concern regarding scan based test methods. Scan allows an arbitrary state to be scanned in. However under normal operation typically many states are not reached. Thus scan based tests tend to operate the circuit in non-functional ways during test. Thus scan based tests often detect many modeled faults that cannot be detected using functionally reachable states only. Concern is that non-functional operation during scan based tests may lead to yield loss by failing good chips (Rearick 2001). Methods to generate tests that avoid non-functional operation are discussed in Section 3.2. Non-functional operation leads to much higher switching activity during test causing supply voltage droops which in turn cause propagation delays in the circuit to increase during test. Several methods to reduce switching activity during scan based test are discussed in Chapter 7. In Section 3.3, DFT methods to increase delay fault coverage and methods to reduce design effort for what is known as launch off shift (LOS) test application method are described.

3.1.1 Basics of Test Generation

Some of the basic terms used in the context of generation of tests for digital logic circuits and terms used in this chapter are defined and illustrated next. For simplicity we assume that the circuits we consider contain primitive gates NOT, NOR, NAND, OR and AND gates and flip-flops. For each gate input we ascribe two signal propagation delays for propagating a falling and a rising transition from an input to the output of a gate. For a multiple input primitive gate cv is called the controlling value if the output of the gate is determined when one of the input value is cv. For example 0(1) is the controlling value for AND and NAND (OR and NOR) gates. Complement of the controlling value is called the non-controlling value ncv. Circuit leads are labeled so that they can be referenced. External inputs to a circuit are called primary inputs (PIs) and the external outputs are called primary outputs (POs). We refer to the outputs of flip-flops driving the combinational core of a sequential circuit as pseudo-primary inputs (PPIs) and the outputs of the combinational circuit driving the inputs of the flip-flops as pseudo-primary outputs (PPOs). A physical path in a combinational circuit is an alternating sequence of circuit leads and is typically represented by a sequence of circuit leads such that two consecutive leads in a path are an input and output of a gate in the circuit and the first lead in the sequence is a PI or PPI and the last lead in the sequence is a PO or PPO. With each circuit

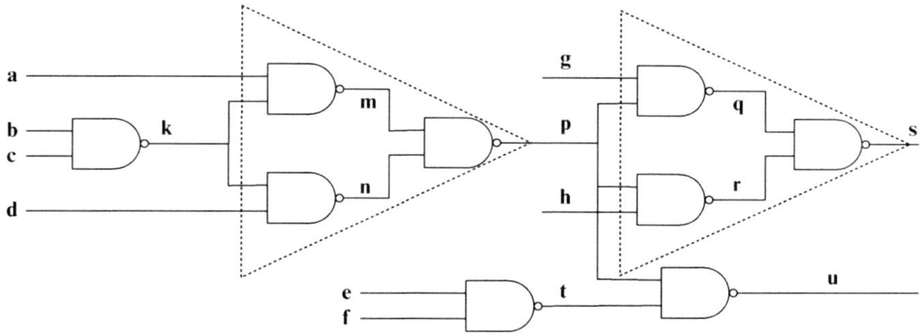

Fig. 3.1 Illustrating some terms used

path we associate two logical paths one with a rising signal transition at the input to the path and one with a falling transition at the input to the path. The circuit leads on a path that are inputs to the gates on the path are called the on-path inputs and the other inputs to the gates on the path are called the side inputs or off-path inputs. As an example consider the circuit shown in Fig. 3.1. The sequence of circuit leads a-m-p-q-s is a physical path with on-path inputs a, m, p and q and side inputs n, r and the upper fan-out branches of k and p. Associated with the physical path a-m-p-q-s we have two logical paths with rising and falling transitions at input a of the path. Each logical path has a delay associated with it which is the sum of the delays on the on-path circuit leads in the path.

In Fig. 3.1 the parts of the circuit enclosed by the dashed triangles are called fan-out-free regions (FFRs). Given any combinational circuit one can uniquely partition the circuit into FFRs. Inputs to a FFR are branches of fan-out stems or circuit inputs without fan-out of >1 and the output of a FFR is either a fan-out stem or a circuit output. For example the inputs to the FFR on the left of Fig. 3.1 are circuit inputs a and d and the two fan-out branches of fan-out stem k. Note that from any input of an FFR there is a unique physical path to the output of the FFR.

In order to detect a modeled fault two steps called fault activation and fault effect propagation are needed. For brevity, fault effect propagation is referred to as fault propagation. Fault propagation requires sensitizing one or more circuit paths or subpaths starting from the fault site to a circuit output. We say that a circuit path or subpath is sensitized if the path/subpath output value changes when the path/subpath input value changes (due to the fault effect). A circuit path/subpath is sensitized by setting the side inputs of the gates along the path/subpath to specific values. Next we illustrate these concepts for the most widely used fault model called the line stuck-at fault model. We say that a line r in a faulty circuit is stuck-at-v, $v = 0$ or 1, if line r is permanently at logic state v. If the circuit has N lines then the number of single line stuck-at faults is $2 \bullet N$ and hence the total number of faults to be considered is linear in the size of the circuit. It is possible to reduce the total of explicitly considered faults for test generation by collapsing the set of faults using what are called fault equivalence and dominance relations (Abramovici et al. 1990).

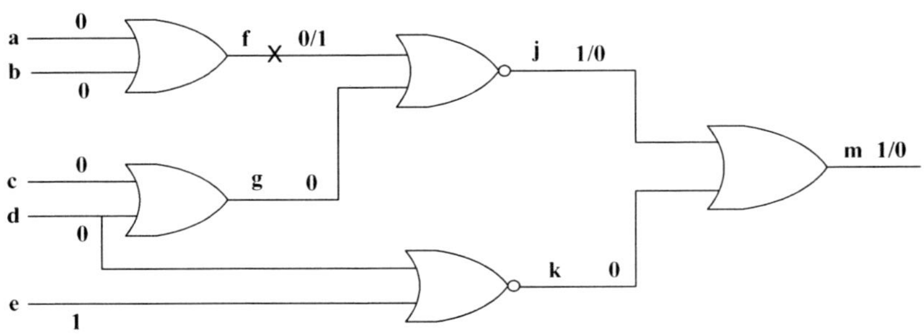

Fig. 3.2 Detecting line f stuck-at-1

A stuck-at-0(1) fault on line r is activated by setting line r to 1(0) in the fault-free circuit. Note that in the faulty circuit, line r remains at the stuck-at value 0(1). To propagate the effect of this fault to a circuit output one must sensitize at least one path starting at line r to the circuit output by setting the side inputs of the gates in the path to non-controlling values. This is illustrated in Fig. 3.2 where a test to detect line f stuck-at-1 is shown. The fault on line f is activated by setting f to 0 in the fault-free circuit and the fault is propagated to the output by setting the side inputs k and g of the unique path from f to the circuit output to the non-controlling value 0. In order to obtain the desired values at the fault site and the side input values the circuit inputs a, b, c, d and e must be set to the values shown in Fig. 3.2. Also, in Fig. 3.2, the signal values shown as p/q are the composite values in the fault-free and faulty circuit with the value above the slash, "/", representing the signal value in the fault-free circuit and below the slash the signal value in the faulty circuit. The example in Fig. 3.2 also illustrates what are called necessary assignments (NA). For example in order to activate the fault f stuck-at 1 line f has to be set to 0 which in turn implies that inputs a and b must be set to 0. Similarly in order to propagate the fault effect to the circuit output line k and g must be set to 0. In order to set line g to 0 inputs c and d must be set to 0. With input d set to 0 in order to set k to zero e must be set to 1. The unique values needed to activate and propagate a fault are called necessary assignments which can be obtained through what are known as forward and backward (unique) implications (Abramovici et al. 1990) and through what are known as static and dynamic learning (Schulz et al. 1988).

3.1.2 Delay Fault Models

Two basic fault models proposed in the literature to model defects that increase circuit delays are gate delay faults (Carter et al. 1987) and path delay faults (Smith 1985). In gate delay fault model we associate with each circuit line a slow to rise (STR) and a slow to fall (STF) fault and a size of the fault which is the additional delay due to a defect added to the modeled delay for propagation of rising and falling transitions, respectively, on the circuit lead. Path delay faults model increases from the modeled delay for the logical paths and hence include the sum of delay increases, if any, anywhere along the path. In a circuit with N lines there are $2 \times N$ single gate delay faults. However one needs to consider a range of delay defect sizes which makes this model more complex than is indicated by the number of faults. Since there could be exponentially large number of paths in a circuit the total number of path delay faults in a circuit could be extremely large. For this reason a fault model called transition delay fault (TDF) or simply transition fault has been proposed (Barzilai et al. 1983). In this model, as in the gate delay fault model, we associate with each circuit line an STF and an STR fault and assume that the defect size is so large that every signal path containing the defect site has a delay higher than the delay permitted by the test clock period (Waicukauski et al. 1987).

3.1.3 Tests to Detect Delay Faults

Methods to generate tests to detect delay faults use the basic steps described in Section 3.1.1. Test generation and fault simulation methods for combinational circuits and scan designs as well as for sequential circuits have been investigated. First some basic procedures are reviewed using combinational circuit examples followed by a short review of work on sequential circuits.

Tests for TDFs can be generated by straight forward modification of procedures for generating tests to detect line stuck-at faults (Waicukauski et al. 1987). For example consider detection of an STR fault on line *d* in the circuit shown in Fig. 3.3. In order to activate this fault we should create a rising transition on line *d*. This implies that the fault can only be activated by applying two consecutive patterns. A pair of such patterns is shown in Fig. 3.3a. In the presence of an STR fault on line *d* the value on *d* will not change to 1 from 0 before the circuit outputs are read due to the assumption that the delay defect size is large enough. Thus when the circuit output is read, line *d* will still be 0. This is shown as the faulty value under the slash in Fig. 3.3b. Thus the STR TDF manifests itself as a stuck-at-0 fault when the second input of the two pattern test is applied. In general, to detect a STR(STF) TDF a two pattern test <t1, t2> must satisfy the following two conditions: (i) the first

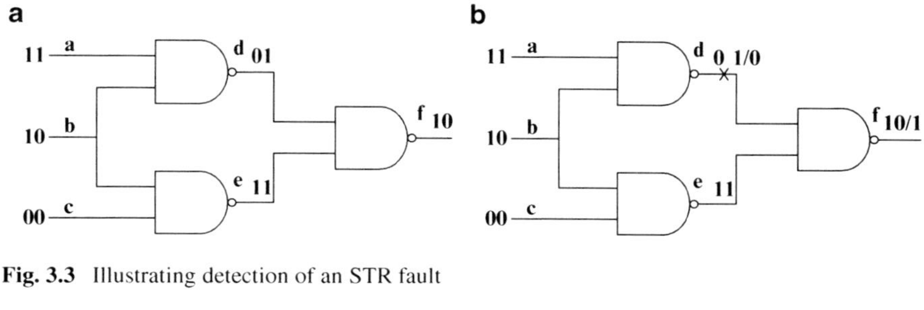

Fig. 3.3 Illustrating detection of an STR fault

pattern t1 should set the value on the faulty line to 0(1) and (ii) the second pattern t2 must detect a stuck-at-1(0) fault on the faulty line. The first pattern t1 is called the initialization pattern that initializes the faulty line to a 0(1) for STR(STF) TDF fault. Thus one can modify test pattern generators and fault simulators for stuck-at faults in a straight forward manner to obtain similar tools for TDFs. This is one of the advantages of the TDF model.

Tests for gate delay faults require accounting for delay defect size. For example if the defect size at a circuit lead r is less than the slack of r the fault may not be detectable by any test. Slack of a circuit line is the difference between the period of the functional clock and the maximum delay of all paths through r. This requires accurate timing models for signal propagation and accommodation of the fact that signal propagation delays can only be modeled as a range between minimum and maximum delays (Iyengar et al. 1990; Pramanick et al. 1997). Additionally propagation delay of a gate input depends on the states of the other inputs to the gates and coupling capacitances to other adjacent lines. Another issue that crops up is the fact that more than one test may be needed to detect all defect sizes at some fault site that may cause malfunction at the desired frequency of operation (Pramanick et al. 1997). Methods to determine a threshold value of defect size above which the fault is detected by a given test have been developed (Iyengar et al. 1990; Pramanick et al. 1997; Dumas et al. 1993). However, such methods pessimistically estimate the defect sizes covered (Pramanick et al. 1997) due to the fact that a given test may detect a range of defect sizes instead of only defects with sizes larger than a threshold. For gate delay faults, as well as transition faults, activation of faults by hazards or glitches also needs to be considered (Pramanick et al. 1997; Brand et al. 1994; Pomeranz et al. 2009b).

Path delay fault model (Smith 1985) is more comprehensive since it can accommodate spot defects and distributed defects. A path delay fault is associated with each logical path and is said to be present if the delay of the logical path exceeds the slack of the path. Tests to detect path delay faults are classified according to the conditions satisfied by the side inputs of the gates in the path. The different types of tests are discussed below.

In Fig. 3.4 a robust test for the logical path c-g-h-k with rising transition at its input is shown. The path delay fault test is called a robust test if it detects the fault independent of the delays in the rest of the circuit (Smith 1985; Lin et al. 1987). In Fig. 3.4 we show the signal values for the two pattern test where S0 represents a signal value that is a glitch free 0 during the application of the two patterns. For the same fault, in Figs. 3.5 and 3.6 non-robust tests called strong non-robust and weak non-robust tests are shown. The signal value H0 represents a signal value that is a 0 in the steady state but may have a hazard or glitch during the transition from the first pattern to the second pattern of the two pattern test. Non-robust tests can be invalidated due to circuit delays that effect signals at the off-path inputs (Konuk 2000; Pomeranz et al. 2008a). For some paths neither robust nor non-robust tests may exist. However such paths may be functional paths and should be tested using functional sensitization (Cheng et al. 1996).

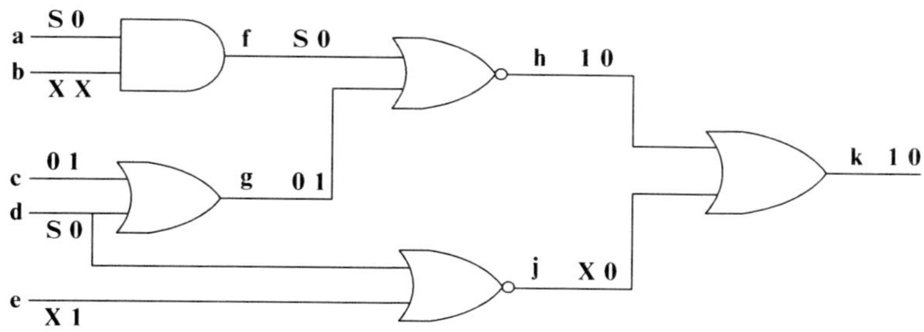

Fig. 3.4 Robust test for path c-g-h-k with rising transition

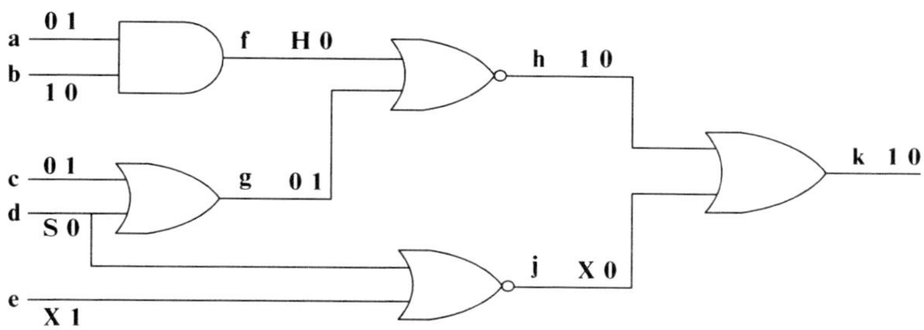

Fig. 3.5 Strong non-robust test for path c-g-h-k with rising transition

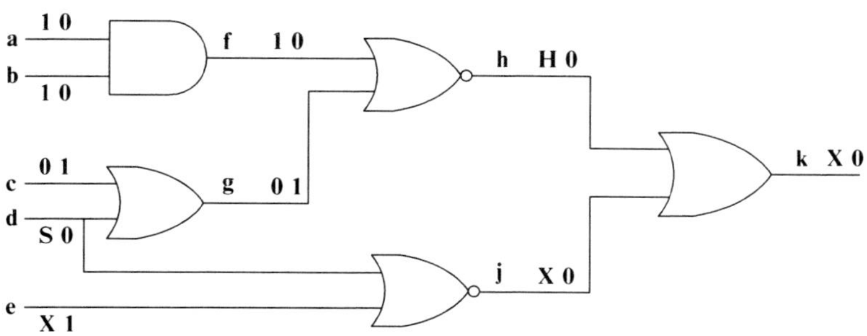

Fig. 3.6 Weak non-robust test for path c-g-h-k with rising transition

Generation of tests for path delay faults and fault simulation have been extensively investigated (Krstic et al. 1998; Pomeranz et al. 1998; Bushnell et al. 2000; Jha et al. 2003). Typically one uses the necessary conditions on off-path inputs of the gates on the path for sensitizing the path to determine additional necessary conditions through implications followed by justifying all the necessary conditions. Path sensitization conditions for different types of two pattern tests are shown in Table 3.1. In Table 3.1 S0(S1) represent hazard free 0(1) during the application of the two pattern test and H0(H1) represent a signal that is a 0(1) in the steady state

Table 3.1 Sensitization conditions on off-path inputs

			Non-robust					
	Robust		Strong		Weak		Functional	
Transition	AND NAND	OR NOR	AND NAND	OR NOR	AND NAND	OR NOR	AND NAND	OR NOR
0 → 1	X1	S0	X1	H0	X1	X0	X1	XX
1 → 0	S1	X0	H1	X0	X1	X0	XX	X0

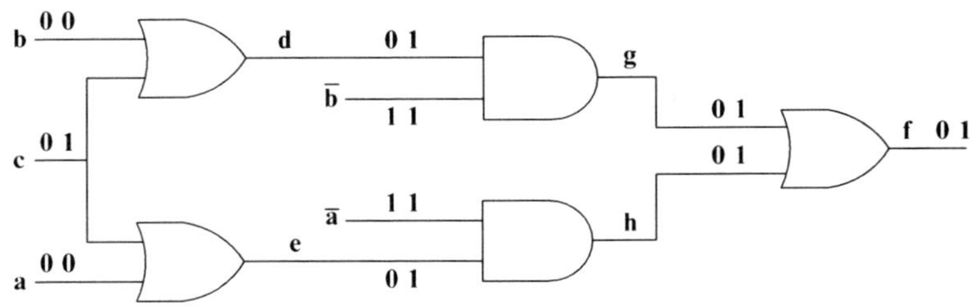

Fig. 3.7 Multipath test (Ke et al. 1995)

but may have a hazard/glitch. It may be necessary to consider subsets of paths or multipaths to generate tests (Ke et al. 1995). In Fig. 3.7 we show an example circuit from Ke et al. (1995), in which the logical paths c-d-g-f and c-e-h-f rising can only be tested together and not separately. In general, it is necessary to consider what are called primitive multipath faults (Ke et al. 1995).

Tests for sequential circuits can be applied using slow and fast clocks (Malaiya et al. 1983), or using a single fast clock (Pomeranz et al. 1992; Cheng 1993). When using slow and fast clocks the initialization pattern is obtained using a slow clock followed by the second vector of a two pattern test. Response to the second vector is captured using a fast clock whose period is the desired clock period during normal operation. The captured error values in flip-flops are then propagated to primary outputs again using a slow clock. It is assumed that when a slow clock is applied the circuit operates as a fault-free circuit. When only the system clock is used for delay faults one has to accommodate the fact that the effect of the delay fault may remain for one or more cycles (Pomeranz et al. 1992; Cheng 1993).

3.1.4 Delay Fault Tests for Scan Designs

Scan is universally used in large industrial designs to facilitate test generation, debug and failure diagnosis of synchronous sequential circuits. In Fig. 3.8 a scan design called MUX-Scan is illustrated. Single pattern tests are applied by shifting/scanning in the state part of the test with scan enable (SE) at logic 1. After loading the scan chain the primary input values for the test are applied. SE is changed to 0 and one

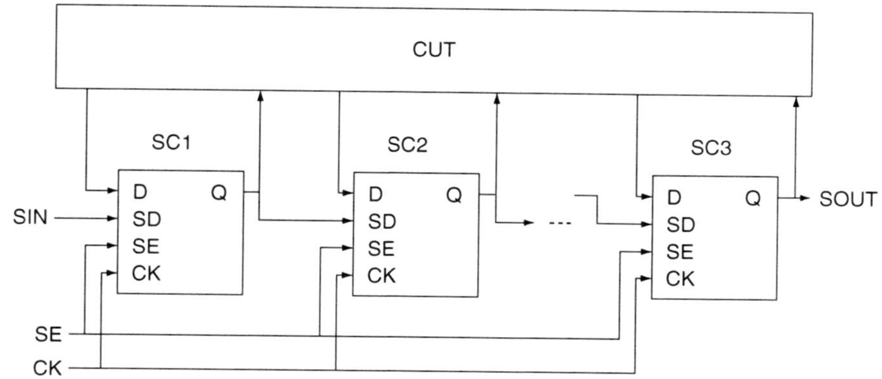

Fig. 3.8 A scan chain

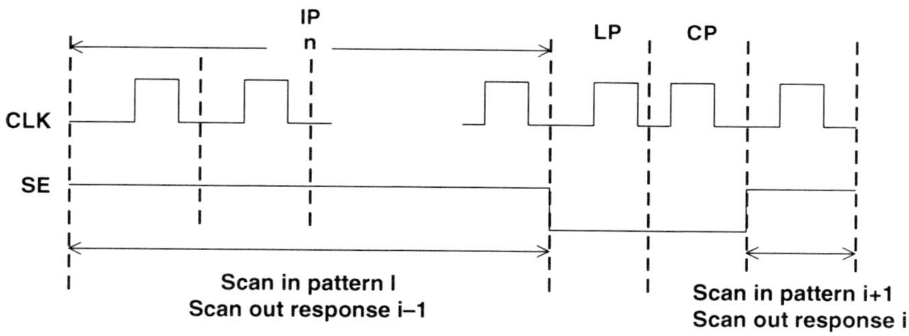

Fig. 3.9 Timing diagram for two pattern tests using LOC test application method

clock cycle is applied to capture the circuit response to the test. SE is changed to 1 and the captured response is scanned out and at the same time the next test is shifted in. For two pattern tests, two methods of test application called skewed-load (Savir et al. 1993) also called launch off shift (LOS) and broadside (Savir et al. 1994) also called launch off capture (LOC) are used. Both methods can be regarded to have three phases. In the first phase, called initialization cycle or initialization phase (IP), the first vector V1 of a two pattern test <V1, V2> is scanned in with SE = 1. The two methods differ in the next phase called the launch phase or launch cycle (LP). In LOS method the second vector V2 is obtained by shifting once with SE staying at 1. Thus V2 is restricted to be a single shift of V1. In LOC test method V2 is obtained through the combinational logic of the circuit by setting SE = 0. Thus in LOC also V2 is obtained as a function of V1. In the third phase, called the capture cycle (CP), in LOS method SE is changed to 0 and the response to the test applied is captured. In LOC method SE is maintained at 0 and the response to the test is captured as for the LOS method. The timing waveforms for the two methods are shown in Figs. 3.9 and 3.10. From the waveforms for the LOS method it can be

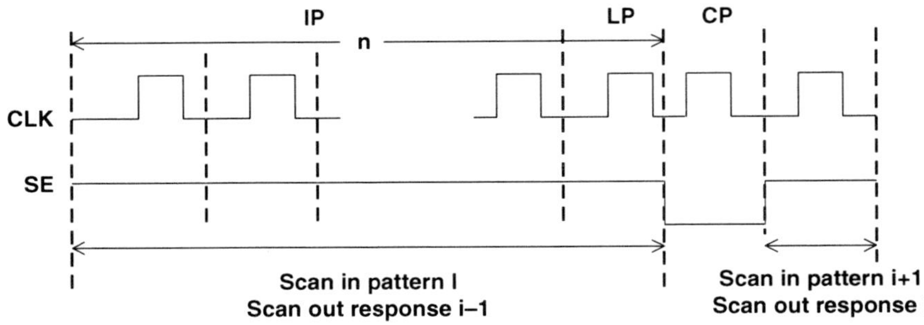

Fig. 3.10 Timing diagram for two pattern tests using LOS test application method

seen that SE has to change fast before the capture cycle. This implies that the SE net must be designed similar to a clock network since it is also distributed to all the scan cells (flip-flops). In LOC method SE has to switch after the initialization cycle and this can happen as slowly as needed by, for example, introducing some idle cycles after the initialization phase. In at-speed test the capture cycle, also referred to as a fast capture cycle, is applied after one clock period of the desired frequency of operation.

In practice, the following advantages and disadvantages of the LOS and LOC test methods have been observed. Test generation times and test set sizes for LOS method are much smaller and achievable fault coverage is higher compared to that for LOC method. Additionally, when multiple scan chains are used as is typical in large industrial designs to reduce test application time, fault coverage using LOS tests increases compared to using a single scan chain (Pomeranz et al. 2002). Fault coverage using LOC tests is independent of the number of scan chains used. However design effort to insure that SE can switch state fast is higher while a fast SE is not needed for LOC test method. LOC tests are often preferred since they are "closer" to the normal functional operation. It should be noted that in one scan design method called Level Sensitive Scan Design (LSSD) (Eichelberger et al. 1978) all scan chain control signals are designed as clocks and hence both LOS and LOC test methods can be used without any additional design effort. Both methods achieve lower fault coverage than if arbitrary two-pattern tests are applicable, for example using enhanced scan (Dasgupta et al. 1981) that has a three latch scan cell to enable storing both patterns of a two pattern test. However using enhanced scan that adds extra hardware overhead may not be acceptable for many designs.

3.1.5 Non-enumerative Procedures and Path Selection Methods

Since the number of paths in a realistic design could be extremely large and so could be the number of tests to detect all detectable path delays, several methods have been developed to address these issues. In order to reduce the impact of the size of the set of path delay faults on fault simulation and test generation non-enumerative procedures were first proposed in Pomeranz et al. (1994), Pomeranz et al. (1995b). Non-enumerative methods do not explicitly consider all path delay faults. These methods have been further developed, for example, in Gharaybeh et al. (1998), Kagaris et al. (2002), Tragoudas et al. (1999). To assess the cost of test application if all path delay faults are targeted, a method to determine a lower bound on the number of tests to detect all path delay faults was proposed in Pomeranz et al. (1996a).

Even though non-enumerative procedures help reduce fault simulation and test generation times for path delay faults, still the number of tests to detect all path delay faults is typically too large. For this reason procedures to select a subset of path delay faults to be targeted for detection have been proposed. These include selecting only paths of maximum delay, selecting paths whose delay is within up to a certain percentage of the maximum and selecting a subset S of paths such that for each circuit lead r there is at least one path in S whose delay is maximum among all paths containing r (Malaiya et al. 1983; Smith 1985). A procedure of polynomial complexity was developed in Li et al. (1989) to select a subset of minimum number of paths S such that S contains at least one path of longest delay, for each circuit lead r, among all paths through r. However, given that many path delay faults in a circuit may not have tests, some or many faults in the selected subset may not be testable. For this reason procedures that efficiently identify untestable paths have been developed (Lam et al. 1993; Cheng et al. 1993; Sparmann et al. 1995; Kajihara et al. 1997; Kajihara et al. 2000; Shao et al. 2001). The methods in Kajihara et al. (1997), Kajihara et al. (2000), Shao et al. (2001) are non-enumerative and the key idea behind these methods is illustrated in Fig. 3.11. The methods find pairs of lines called (b,f) pairs (Kajihara et al. 1997) such that there are paths between line b and line f of the circuit and any path fault containing the two lines is untestable. The lines b and f are logical lines which are physical lines with which a rising or falling transition is associated. The pairs of lines considered are inputs to FFRs. Consider inputs a and c of the two FFRs shown in Fig. 3.11. There are unique subpaths from a to b and c to d in the two FFRs. If the necessary assignments to sensitize these two subpaths cannot be justified simultaneously then all the path delay faults containing lines a and c are untestable. In Murakami et al. (2000) a subset of path delay faults were selected that

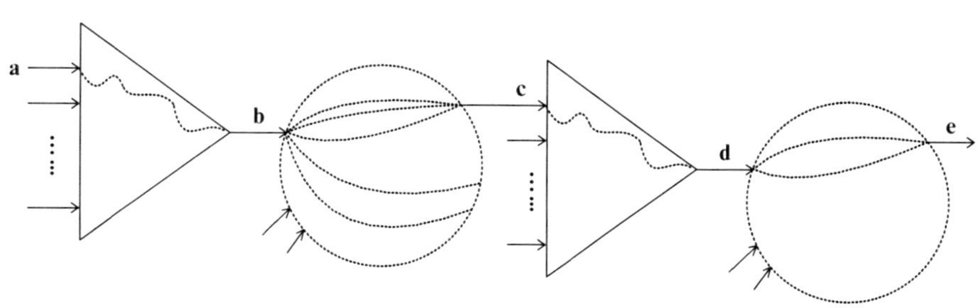

Fig. 3.11 Determining untestable paths

avoid (b,f) pairs of lines and such that the subset contains at least one path delay fault for each circuit lead with maximum delay among all paths containing the lead. Over 90% of the path delay faults in such subsets were found to be testable.

3.1.6 Additional Delay Fault Models

In addition to the basic delay fault models, gate and path delay faults, several other fault models have been proposed. These include double and multiple transition fault model (Pomeranz et al. 1996b) and the segment fault model (Heragu et al. 1996). These fault models require tests that robustly propagate transitions through subpaths containing pairs or multiple lines of circuits. Segment fault model considers a set of two or more consecutive circuit lines. These fault models are more complex than TDF model but are less demanding than path delay fault model.

Even though path delay faults model the effect of accumulated delays along the circuit lines on the path a non-robust test for a path delay fault may not detect extra delay in a lead or a subpath of the path. This is illustrated in Fig. 3.12. The two pattern test shown in Fig. 3.12 is a non-robust test for the path b-d-f with a rising transition at b. However, this test does not detect the STR fault on line b as shown by the faulty circuit values under "/". This means that if the circuit shown in Fig. 3.12 is part of a larger circuit a non-robust test for a path that contains the subpath b-d-f may not detect accumulated excess delay up to line b. Given that many if not most of the path delay faults can only be detected by non-robust tests, methods to generate non-robust tests to address this weakness were investigated. Towards this goal, in Pomeranz et al. (2008a) a fault model called Transition Path Delay Faults was proposed. This model requires that a test that detects a path delay fault also detects appropriate transition delay faults on each on-path line.

In many designs handcrafted custom blocks are used for which accurate or even any gate level descriptions may not be available. Tests for delay faults for such designs need to consider them as black boxes. For such designs functional test methods were proposed in Underwood et al. (1994) and Pomeranz et al. (1995a).

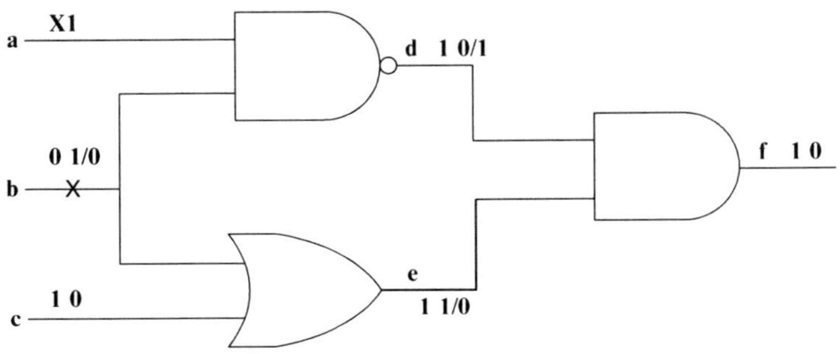

Fig. 3.12 Invalidation of a non-robust test

Resistive interconnect opens are one cause for delay defects. Noting that a resistive open slows down both the rising and falling transition on the defective line, Inline Resistance Fault model was proposed in Benware et al. (2004). An inline resistance fault on line r is detected if either a slow to rise or a slow to fall fault is detected on line r. Inline resistance fault model allows reduction in test patterns compared to TDF fault model.

When determining TDF coverage by a given sequence for a non-scan sequential circuit it is necessary to consider persistence of fault effects over more than one clock cycle (Cheng 1993). This requires simulating the sequence several times with different numbers of fault effect persistence cycles. In Pomeranz et al. (2008b) a transition delay fault model called Unspecified Transition Fault model was proposed which allows a one pass simulation of the given sequence.

3.2 Test Generation for TDFs and Small Delay Defects

In delay fault testing two conflicting goals need to be considered. One is achieving as high defect coverage as possible and the other is to avoid over testing. Over testing occurs due to non-functional operation during scan based test application (Rearick 2001). In this section we review some of the recent works related to both these issues.

As discussed above defects that increase circuit delays are modeled by gate delay faults, transition delay faults (TDFs) and path delay faults. Application of tests to detect all path delay faults is impractical and gate delay faults require accurate timing models. For these reasons for the detection of delay defects in industrial designs typically tests for TDFs are used together with tests for selected critical paths. However tests for TDFs may not provide adequate coverage of delay defects that are of small size. This can be seen by the example in Fig. 3.13. A TDF on line a can be propagated either through path a-f-g-j or through a-f-k. Typically test pattern generation tools propagate tests through easier to sensitize paths and hence the

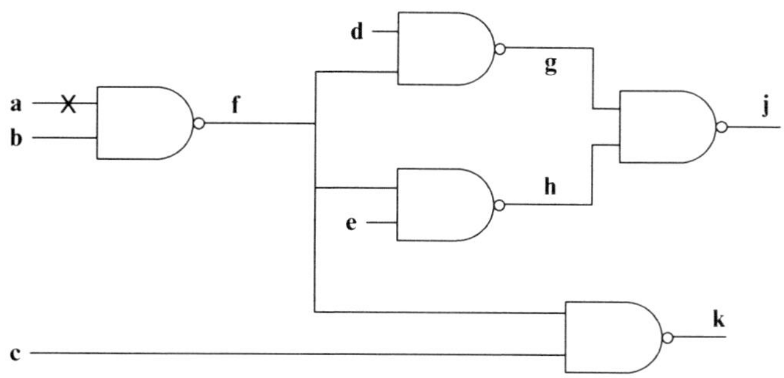

Fig. 3.13 Gate delay faults

fault may be propagated through the shorter path a-f-k. In this case the delay defect size needs to be larger for it to be detected. However a defect of a smaller size than detectable by the test will affect circuit operation when the transition on *a* is propagated through the longer path under normal operation. For this reason methods to activate and propagate TDFs through longest delay paths have been proposed. We review some of the recent works on generating TDF tests to detect small delay defects.

3.2.1 Functional Broadside Tests

There are two reasons for non-functional operation in scan based tests. One is the very fact that tests are shifted in to scan chains which is not a functional operation and the states of the circuit under tests go through many states that are not functional. The other is during launch and capture cycles of the application of two pattern tests non-functional operation may cause excessive switching activity that may cause supply voltage droops and higher heat dissipation. Voltage droops cause increase in circuit delays which may fail good chips (Saxena et al. 2003). Additionally tests using non-functional operation may propagate faults along non-functional paths potentially failing good chips even if the switching activity during test is not excessive. In this section we discuss recently developed methods to address the issue of non-functional operation during launch and capture cycles.

An LOC or broadside test can be represented by <s1,a,b>, where s1 is the state scanned in and a and b are the primary input values. The state part of the second pattern of the two pattern test is obtained through the functional logic. Hence if s1 is a state that can be reached during normal functional operation then the circuit will only operate within normal functional operation during test also. Observing this, Functional Broadside Tests were proposed in Pomeranz et al. (2006). In a functional broadside test the shifted in state s1 is a reachable state. A reachable state is a state that can be reached from the state of the circuit after it is synchronized. Any state reached after synchronization is a state that can occur during the normal operation of the circuit. Functional broadside tests insure that switching activity and supply current demands during launch and capture cycles are within those during normal operation. Additionally no non-functional paths will be activated. In Table 3.2 the numbers of TDFs detected by functional broadside tests (Lee et al. 2008) are compared with the numbers of faults detected using arbitrary broadside tests in full scan ISCAS-89 circuits. In Table 3.2, after the circuit name the numbers of collapsed TDFs are given followed by the numbers of faults detected by functional broadside and arbitrary broadside tests. From this data one can observe that numbers of faults detected by the functional broadside tests are sometimes smaller as can be expected. However, overall the numbers of detected faults are similar in most circuits. Expanding the functional operation to include the state transitions encountered during the application of a synchronizing sequence permits additional tests called Synchronization Broadside Tests (Pomeranz et al. 2009a). These tests may shift in

Table 3.2 TDFs detected by functional broadside tests

Circuit	# Faults	# Func. det	# Arb. det
S298	508	403	403
S344	552	522	522
S349	566	505	530
S382	646	488	500
S386	690	505	530
S444	764	554	568
S526	948	571	590
S641	734	575	699
S713	918	648	777
S820	1,574	1,281	1,283
S832	1,614	1,290	1,290
S1196	2,110	2,108	2,108
S1238	2,316	2,234	2,234
S1423	2,512	2,207	2,239
S1488	2,770	2,529	2,529
S1494	2,810	2,548	2,548
S5378	7,040	5,353	6,412
S35932	63,502	54,599	54,599

unreachable states, however they are restricted to state transitions that occur during synchronization of the circuit. Additional fault coverage beyond that obtained by functional broadside tests can be obtained using the synchronization broadside tests (Pomeranz et al. 2009a).

3.2.2 Pseudo-Functional Tests

Functional broadside tests require scanning in a reachable state. An alternate approach is to avoid shifting in an unreachable state. Unreachable states can be avoided by implications learned from the sequential circuit. Several earlier works, for example, Lin et al. (1998), Chen et al. (2003) used sequential static learning to identify untestable stuck-at and TDF faults. These learned implications help in insuring that the shifted in state of a broadside test is not an unreachable state. However, they do not guarantee that a state that does not violate the learned implications is indeed a reachable state. For this reason tests generated using sequential learning are called pseudo-functional tests (Lin et al. 2005). Several works have investigated methods to generate pseudo-functional tests (Lin et al. 2005; Zhang et al. 2005; Syal et al. 2006). In Table 3.3, the numbers of TDFs detected by pseudo-functional broadside tests in larger ISACAS-89 benchmark circuits are given from Zhang et al. (2005). As expected the sets of faults detected by pseudo-functional broadside tests are smaller and proper subsets of the faults detected by arbitrary broadside tests. Also pseudo-functional tests cause less switching activity during launch and capture cycles (Zhang et al. 2005). Another observation regarding the faults detected

Table 3.3 TDFs detected by pseudo-functional tests

Circuit	# Det-pseudo	# Det-arb
S3330	3,302	3,937
S5378	5,404	6,412
S9234	4,819	9,505
S13207	9,658	12,489
S15850	11,738	13,535
S38417	46,926	48,761
S35932	54,599	54,599
S38584	53,349	55,123

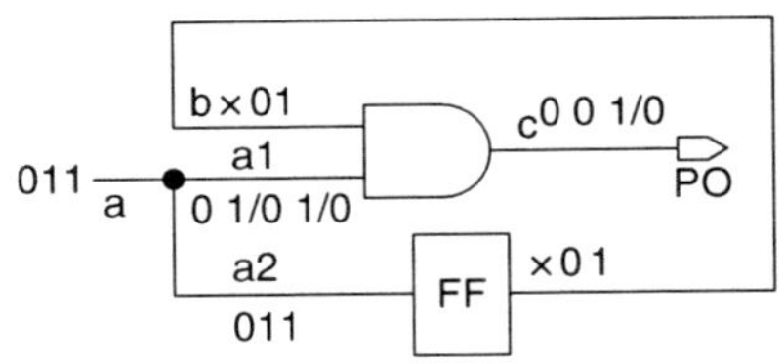

Fig. 3.14 A sequential circuit with STR fault et al.

by pseudo-functional tests is that even though in general LOS tests detect more faults than LOC (broadside) tests many faults that are detected by functional and pseudo-functional tests are not detected by LOS tests (Zhang et al. 2007a). Thus LOS tests may cause test escapes that cause malfunction of circuits in normal operation.

3.2.3 Tests with Multiple Activation Cycles

Tests to detect delay faults described so far used one launch cycle and one capture cycle. The launch cycle activates and propagates the fault. However some delay faults require multiple activation cycles for detection (Brand et al. 1994, Zhang et al. 2006a, Abraham et al. 2006). This is illustrated using an example from Zhang et al. (2006a).

Consider the sequential circuit shown in Fig. 3.14. Assume a slow to rise (STR) TDF on line a1. By definition a transition fault represents a delay fault of large (infinite) size. Consider a sequence of inputs 011 applied to a in three consecutive clock cycles. The values on all the signal lines in the circuit are shown using the standard notation of p/q to represent fault-free/faulty values on a signal line. It can be seen that the TDF on a1 affects the circuit performance in the sense that in its presence the circuit malfunctions when the input sequence 011 is applied. Now consider generating a test to detect the STR fault on a1 using a standard single activation cycle LOC test. Generation of such tests use an iterative logic array of two time frames as illustrated in Fig. 3.15a. Clearly the STR fault at a1 is not detectable since the fault effect is not propagated to the primary output or the flip-flop. A three cycles test, which uses an ILA of three time frames, is illustrated in Fig. 3.15b.

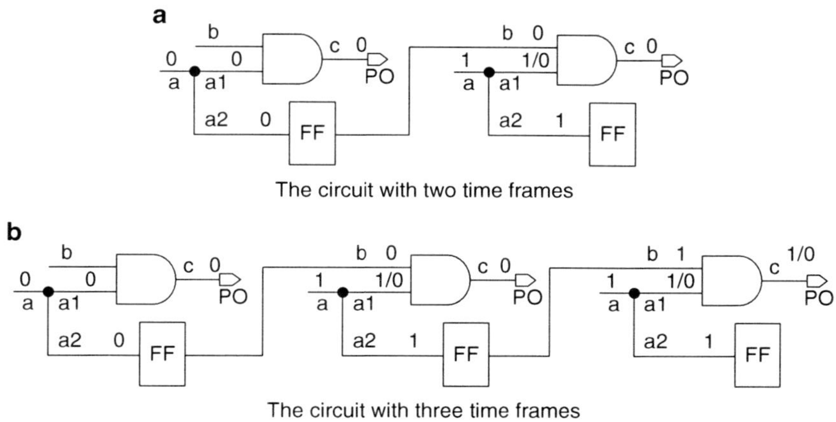

Fig. 3.15 LOC test for the circuit in Fig. 3.14

Using a three cycle test, with two activation cycles, the STR fault on a1 is detected as shown by the 1/0 on output c in time frame 3. This example shows that TDFs at some fault sites may not be detectable using LOC tests that use only one fault activation cycle but may be detectable using tests with more than one activation cycles. Similarly some TDFs not detected by two pattern LOS tests are detected by LOS tests with multiple activation cycles (Zhang et al. 2006a).

In Table 3.4 results on TDF detection using multiple fault activation cycles are given for ISCAS-89 circuits. After the circuit name the numbers of TDFs that can be detected using enhanced scan are given. This is the maximum number of TDFs that can be detected by any scan based tests. Next the numbers of faults detected by single activation cycle using LOC, LOS and jointly by LOC and LOS tests are given. Finally similar numbers are given when multiple activation cycles up to 11 are used. It can be seen that using multiple activation cycles and both LOC and LOS test methods, for most of the benchmark circuits, the same fault coverage as that achievable using enhanced scan can be achieved.

3.2.4 *Tests for Small Delay Defects*

In order to improve delay defect coverage whilst keeping the advantages of TDF model, it has been proposed to use tests that activate and propagate TDF faults through longest paths (Pramanick et al. 1989, Majhi et al. 1996, Shao et al. 2002). Following is a brief review of one of these works (Shao et al. 2002).

Tests for TDFs can be classified in to six types, shown in Table 3.5, based on how faults are activated and how they are propagated to observed outputs. In Table 3.5 SNRB and WNRB stand for strong non-robust and weak non-robust. Note that in this classification robust activation and propagation are considered as contained in

Table 3.4 TDFs detected by multi-cycle tests

Circuit	Max. Det.	Method	Det. Sngl.	Det. Mult.
S1423	2,488	LOC	2,239	2,450
		LOS	2,412	2,488
		LOC/S	2,476	2,488
S1488	2,770	LOC	2,529	2,728
		LOS	2,211	2,770
		LOC/S	2,694	2,770
S1494	2,794	LOC	2,548	2,753
		LOS	2,225	2,794
		LOC/S	2,718	2,794
S5378	6,961	LOC	6,412	6,428
		LOS	6,522	6,960
		LOC/S	6,899	6,961
S9234	10,698	LOC	9,517	9,687
		LOS	9,882	10,698
		LOC/S	10,608	10,698
S13207	15,379	LOC	12,489	13,193
		LOS	13,377	15,333
		LOC/S	14,895	15,367
S15850	18,403	LOC	13,535	14,920
		LOS	17,176	18,343
		LOC/S	17,752	18,385
S35932	56,446	LOC	54,599	56,257
		LOS	56,446	56,446
		LOC/S	56,446	56,446
S38417	49,544	LOC	48,761	49,039
		LOS	48,560	49,544
		LOC/S	49,487	49,544
S38584	58,979	LOC	55,129	56,811
		LOS	56,118	58,963
		LOC/S	58,060	58,979

Table 3.5 Six types of TDF tests

	Activation		Propagation	
	Type of path	Sensitization	Type of path	Sensitization
Type-I	Single	SNRB	Single	SNRB
Type-II	Multiple	Functional	*Single*	*SNRB*
Type-III	*Single/multi*	*Functional*	*Multi*	*SNRB*
Type-IV	*Single*	*SNRB*	*Single*	*WNRB*
Type-V	*Multi*	*Functional*	*Single*	*WNRB*
Type-VI	*Single/multi*	*Functional*	*Multi*	*WNRB*

the strong non-robust activation and propagation. In Figs. 3.16 to 3.18 three of the six types of tests are illustrated. Testable paths of largest delay are constructed by extending subpath(s) containing the fault site towards circuit inputs and circuit outputs.

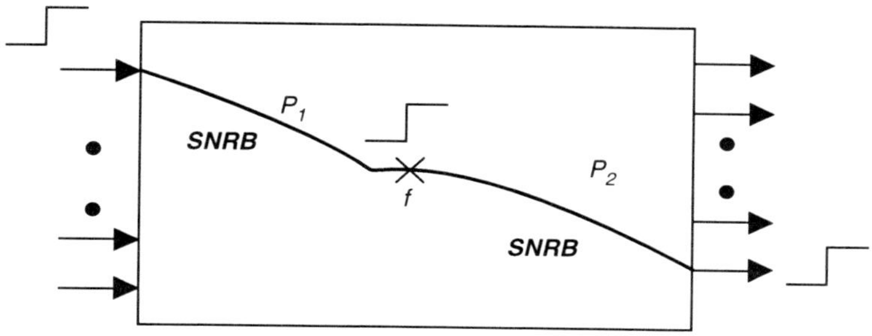

Fig. 3.16 Type-I tests

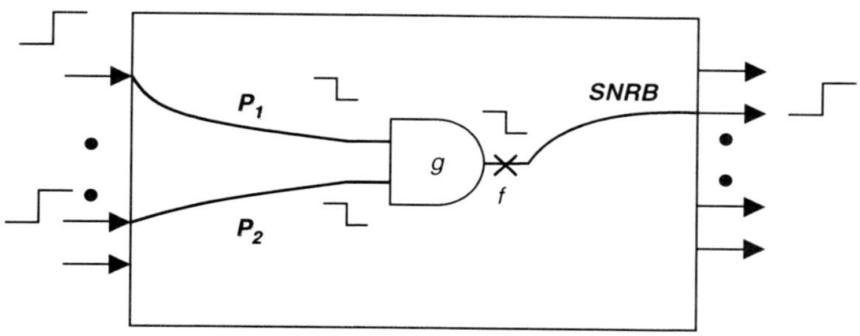

Fig. 3.17 Type-II tests

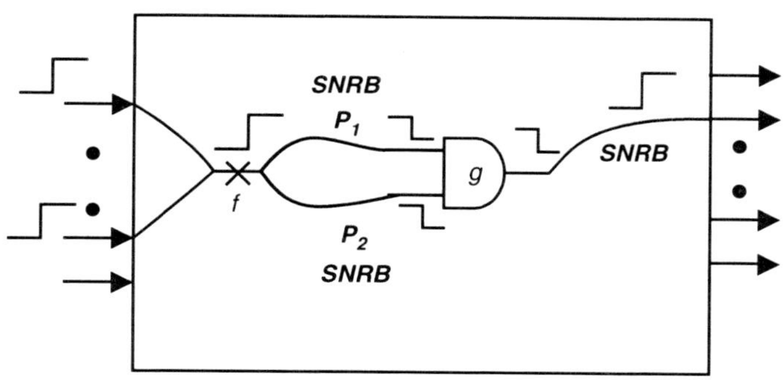

Fig. 3.18 Type-III tests

During the process of path extension information on (b,f) pairs is used to guide the path extension. Recall that all paths through a (b,f) pair are untestable. This is illustrated in Fig. 3.19 for Type I and Type IV tests. Each time the current subpath is extended a unique subpath through a FFR is chosen such that the extended path does not contain any (b,f) pairs.

Methods to generate compact test sets that attempt to activate and detect TDFs through largest delay paths have been proposed and a sketch of the method in Wang et al. (2008b) is given next. The method in Wang et al. (2008b) first finds a testable

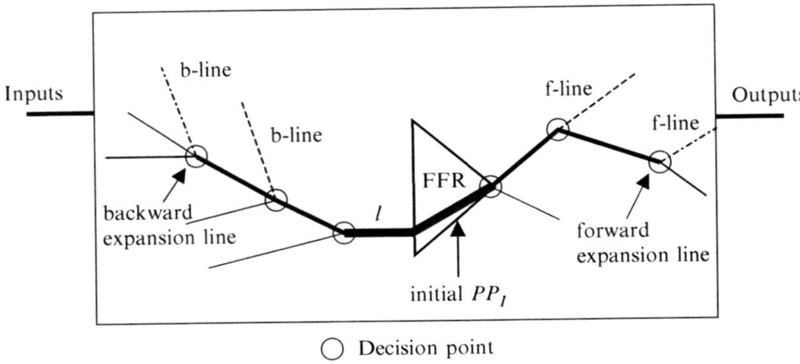

Fig. 3.19 Step-wise path expansion

path of largest delay for each TDF. For each such path necessary assignments to sensitize the path are found. Next clusters of TDFs such that the necessary assignments of the largest delay paths for any pair of faults do not conflict are determined together with the union of the necessary assignments for the largest delay paths of the faults in the cluster. Let the union of the necessary assignments for a cluster be CNA. Tests that satisfy all the necessary assignments in a CNA detect all the faults in the cluster through largest delay paths.

A method proposed to improve delay defect coverage is to reduce the test clock period to even less than the system clock period (Pramanick et al. 1990, Iyengar et al. 1992). This will allow detection of delay defects smaller than the slacks of the faults.

It is common to have device specifications that cover a range of supply voltage and temperatures. It is also now common that devices are operated at different power supply voltages to save power by dynamic supply voltage switching during run time (Cai et al. 2007). For such devices methods to generate TDF tests through longest paths need to consider the supply voltages and the range of operating conditions. The longest paths through target fault sites change with supply voltage and operating conditions such as temperature (Seshadri et al. 2005). However testing at several supply voltages and temperatures may be costly in test time. One solution suggested is to test at one or a minimal number of operating conditions and generate tests to detect the target faults N times through longest paths, for a small value of N (Seshadri et al. 2005). Testing more than one path may also be necessary to address variations in the delay of paths through a circuit lead due to process variations.

Several metrics to evaluate the effectiveness of a given TDF test set S to detect small defects (Pramanick et al. 1989; Park et al. 1989; Shao et al. 2002; Sato et al. 2005; Lin et al. 2006) and to determine probability of test escapes and defect levels have been proposed (Park et al. 1989; Sato et al. 2005).

Assume that the clock period of the capture cycle for tests used is Tc and the system clock period is Ts. Consider a TDF on a logical line r of the circuit. Let the maximum of the delays of all sensitizable paths (functional paths) through r be

Max(dr) and let the maximum of the delays of the paths through r used to detect the fault by tests in S be Max(tr). The slack of line r slack(r) = Ts – Max(dr) and let the test slack of line $r$be testslack(r) = Tc – Max(tr). Typically Tc is larger than Ts. A delay defect of size greater than or equal to slack(r) is detectable while the given test only detects delay defects of size greater than or equal to testslack(r). If the fault is not detected testslack(r) is defined to be infinity. Let the probability density function of defect sizes on line r be P_r (s), where s is the size of the defect. One can compute the coverage of defects on line r, C_r, by tests in S as given below:

$$C_r = \int_{testslack(r)}^{\infty} P_r(s)ds \bigg/ \int_{testslack(r)}^{\infty} P_r(s)ds \tag{3.1}$$

Note that if Ts = Tc and testslack(r) = slack(r) all defect sizes that are detectable at r are detected by the tests in S and in this case C_r is also 1. If the number of faults in the set of faults F is N then the coverage of delay defects in the entire circuit, DDC, can be computed as:

$$DDC = \sum_{r \in F} C_r / N \tag{3.2}$$

Note that DDC will be equal to 1 if for every fault p in the circuit coverage C_p is 1. Thus DDC is similar to the fault coverage metric typically used to report the effectiveness of covering modeled faults by a test set. Equations 3.1 and 3.2 are obtained from the statistical delay fault coverage (SDFC) metric proposed in Park et al. (1989) assuming that Tc and Ts can be different as assumed in Sato et al. (2005) and that the circuit path delays are constants.

A difficulty in using the coverage metric in Eqs 3.1 and 3.2 is the need to know the delays of sensitizable paths to compute slacks of fault sites. Instead one can use maximum delay of the structural paths through circuit leads (Park et al. 1989) which are typically higher than sensitizable path delays. A tighter estimate of Max(dr) could be obtained by determining the longest delay functionally sensitizable paths that do not contain any (b,f) pairs discussed in the last section. Another metric, used in the statistical delay quality model (SDQM), proposed in Sato et al. (2005) measures probability of not detecting delay faults using a given test set. Delay quality of a test set for a given fault r is defined as:

$$DQ(r) = \int_{testslack(r)}^{slack(r)} P_r\,(s)\,ds \tag{3.3}$$

The delay quality of a circuit with respect to a given test set is:

$$DQ = \sum_{r \in F} DQ\,(r)/N \tag{3.4}$$

DQ is an estimate of the probability that a chip that passed a given test set is defective. The metrics given in Eqs 3.1 through 3.4 require knowledge of the probability density function of defects. Metrics that can be used without the knowledge of the probability density functions have also been proposed (Pramanick et al. 1992; Shao et al. 2002; Lin et al. 2006). A metric proposed for the case where Pr (s) is not known is called delay test coverage (DTC) (Lin et al. 2006) given in Eq. 3.5 given below. DTC can also be used with any test clock period Tc.

$$DTC = \left(\sum_{r \in F} Max\,(tr)/Max\,(dr) \right) \Big/ N \tag{3.5}$$

3.3 DFT Techniques

In this section design for test (DFT) techniques proposed recently to reduce design effort for LOS test method and methods to increase delay fault coverage are discussed. In Section 3.3.1 two methods to reduce the design effort for LOS designs are discussed. Both these methods do not require any additional global signals and use only the signals already available in MUX-scan designs. In Sections 3.3.2 a method to increase delay fault coverage using multiple scan enable signals is outlined. In Section 3.3.3 achieving higher delay fault coverage using segmented scan designs is discussed.

3.3.1 LOS Testing Using Slow Scan Enable

As pointed out in Section 3.1.4, for MUX scan LOS tests require scan enable line to switch fast from 1 to 0. This is typically achieved by pipeline design for distributing scan enable line which has high design time overhead and area overhead. In Ahmed et al. (2007) a method to locally generate fast scan enable signal from a slow scan enable signal has been proposed. The method adds one or more additional cells called LTG cell, shown in Fig. 3.20, to each scan chain as illustrated in Fig. 3.21. In Fig. 3.20 SD is the scan data, GSEN is the global slow scan enable, and LSEN is the fast scan enable signal. Each LSEN drives the scan enable signals of a subset of scan cells which are close to it. When the initialization pattern is scanned in, the flip-flops in the LTG cell, which are part of the scan chain, are loaded with 01. GSEN is changed to 0 after initialization phase as for LOC test. However the LSEN signals which drive the scan cells changes only on the leading edge of the launch cycle. Thus during the launch cycle LSEN is 1 thus the second pattern of the test is obtained by a shift of the first pattern as required for LOS tests.

An alternate method to generate a fast scan enable signal proposed in Xu et al. (2007) replaces each scan cell by, what is called, a DTS flip-flop shown in Fig. 3.22. In Fig. 3.22 the select input of the multiplexer in the scan cell is driven by the Timed Multiplexer Control (TMC) signal. TMC is the fast scan enable signal in this design. The timing waveform for the operation of DTS flip-flop is shown

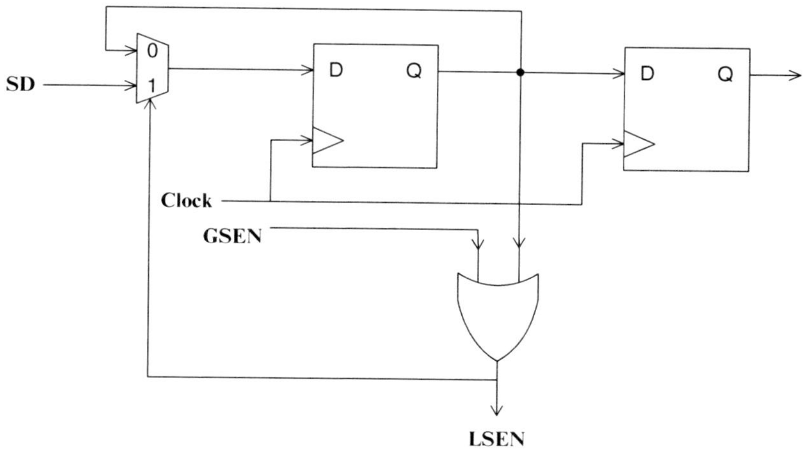

Fig. 3.20 LTG cell

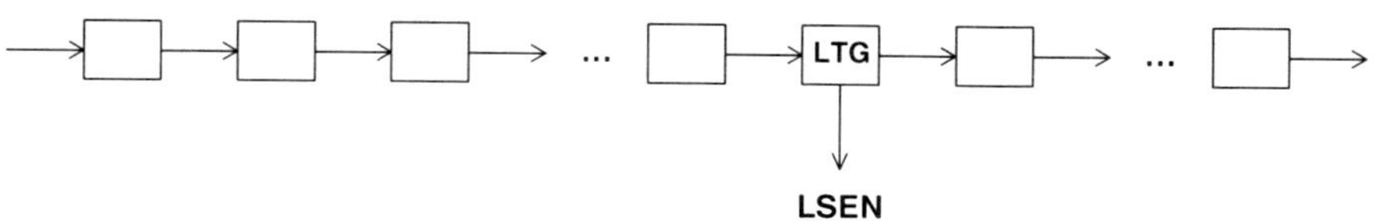

Fig. 3.21 A scan chain with an LTG cell inserted

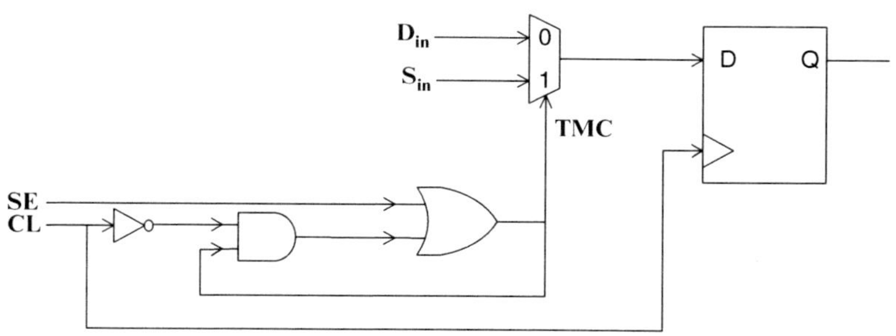

Fig. 3.22 DTS flip-flop

in Fig. 3.23. The global scan enable signal SE changes to 0 after the initialization cycle. However the local scan enable signal TMC of each scan cell changes to 0 on the leading edge of the launch cycle. Thus during the launch cycle the scan chain is shifted to generate the second pattern of the test.

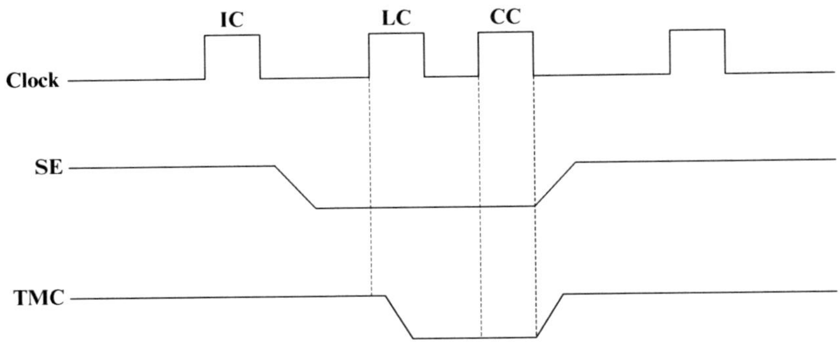

Fig. 3.23 DTS flip-flop operation

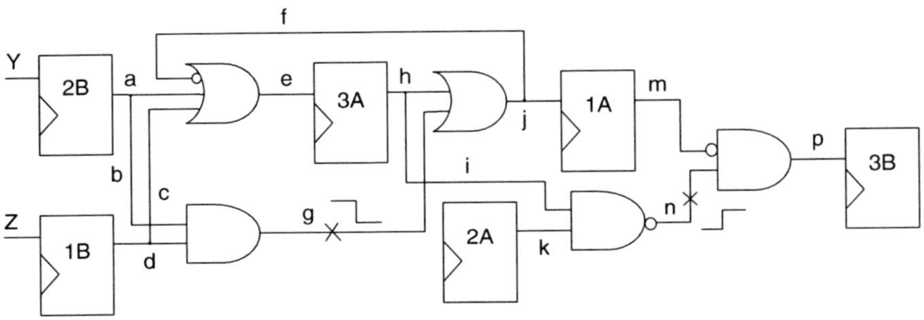

Fig. 3.24 An example to illustrate higher fault coverage using multiple scan enables

3.3.2 *Multiple Scan Enable Signals*

As noted earlier delay fault coverage using LOC test method are typically lower than the coverage obtained using LOS test method. In order to increase fault coverage using LOC tests use of multiple scan enable signals was investigated in Devtaprasanna et al. (2005). The following example from Devtaprasanna et al. (2005) illustrates how some faults not detected using a single scan enable are detected using multiple scan enable signals.

Consider the circuit shown in Fig. 3.24. Consider the line g STF TDF. This fault is untestable using the LOC test method since the initialization condition a $=$ d $=$ 1 implies h $=$ 1 during the launch cycle. Thus the fault effect is blocked from being propagated to flip-flop 1A during the capture cycle. Similarly, the line n STR TDF is LOC untestable. Next, assume that there are two scan enable signals SEN_1 and SEN_2 with SEN_1 connected to flip-flops 1A, 2A, 1B and 2B and SEN_2 connected to flip-flops 3A and 3B as shown in Fig. 3.25a. SIN_A and SIN_B are the two scan-in inputs. Figure 3.25b shows the timing diagram for a test in which SEN1 is 0 and SEN2 is 1 during launch and capture. Figure 3.25c shows the contents of the flip-flops in the two scan chains during initialization (IC), launch (LC) and capture (CC)

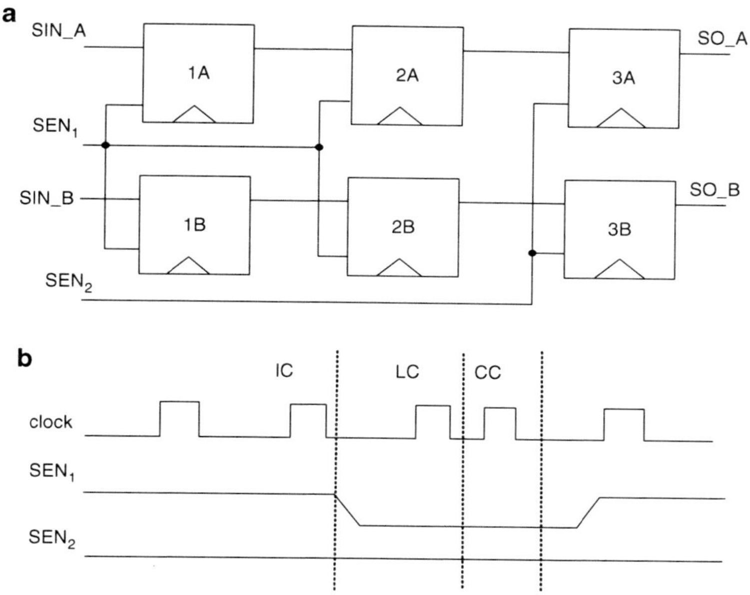

c

	1A	2A	3A	1B	2B	3B
IC	X	0	X	1	1	X
LC	1	X	0	0	0	1
CC	0/1	X	X	0	0	0

Fig. 3.25 Circuit of Fig. 3.24 with two scan enable signals

cycles required to test line g STF fault. Initialization vector (1A, 2A, 3A, 1B, 2B, 3B) = (X, 0, X, 1, 1, X) (X = don't care) is scanned in with both the scan enables SEN1 and SEN2 set to 1. Both flip-flops 1B and 2B are initialized to 1 to set line g to 1. Then the scan enable signal SEN1 is switched to 0 before the launch and capture clocks are applied while SEN2 is held at 1 throughout this test. Assume that the circuit inputs Y and Z (cf. Fig. 3.24) are both set to 0 during launch and capture cycles. During the launch cycle flip-flops 1B and 2B are set to 0 and a 1 $\rightarrow$ 0 transition is launched at the fault site. If an STF fault exists on g, then the value of line g will be 1 during the capture cycle instead of 0. Since SEN2 = 1, flip-flop 3A receives its launch cycle value (0) from flip-flop 2A instead of through its functional data input. Therefore the fault effect is propagated to flip-flop 1A during capture cycle and captured since SEN1 is 0. Similarly line n STR fault can be detected if SEN1 = 0 and SEN2 = 1 during the launch and capture cycles. Thus both the LOC untestable TDFs can be tested using two independent scan enable signals instead of one scan enable signal.

In the test discussed above the scan enable signals are held constant at 1 or 0 during the launch and capture cycles and hence they do not need to switch fast. Also the scan cells connected to the scan enable signals that are at 1 shift during launch and capture cycles and hence do not capture test responses and the achievable fault coverage with this method can be expected to strongly depend on the grouping of scan cells into subsets driven by different scan enable signals. In Devtaprasanna et al. (2005) two different methods to group scan cells were investigated. It is important to note that when multiple independent scan enable signals are used standard LOC tests can still be applied with all scan enables set to 0 during launch and capture cycles. Thus the fault coverage using multiple scan enables is not lower than that by LOC test method using a single scan enable signal. Additionally, it was shown that multiple scan enable lines facilitate generation and application of tests with reduced switching activity during scan shift and capture cycles (Wang et al. 2007). Thus use of multiple independent scan enable signals facilitates simultaneous achievement of higher delay fault coverage and reduced switching activity during test.

3.3.3 Higher Delay Fault Coverage Using Segmented Scan Designs

Segmented scan design was proposed to reduce switching activity during loading of scan based tests (Whetsel 2000). Segmented scan design is illustrated in Fig. 3.26 taken from Zhang et al. (2007b) which shows a scan chain divided in to three

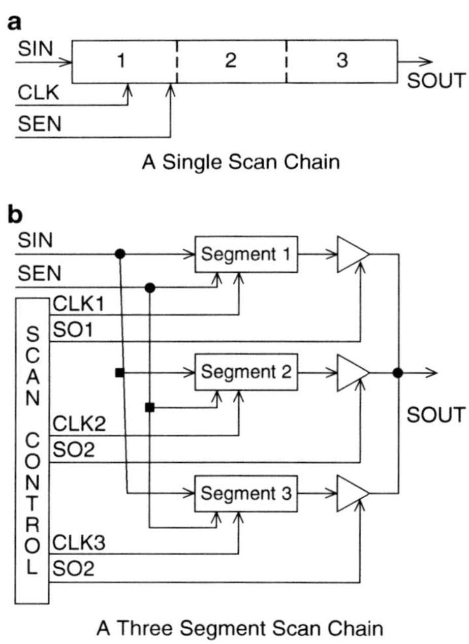

Fig. 3.26 A segmented scan design

segments. All segments share one scan-in and one scan-out lines. Each segment can thus be loaded and unloaded independently while the other segments are inactive. This reduces the switching activity during scan loads and unloads. Also if each segment can be independently clocked each segment can independently capture thus permitting reduced switching activity in capture cycles (Rosinger et al. 2004). The clocks and the tri-state buffers at scan-outs are controlled through additional logic as shown in Fig. 3.26. In Lee et al. (2004) and Rosinger et al. (2004) it was shown that stuck-at fault coverage for segmented scan designs is the same as for the corresponding unsegmented scan designs. However the number of test patterns could be lower for segmented designs using appropriate test generation procedures (Zhang 2006b). For some circuits segmented designs may require fewer stuck-at patterns than the minimum possible number of test patterns for the corresponding unsegmented design (Zhang 2006b). As for delay faults, segmented scan designs may have higher fault coverage, as discussed below, due to the fact that many different combinations of launch and capture in different segments can be used.

Consider LOC tests in which the first vector V1 of a two pattern test <V1, V2> is scanned in. In a design with two segments the first pattern is shifted in to the segments one at a time as shown in Fig. 3.26 where the scan load cycles are shown under SC. In non-segmented designs after scanning in V1 two capture cycles launch and capture cycles are applied to the scan chain. For a two segment scan design one can apply a capture cycle to segment 1 followed by a capture cycle to segment 2 as shown in Fig. 3.27a. In this case the capture cycle applied to segment 1 launches the second vector V2 of the two pattern test which is comprised of the captured values in segment 1 and the shifted in values in segment 2 of V1. The capture cycle applied to segment 2 captures the test response and no test response is captured in segment 1. Other possibilities of launch and capture for two pattern tests are shown in Fig. 3.27. In Fig. 3.27e both segments are simultaneously applied launch and capture cycles which amounts to be the same as in the case of unsegmented scan. Since there are many different launch and capture scenarios which increase in number with increasing numbers of segments many tests not possible using unsegmented can be applied. Thus, higher delay fault coverage can be obtained in segmented scan designs. The following example from Zhang et al. (2007b) illustrates this.

Consider the sequential circuit with two flip-flops shown in Fig. 3.28a. The set of TDFs in this circuit are shown on the right in Fig. 3.28a. The two copies of the circuit shown in Fig. 3.28b through d represent the two time frame iterative logic array used to generate two pattern LOC tests. If a single unsegmented scan chain is used, the four TDFs on the right of Fig. 3.28b cannot be detected. However, if a two segment design is used with SC_i in segment i, i = 1, 2, then the following faults are not detected. Using launch off segment 1 and capturing in both the segments the three faults shown on the right of Fig. 3.28c are not detectable. Using launch off segment 2 and capturing test response in both the segments one fault shown on the right of Fig. 3.27d is not detected. It should be noted that the fault coverage with a given launch method can be achieved either capturing in both segments or in a single segment as long as responses are captured in both segments by the tests used.

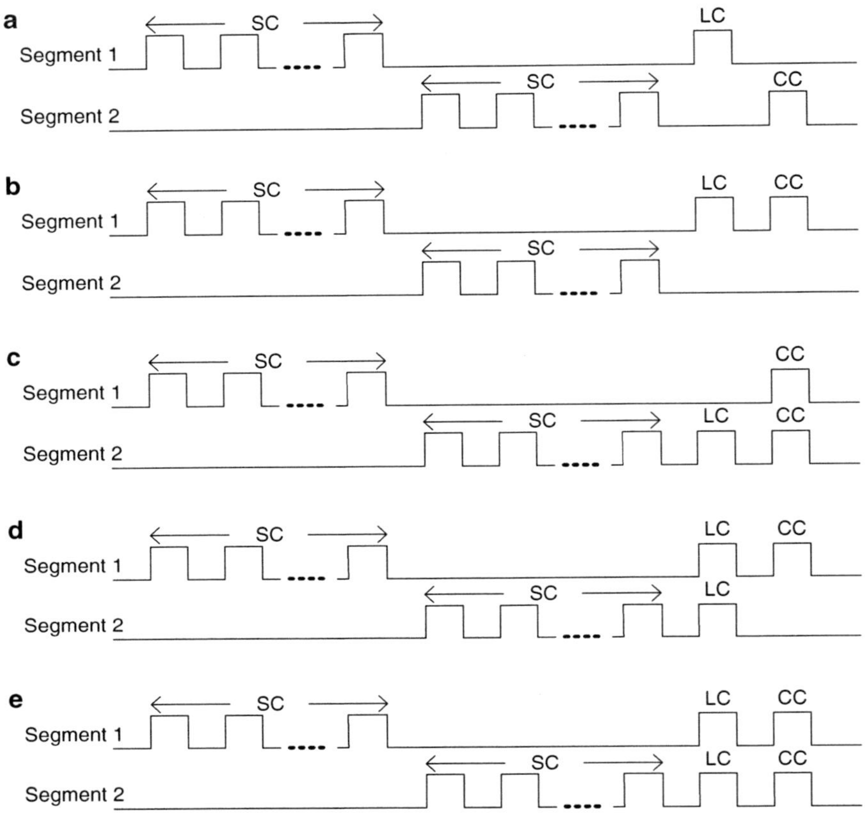

Fig. 3.27 Various ways to launch and capture in a segmented scan chain design

This is in general true independent of the number of segments. That is, the fault coverage using any selected launch method is the same independent of the capture scheme used as long as capturing in every segment is considered.

From Fig. 3.28b through d it can be noted that using LOC tests that launch off a single segment achieves better coverage than launching off both segments as done using unsegmented design. It was found that this is true in most of the ISCAS-89 benchmark circuits (Zhang et al. 2007b). Additionally, it can be seen that if the normal LOC that launches all segments is used together with launching off the second segment 100% TDF coverage can be obtained for the circuit in Fig. 3.28. Thus, using a combination of launching schemes it is possible to achieve much higher TDF coverage in segmented scan designs in addition to reduced switching activity during test. Finally, even though the discussion above used LOC tests for TDFs similar observations are valid for LOS tests and for other delay fault models.

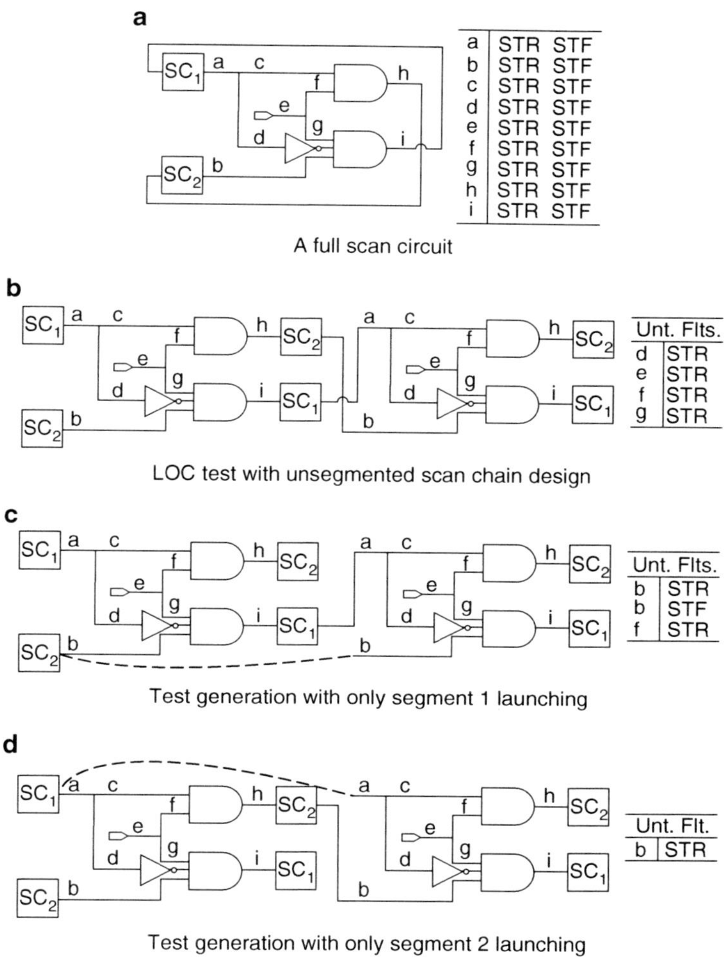

Fig. 3.28 An example illustrating higher TDF coverage in segmented scan designs

3.4 Summary

In this chapter basic fault models for defects that increase signal propagation delays in digital logic circuits were presented together with methods to generate and apply tests to detect modeled faults. Methods to detect small delay faults and design for test methods to improve delay fault coverage and reduce design effort were discussed. There is a vast amount of literature on delay faults that could not be reviewed in the chapter. Among the topics not discussed in the chapter are built-in-self-test methods and fault diagnosis. Additional material and in-depth treatment can be obtained from several books, book chapters and review articles that cover delay fault testing in greater detail (Krstic et al. 1998; Pomeranz et al. 1998; Bushnell et al. 2000; Jha et al. 2003; Wang et al. 2008a).

References

Abraham J, Goel U, Kumar A (Apr 2006) Multi-cycle sensitizatizable transition delay faults. Proceedings of VLSI test symposium, pp 308–313

Abramovici M, Breuer M, Friedman AD (1990) Digital systems testing and testable design. IEEE Press

Ahmed N, Tehranipoor M, Ravikumar CP, Butler KM (May 2007) Local at-speed scan enable generation for transition fault testing using low-cost testers. IEEE Trans Comput-Aided Des Integrat Circuits Sys 26:896–906

Barzilai Z, Rosen B (Sep 1983) Comparison of AC self-testing procedures. Proceedings of international test conference, pp 89–94

Benware B, Liu C, Van Slyke J, Krishnamurthy P, Madge R, Keim M, Kassab M, Rajski J (Oct 2004) Affordable and effective screening of delay defects in ASICS using the inline resistance fault model. Proceedings of international test conference, pp 1285–1294

Brand D, Iyengar VS (Oct 1994) Identification of redundant delay faults. IEEE Trans Comput-Aided Des Integrat Circuits Sys 13:553–565

Breuer MA (Oct 197) The effects of races, delays, and delay faults on test generation. IEEE Trans Comput C-23:1078–1092

Bushnell M, Agrawal VD (2000) Essentials of electronic testing for digital, memory, and mixed signal circuits. Frontiers in electronic testing volume 17. Springer

Cai Y, Schmitz MT, Al-Hashimi BM, Reddy SM (Jan 2007) Workload-ahead-driven online energy minimization techniques for battery-powered systems with time-constraint. ACM transaction on design automation of electronic systems, vol 12

Carter JL, Iyengar VS, Rosen BK (Sep 1987) Efficient test coverage determination for delay faults. Proceedings of international test conference, pp 418–427

Chen G, Reddy SM, Pomeranz I (Oct 2003) Procedures for identifying untestable and redundant transition faults in synchronous sequential circuits. Proceedings of international conference on computer design: VLSI in computers and processors, pp 36–41

Cheng K-T, Chen H-C (Sep 1993) Delay testing for non-robust untestable circuits. Proceedings of international test conference, pp 954–961

Cheng K-T (Dec 1993) Transition fault testing for sequential circuits. IEEE Trans Comput-Aided Des Integrat Circuits Sys 12:1971–1983

Cheng K-T, Chen H-C (Aug 1996) Classification and identification of nonrobust untestable path delay faults. IEEE Trans Comput-Aided Des Integrat Circuits Sys 15:845–853

Dasgupta S, Walthers RG, Williams TW, Eichelberger EB (Jun 1981) An enhancement to LSSD and some applications of LSSD in reliability, availability and serviceability. Proceedings of international symposium on fault-tolrant computing, pp 880–885

Devtaprasanna N, Gunda A, Krsihnamurthy P, Reddy SM, Pomeranz I (Oct 2005) A novel method of improving transition delay fault coverage using multiple scan enable signals. Proceedings of international conference on computer design: VLSI in computers and processors, pp 471–474

Dumas D, Girard P, Landrault C, Pravossoudovitch S (Oct 1993) An implicit delay fault simulation method with approximate detection threshold calculation. Proceedings of international test conference, pp 705–713

Eichelberger EB, Williams TW (1978) A logic design structure for LSI testability. J Des Automation Fault-Tolerant Comput 2:165–178

Gharaybeh MA, Bushnell ML, Agrawal VD (Apr 1998) The path-status graph with applications to delay fault simulation. IEEE Trans Comput-Aided Des Integrat Circuits Sys 17:324–332

Guo R, Venkataraman S (Sep 2006) An algorithmic technique for diagnosis of faulty scan chains. IEEE Trans Comput-Aided Des Integrat Circuits Sys 25:1861–1868

Heragu K, Patel JH, Agrawal VD (Jan 1996) Segment delay faults: a new fault model. Proceedings of international conference on VLSI design conference, pp 32–39

Iyengar VS, Rosen BK, Waicukauski JA (Mar 1990) On computing the sizes of detected delay faults. IEEE Trans Comput-Aided Des Integrat Circuits Sys 9:299–312

Iyengar VS, Vijayan G (Nov 1992) Optimized test application timing for AC testing. IEEE Trans Comput-Aided Des Integrat Circuits Sys 11:1439–1449

Jha NK, Gupta S (2003) Testing of digital systems. Cambridge University Press

Kagaris D, Tragoudas S (Sep 2002) On the nonenmerative path delay fault simulation problem. IEEE Trans Comput-Aided Des Integrat Circuits Sys 21:1095–1101

Kajihara S, Kinoshita K, Pomeranz I, Reddy SM (Jan 1997) A method for identifying robust dependent and functionally unsensitizable paths. Proceedings of international conference on VLSI design conference, pp 82–87

Kajihara S, Shimono T, Pomeranz I, Reddy SM (Dec 2000) Enhanced untestable path analysis using edge graphs. Proceedings of Asian test symposium, pp 139–144

Ke W, Menon PR (Feb 1995) Synthesis of delay-verifiable combinational circuits. IEEE Trans Comput 44:213–222

Konuk H (Oct 2000) On invalidation mechanisms for non-robust delay tests. Proceedings of international test conference, pp 393–399

Krstic A, Cheng K-T (1998) Delay fault testing for VLSI circuits. Frontiers in electronic testing, Springer

Lam WK, Saldanha A, Brayton RK, Sangiovanni-Vincentelli AL (Jun 1993) Delay fault coverage and performance tradeoffs. Proceedings of design automation conference, pp 446–451

Lee K-J, Hsu S-J, Ho C-M (Nov 2004) Test power reduction with multiple capture orders. Proceedings of Asian test symposium, pp 26–31

Lee H, Pomeranz I, Reddy SM (Mar 2008) On complete functional broadside tests for transition faults. IEEE Trans Comput-Aided Des Integrat Circuits Sys 27:583–587

Li WN, Reddy SM, Sahni S (Jan 1989) On path selection in combinational logic circuits. IEEE Trans Comput-Aided Des Integrat Circuits Sys 8:56–63

Lin CJ, Reddy SM (Sep 1987) On delay fault testing in logic circuits. IEEE Trans Comput-Aided Des Integrat Circuits Sys 6:694–703

Lin X, Pomeranz I, Reddy SM (Oct 1998) On finding undetectable and redundant faults in synchronous sequential circuits. Proceedings of international conference on computer design: VLSI in computers and processors, pp 498–503

Lin Y-C, Lu F, Yang K, Cheng K-T (Jan 2005) Constraint extraction for pseudo-functional scan-based delay testing. Proceedings of Asia and South Pacific design automation conference, pp 166–171

Lin X, Tsai K-H, Kassab M, Rajski J, Kobayashi T, Klingenberg R, Sato Y, Hamada S, Aikyo T (Nov 2006) Timing-aware ATPG for high quality at-speed testing of small delay defects. Proceedings of Asian test symposium, pp 139–146

Majhi AK, Jacob J, Patnaik LM, Agrawal VD (Jan 1996) On test coverage of path delay faults. Proceedings of international conference on VLSI design conference, pp 418–421

Malaiya YK, Narayanaswamy R (Oct 1983) Testing for timing faults in synchronous sequential circuits. Proceedings of international test conference, pp 560–571

Murakami A, Kajihara S, Sasao T, Pomeranz I, Reddy SM (Oct 2000) Selection of potentially testable path delay faults for test generation. Proceedings of international test conference, pp 376–384

Park ES, Mercer MR, Williams TW (Feb 1989) A statistical model for delay-fault testing. IEEE Des Test Comput 6:45–55

Pomeranz I, Reddy SM (Jun 1992) At-speed delay testing of synchronous sequential circuits. Proceedings of ACM/IEEE design automation conference, pp 177–181

Pomeranz I, Reddy SM (Feb 1994) An efficient non-enumerative method to estimate the path delay fault coverage in combinational circuits. IEEE Trans Comput-Aided Des Integrat Circuits Sys 13:240–250

Pomeranz I, Reddy SM (Nov 1995) Functional test generation for delay faults in combinational circuits. Proceedings of international conference on computer-aided design, pp 687–694

Pomeranz I, Reddy SM, Uppaluri P (Dec 1995) NEST: a non-enumerative test generation method for path delay faults in combinational circuits. IEEE Trans Comput-Aided Des Integrat Circuits Sys 14:1505–1515

Pomeranz I, Reddy SM (Jan 1996a) On the number of tests to detect all path delay faults in combinational logic circuits. IEEE Trans Comput-Aided Des Integrat Circuits Sys 15:50–62

Pomeranz I, Reddy SM, Patel JH (Mar 1996b) On double transition faults as a delay fault model. Proceedings of Great Lakes symposium on VLSI, pp 282–287

Pomeranz I, Reddy SM (1998) Delay fault models for VLSI circuits. Integrat VLSI J 26:21–40

Pomeranz I, Reddy SM (Sep 2002) On the coverage of delay faults in scan designs with multiple scan chains. Proceedings of international conference on computer design: VLSI in computers and processors, pp 206–209

Pomeranz I, Reddy SM (Oct 2006) Generation of functional broadside tests for transition faults. IEEE Trans Comput-Aided Des Integrat Circuits Sys 25:2207–2218

Pomeranz I, Reddy SM (Jan 2008) Transition path delay faults: a new path delay fault model for small and large delay defects. IEEE Trans VLSI Sys 16:98–107

Pomeranz I, Reddy SM (Jan 2008) Unspecified transition fault model: a transition fault model for at-speed fault simulation and test generation. IEEE Trans Comput-Aided Des Integrat Circuits Sys 27:137–146

Pomeranz I, Reddy SM (Jan 2009) Functional broadside tests under an expanded definition of functional operation conditions. IEEE Trans Comput-Aided Des Integrat Circuits Sys 28:121–129

Pomeranz I, Reddy SM (2009) Hazard-based detection conditions for improved transition fault coverage of scan-based tests. IEEE Trans VLSI Sys 17

Pramanick AK, Reddy SM (Sep 1989) On the detection of delay faults. Proceedings of international test conference, pp 680–687

Pramanick AK, Reddy SM (Mar 1990) On the fault coverage of gate delay fault detecting tests. Proceedings of European Design Automation Conference (EDAC), pp 334–338

Pramanick AK, Reddy SM (Jan 1997) On the fault coverage of gate delay fault detecting tests. IEEE Trans Comput-Aided Des Integrat Circuits Sys 16:78–94

Rearick J (Oct 2001) Too much delay fault coverage is a bad thing. Proceedings of international test conference, pp 624–633

Reddy SM, Pomeranz I, Kajihara S, Murakami A, Takeoka S, Ohta M (Oct 2000) On validating data hold times for flip-flops in sequential circuits. Proceedings of international test conference, pp 317–325

Rosinger P, Al-Hashimi BM, Nicolici N (Jul 2004) Scan architecture with mutually exclusive scan segment activation for shift-and-capture-power reduction. IEEE Trans Comput-Aided Des Integrat Circuits Sys 23:1142–1153

Sato Y, Hamada S, Maeda T, Takatori A, Kajihara S (Jan 2005) Evaluation of the statistical delay quality model. Proceedings of Asia and South Pacific design automation conference, pp 305–310

Savir V, Patil S (Aug 1993) Scan-based transition test. Trans Comput-Aided Des Integrat Circuits Sys 12:1232–1241

Savir J, Patil S (Aug 1994) Broad-side delay test. Trans Comput-Aided Des Integrat Circuits Sys 13:1057–1064

Saxena J, Butler KM, Jayaram VB, Kundu S, Arvind NV, Sreeprakash P, Hachinger M (Sep 2003) A case study of ir-drop in structured at-speed testing. Proceedings of international test conference, pp 1098–1104

Seshadri B, Pomeranz I, Reddy SM (May 2005) Path-oriented transition fault test generation considering operating conditions. Proceedings of European test symposium, pp 54–59

Schulz MH, Trischler E, Sarfert TM (Jan 1988) SOCRATES: a highly efficient automatic test pattern generation system. IEEE Trans Comput-Aided Des Integrat Circuits Sys 7:126–137

Shao Y, Reddy SM, Kajihara S, Pomeranz I (Nov 2001) An efficient method to identify untestable path delay faults. Proceedings of Asian test symposium, pp 233–238

Shao Y, Pomeranz I, Reddy SM (Nov 2002) On generating high quality tests for transition faults. Proceedings of Asian test symposium, pp 1–8

Sinanoglu O, Schremmer P (Apr. 2007) Diagnosis, modeling and tolerance of scan chain hold-time violations. Proceedings of design automation and test in Europe conference

Smith GL (Sep 1985) Model for delay faults based upon paths. Proceedings of international test conference, pp 342–349

Sparmann U, Luxenburger D, Cheng K-T, Reddy SM (Jun 1995) Fast identification of robust dependent path delay faults. Proceedings of design automation conference, pp 119–125

Syal M, Chandrasekar K, Vimjam V, Hsiao MS, Chang Y-S, Chakravarty S (Oct 2006) A study of implication based pseudo functional testing. Proceedings of international test conference

Tragoudas S, Karayiannis D (Jul 1999) A fast nonenumerative automatic test pattern generator for path delay faults. IEEE Trans Comput-Aided Des Integr Circuits Sys 18:1050–1057

Underwood B, Law W-O, Kang S, Konuk H (Oct 1994) Fastpath: a path-delay test generator for standard scan designs. Proceedings of international test conference, pp 154–163

Waicukauski J, Lindbloom E, Rosen B, Iyengar V (Apr 1987) Transition fault simulation. IEEE design and test, pp 32–38

Wang S, Wei W (Jan 2007) A technique to reduce peak current and average power dissipation in scan designs by limited capture. Proceedings of Asia and South Pacific design automation conference, pp 810–816

Wang LT, Stroud C, Touba N (2008) System on chip test architectures. Morgan Kaufmann Publishers

Wang Z, Walker DMH (May 2008) Dynamic compaction for high quality delay test. Proceedings of VLSI test symposium, pp 243–248

Whetsel L (Oct 2000) Adapting scan architecture for low power operation. Proceedings of international test conference, pp 863–872

Xu G, Singh AD (May 2007) Scan cell design for launch-on-shift delay tests with slow scan enable. IET Comput Dig Tech 1:213–219

Zhang Z, Reddy SM, Pomeranz I (Oct 2005) On generating pseudo-functional delay fault tests for scan designs. Proceedings of IEEE international symposium on defect and fault tolerance in VLSI systems, pp 398–405

Zhang Z, Reddy SM, Pomeranz I, Lin X, Rajski J (Apr 2006) Scan tests with multiple fault activation cycles for delay faults. Proceedings of VLSI test symposium, pp 343–348

Zhang Z (Dec 2006) New test generation methods for transition delay faults in scan designs. Ph.D. Thesis, University of Iowa, Iowa City, Iowa, USA

Zhang Z, Reddy SM, Pomeranz I (Jan 2007) Warning: launch off Shist tests for delay faults may contribute to test escapes. Proceedings of Asia and South Pacific design automation conference, pp 817–822

Zhang Z, Reddy SM, Pomeranz I, Rajski J, Al-Hashimi BM (May 2007) Enhancing delay fault coverage through low-power segmented scan. IET computers and digital techniques, pp 220–229

Chapter 4
Fault Modeling for Simulation and ATPG

Bernd Becker and Ilia Polian

Abstract Fault simulation and ATPG are core algorithms in the context of digital hardware test. Their deployment for resistive fault models is challenging as the behavior of the defective circuit depends on the defect resistance and the number of possible resistances is infinite. In this chapter, we show that efficient fault simulation and ATPG algorithms are feasible for resistive bridging faults. Application of the algorithms to multi-million gate industrial circuits without sacrificing accuracy is demonstrated. We cover in detail the abstraction mechanisms required for the algorithms, the algorithms themselves and their optimizations. We also indicate how a resistive bridging fault framework can be employed for problems outside the classical test development.

Keywords Resistive defects · Fault simulation · Automatic test pattern generation

Fault simulation and automatic test pattern generation (ATPG) are essential steps of test preparation for digital integrated circuits (ICs). Fault simulation is used to estimate the quality of an existing collection of test patterns (test set), i.e., its ability to identify a failing chip. Fault simulation is also useful to evaluate the efficiency of fault-tolerance mechanisms in dependable computer architectures described in Chapter 8 of this book. ATPG is used to produce a high-quality test set. Typical ATPG algorithms work iteratively: they generate patterns, run fault simulation to determine whether the quality of the test set obtained so far is sufficient, and produce more patterns if required.

Both fault simulation and ATPG are defined with respect to a *fault model*. A *fault* is a *model* of a defect which could have occurred during the manufacturing of an IC. An instance of a defect is a conducting particle leading to a bridge between an interconnect and a neighboring power rail. An instance of a fault is a logic signal line connected to an output of a logic gate being permanently set to logic-1 regardless of the value which the gate attempts to impose on the line (*stuck-at-1* fault). *Stuck-at-0* faults are defined symmetrically.

B. Becker (✉) and I. Polian
Albert-Ludwigs-University of Freiburg, Germany
e-mail: becker@informatik.uni-freiburg.de

H.-J. Wunderlich (ed.), *Models in Hardware Testing: Lecture Notes of the Forum in Honor of Christian Landrault*, Frontiers in Electronic Testing 43,
DOI 10.1007/978-90-481-3282-9_4,

Given a circuit and a simple fault model such as the stuck-at fault model, it is possible to construct a *fault list* which consists of all possible faults in the circuit under the fault model. For example, a fault list under the stuck-at fault model consists of stuck-at-1 and stuck-at-0 faults on all signal lines of a circuit. Fault simulation would decide, for each fault in the fault list, whether any of the test patterns in the test set detects the faults, i.e., the circuit with the pattern applied to its inputs produces different responses when the fault is present and absent, respectively. The fraction of detected faults among all faults is called *fault coverage*. Higher fault coverage indicates a higher quality of the test set; fault coverage of 100% implies that all faults have been detected. ATPG would be invoked for a fault which is not detected yet; it would generate a test pattern which detects the fault and, possibly, other faults, thus enhancing coverage.

It turns out that some faults in a circuit might be redundant, i.e., undetectable. An instance is a stuck-at-1 fault at a line which assumes the logical value of 1 under all input vectors. ATPG will not be able to produce a pattern detecting a redundant fault (however, it could mathematically prove that a fault is redundant). If a circuit has redundant faults, fault coverage of 100% cannot be achieved. Hence, the metric called *fault efficacy* (sometimes also called fault efficiency) is used instead: for a fault list of N faults, where D faults are detected and R faults are known to be redundant, fault efficacy is defined as $D/(N - R)$. Fault efficacy of 100% is the maximal quality a test set could have under the considered fault model.

The complexity of fault simulation is polynomial in the size of the circuit and the test set. The trivial fault simulation algorithm would, for every test pattern, simulate the circuit in absence of faults and then take each fault from the fault list, simulate the circuit in presence of that fault and compare the results. ATPG is NP-complete and proving fault redundancy is co-NP-complete, meaning that there is probably no algorithm which is guaranteed to generate a test pattern for any given fault in polynomial time. State-of-the-art ATPG methods are often successful in generating patterns for most faults even in very large practical circuits. Interestingly, calculating fault efficacy cannot be done in polynomial time since undetected faults must be checked for redundancy.

It is tempting to believe that achieving fault efficacy of 100% will yield perfect test quality, i.e., all manufactured ICs having a defect will be identified during the test process. Unfortunately, this is not the case. As mentioned above, a circuit with a fault is just a model of a manufactured IC having a defect. The modeling often implies abstraction, i.e., important behavioral details of low-level defective circuit behavior are not considered to reduce the complexity of fault simulation and ATPG. Different fault models are supposed to model different classes of actual defects with different degrees of accuracy. Furthermore, there are defects which lead to a circuit behavior which is so complex that there is no fault model that could represent it (Khare 1996) (unmodeled defects). See Fig. 8.3a in Chapter 8 for an example of such a defect.

The ultimate quality of a test strategy is the *quality level*, measured in defective parts per million (DPPM), indicating the number of defective chips which passed the test, were delivered and resulted in a customer return. There is a non-trivial and poorly understood relationship between fault coverage (or fault efficacy) and the DPPM level. In general, a higher fault coverage implies a lower DPPM level. However, it is difficult to predict the DPPM level from known fault coverage. For products with stringent DPPM targets of a few tens and less, focusing on fault efficacy of 100% for a simple fault model such as the stuck-at fault model is often associated with an unacceptable risk of missing the target, resulting in financial and reputation impact. As a consequence, several strategies to improve the quality of the test process have been devised:

- Defect-based test (DBT): in addition to conventional fault models, more accurate models are used to better capture the actual physical defect mechanisms and the low-level behavior of the defective circuits. Instances of such models can be found in Chapters 1 through 3 of this book.
- N-detect: simple fault models are used, but every fault is required to be detected several times by different test patterns, thus increasing the probability of incidentally detecting an unmodeled defect.
- Non-nominal test: the test is applied under stress conditions outside of the IC's specification, e.g., too high or too low supply voltage or temperature. Non-functional behavior of the chip is observed, e.g., the current consumption during test (I_{DDQ} test). Chips which passed the test may be rejected based on statistical reasoning, e.g., dies on the wafer surrounded by a large number of dies which have failed the test are rejected also.

At this time, it is impossible to say which of the strategies is best; often a mix of strategies appears to be optimal. DBT is a systematic, pinpointed approach which requires a good understanding of the failure mechanisms. N-detect is attractive because necessary adjustments to existing tools, including fault simulators and automatic test pattern generators, are limited. Non-nominal test methods often require costly test equipment and longer test times. Furthermore, chips good under nominal conditions could fail under stress conditions and thus be rejected (yield loss).

This chapter discusses modeling resistive faults discussed in Chapter 2 in a way which enables efficient fault simulation and ATPG algorithms. Although this approach clearly belongs to the class of DBT strategies, it can be leveraged to evaluate non-nominal test methods, in particular low-voltage and low-temperature. The difficulty in handling resistive defects arises from the defect resistance being a continuous parameter affecting the behavior of a faulty circuit. A bridge between two circuit lines could have an infinite number of bridge resistances. The bridge could be detected for some values of these resistances and remain undetected for other values. Hence, the notion of fault list in the sense introduced above is no longer well-defined. The conventional understanding of fault coverage must thus be replaced by a statistical definition. Consequently, fault simulation and ATPG must be based on different principles than the standard algorithms.

The next section introduces basic concepts used in this chapter and defines various fault coverage metrics. In Section 4.2, a fault simulation algorithm is presented. Section 4.3 describes a high-performance resistive bridging fault simulator which leverages some of the speed-up techniques known for stuck-at fault simulation.

An ATPG algorithm for resistive bridging faults is presented in Section 4.4. Section 4.5 briefly describes extensions of the model to sequential circuits, feedback faults, non-nominal test conditions, and dynamic effects.

4.1 Resistive Fault Coverage

Under a simple fault model such as the stuck-at fault model, a fault is either detected with certainty (i.e., its detection probability is 100%), or definitely undetected (its detection probability is 0%). If we assume that all faults are equally likely to occur, fault coverage can be interpreted as the average of individual detection probabilities of all faults, i.e., the probability that a fault in the circuit will be detected by a test set. Resistive fault coverage should provide the same information for resistive faults.

Resistive fault models incorporate the defect resistance as a continuous parameter. For clarity, we employ the following terminology throughout this chapter (we restrict ourselves to two-node non-feedback resistive bridging (RBF) defects although the concepts can be readily extended to other types of bridges and other defect classes such as resistive interconnect opens). We denote by a *bridging fault* the pair of circuit lines involved in the bridge. We call the pair of circuit lines and the bridge resistance R_{sh} a *bridging defect*. A bridging fault corresponds to an infinite number of bridging defects with different resistances.

A test set could detect some of the bridging defects belonging to a bridging fault and miss other defects. Some of the missed defects could be detectable by test patterns not included in the test set while other missed defects could be redundant, i.e., not detectable by any possible test pattern. Moreover, the probability that a bridging defect occurs generally depends on R_{sh}. We assume that the probability density function of the bridging defect resistance ρ is known. $\rho\,(R_{sh})$ gives the probability of occurrence of a bridging defect with resistance R_{sh}; it can be extracted from manufacturing process data (Renovell 1996).

Resistive fault coverage of a test set is first defined with respect to one resistive bridging fault f. Let the set of all resistances for which the bridge defect corresponding to f is detected be called the *covered analogue detectability interval* (C-ADI) and denoted $C(f)$. As the name suggests, $C(f)$ is often (though not always) an interval of shape $[0, R]$ for some R. This means that fault f is detected if the defect resistance R_{sh} does not exceed R.

The first resistive fault coverage metric is called *pessimistic fault coverage P-FC* [Renovell 1995]. It is defined as follows:

$$P - FC(f) = 100\% \cdot \frac{\int_{C(f)} \rho(r)dr}{\int_0^{\infty} \rho(r)dr}.$$

The "fraction" of R_{sh} values for which fault f is detected among all possible resistance values from 0 to ∞ is weighted by ρ. ρ is often normalized such that the integral in the denominator evaluates to 1. For a complete fault list $f_1, \ldots, f_N$, *P-FC* is obtained by averaging:

$$P - FC = \frac{1}{N} \cdot \sum_{i=1}^{N} P - FC(f_i).$$

Hence, to calculate *P-FC*, the fault simulation algorithm must calculate $C(f)$ for each fault f. The disadvantage of *P-FC* is the uniform handling of uncovered bridge resistances: the integral in the denominator contains all possible R_{sh} values, including ones corresponding to redundant defects. Hence, *P-FC* is typically significantly below 100%, even for exhaustive test sets which detect all detectable defects. This is the reason why *P-FC* is called pessimistic. *P-FC* is conceptually similar to fault coverage for simple fault models because the effect of redundant faults and defects is ignored by both definitions. However, redundant faults with respect to simple fault models are rare in practical circuits. On the other hand, every resistive bridging fault does have a range of very high resistances for which it is undetectable. Hence, the error introduced by the inaccurate handling of redundancy tends to be prohibitive when modeling resistive faults.

This is remediated by calculating the range $G(f)$ of all bridge resistances for which fault f is irredundant, i.e., is detected by at least one test pattern. $G(f)$ is called *global analogue detectability interval* (G-ADI). For circuits with a very small number of inputs, $G(f)$ can be obtained directly by simulating the exhaustive test set, i.e., determining $C(f)$ of that test set. $G(f)$ can also be calculated by the ATPG method described in Section 4.4. The *global fault coverage* G-FC is defined as follows:

$$G - FC(f) = 100\% \cdot \frac{\int_{C(f)} \rho(r)dr}{\int_{G(f)} \rho(r)dr}.$$

G-FC is the accurate fault coverage metric for resistive faults: *G-FC* of 100% indicates that every detectable defect has been covered by the test set. In this sense, *G-FC* is similar to fault efficacy for simple fault models. On the other hand, the calculation of *G-FC* requires G-ADI. It could be proven that a polynomial algorithm to calculate G-ADI would imply $P = NP$. Hence, the determination of G-ADI is as unlikely to be done efficiently as the redundancy proof for simple fault models. As a consequence, it has been suggested to approximate G-ADI if it cannot be calculated in practical time by available methods[0]. The approximation is based on calculating an upper bound R_{max} for any resistance contained in G-ADI. The interval $[0, R_{max}]$ is used instead of its accurate subset G-ADI, resulting in a new fault coverage metric *E-FC*:

$$E - FC(f) = 100\% \cdot \frac{\int_{C(f)} \rho(r)dr}{\int_0^{R_{max}} \rho(r)dr}.$$

Similar to *P-FC*, *E-FC* is an under-approximation of the accurate metric *G-FC*, but the inaccuracy is much smaller for *E-FC* compared with *P-FC*. The final metric is called *optimistic fault coverage* or *O-FC*. For a fault f which is detected for at least one defect resistance, $O\text{-}FC(f)$ is set to 100%. As in the case of $P\text{-}FC$, to calculate $G\text{-}FC$, $E\text{-}FC$ and $O\text{-}FC$ of a fault list, the values for individual faults are averaged. It is obvious that

$$P - FC \leq E - FC \leq G - FC \leq O - FC$$

holds. This means that $E\text{-}FC$ and $O\text{-}FC$ can be used as lower and upper bounds of the exact fault coverage $G\text{-}FC$ for large circuits for which $G\text{-}FC$ cannot be computed.

The subsequent sections will provide more details on algorithms for resistive fault simulation and ATPG. Fault simulation computes fault coverages with respect to the definitions given above. The main part of a fault simulation procedure is to obtain C-ADI of a fault. ATPG attempts to find a test pattern for a specific defect or prove that this defect is redundant. If done consequently, ATPG yields G-ADI as a by-product and allows the calculation of $G\text{-}FC$.

4.2 Interval-Based Fault Simulation

Interval-based fault simulation is the simplest algorithm to determine the resistive bridging fault (RBF) coverage of a test set. It is based on an electrical analysis and construction of *analogue detection intervals* (ADIs) at fault site and the propagation of the ADIs to the outputs of the circuit. C-ADI of a fault is obtained by aggregating the ADIs at different outputs for all test patterns in a test set. Fault coverage is then calculated as outlined in the previous section.

Figure 4.1 shows the pseudo code of the fault simulation procedure **RBF_FSIM**. It takes the circuit and the technology parameters needed for electrical analysis at the

```
Procedure RBF_FSIM
Input: Circuit CKT, technology parameters,
      fault list F = {f1, f2, ...}, test set T = {t1, t2, ...},.
Optional input: G(fi) for all fi ∈ F.
Output: C(fi) for all fi ∈ F, P-FC, E-FC, O-FC; G-FC if G-ADIs provided
  for each fi ∈ F C(fi) := ∅;
  for each t ∈ T begin
     for each fi ∈ F begin
        perform good-simulation;
        determine critical resistances R1, ..., Rm at fault site;
        construct local ADIs [0, Ri];
        propagate ADIs to circuit output;
        for each circuit output with ADI A
           C(fi) := C(fi) ∪ A;
     end for
  end for
  calculate and return P-FC, E-FC, G-FC, O-FC;
end RBF_FSIM
```

Fig. 4.1 Fault simulation algorithm for resistive bridging faults

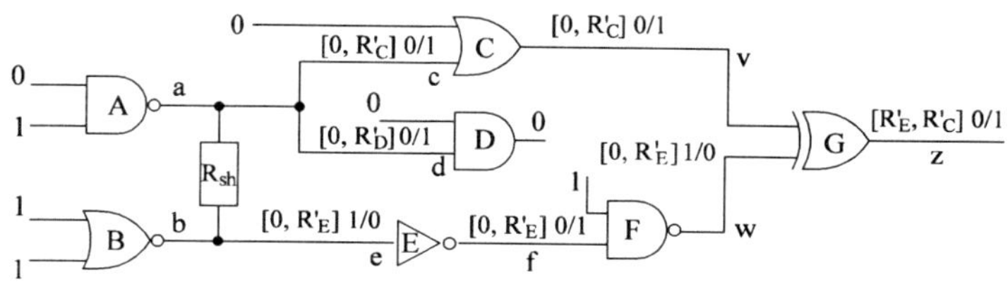

Fig. 4.2 Example circuit

fault site as inputs. Furthermore, the test set and the fault list must be provided. The fault list could include all bridging faults in the circuit or a selection of faults which are most likely to occur (*realistic faults*). Techniques such as *inductive fault analysis* (Ferguson 1988) or *inductive contamination analysis* (Khare 1996) are often employed to determine realistic faults: the proximity of interconnects in the physical layout of the circuit is evaluated and the probability that a particle of certain size will bridge two interconnects is calculated. Interconnect pairs for which this probability is sufficiently high are considered as candidates for realistic bridging faults.

Procedure **RBF_FSIM** calculates C-ADI of each fault and aggregates it to fault coverage metrics introduced above (G-ADI information must be provided to obtain *G-FC*). C-ADI of each fault is initially set to empty in Line (1). In Lines (2) through (11), the procedure determines, for each test vector and each fault f_i, resistance ranges (ADIs) in which the fault is detected and adds these ranges to C-ADI (Line 9). The calculation of the ADIs in Lines (5) through (7) is the core of the algorithm. These computations are explained in more detail using the bridging fault between signal lines a and b in the circuit in Fig. 4.2 as an example. The description avoids in-depth discussions on electrical modeling issues. Only concepts essential for understanding the algorithm, such as critical resistances, are introduced. Refer to Chapter 2 for more information on electrical modeling.

4.2.1 Local Electrical Analysis

Consider the circuit in Fig. 4.2. We call the logical values applied to the inputs of the gates which drive the bridged signal lines *fault-site input combination* (FSIC). Note that in Fig. 4.2, these lines are primary inputs of the circuit, while in general they could also be located within a larger circuit. In a combinational circuit, the FSIC is induced by the input vector. Assume FSICs 0011 and 0111. Good-simulation in Line (4) of Procedure **RBF_FSIM** will report the logic values of 1 and 0 at signal lines a and b, respectively, for both FSICs. In absence of the bridge, or for a bridge of infinite resistance, the voltage on a will equal V_{DD} and the voltage on b will equal 0V. If the bridge resistance R_{sh} equals 0 Ω, both a and b will assume some

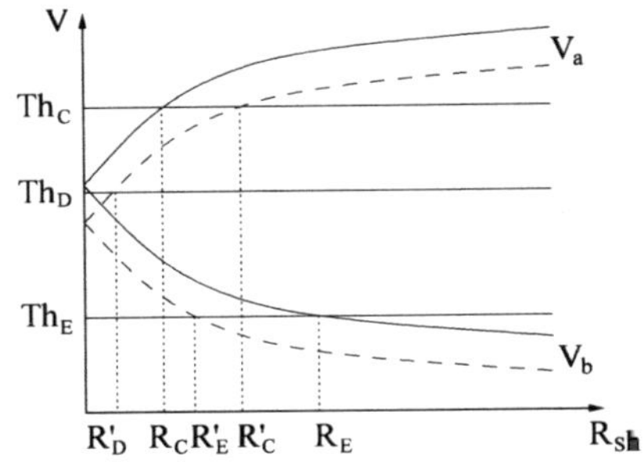

Fig. 4.3 Critical resistances in circuit from Fig. 4.2 for fault-site input combinations 0011 (*solid lines*) and 0111 (*dashed lines*)

intermediate voltage V_0. This voltage will be lower under FSIC 0111 compared to FSIC 0011, because only one *p*-transistor in the NAND gate *A* is pulling up the voltage to V_{DD}. (Speaking colloquially, one could say that the logic-1 value on *a* is driven with less strength.)

A bridging defect with non-zero resistance leads to voltages V_a and V_b on lines *a* and *b* with $V_a > V_b$. The difference $V_a > V_b$ is larger for larger values of R_{sh}. Possible voltage characteristics $V_a(R_{sh})$ and $V_b(R_{sh})$ are indicated in Fig. 4.3. Note that the characteristics for FSIC 0011 (solid lines) are located above their counterparts for vector 0111 (dashed lines), due to the different numbers of the active transistors in gate *A*.

The intermediate voltages are interpreted by subsequent logic gates as either logic-1 or logic-0, depending on the *logic thresholds* of these gates. (It is also possible to consider an intermediate voltage region in which no definite logic value is interpreted (Cheung 2007).) Thresholds Th_C, Th_D and Th_E of gates C, D and E driven by the bridged lines *a* and *b* are shown in Fig. 4.3 as horizontal lines because they are independent of the bridge resistance R_{sh}. In general, a gate will interpret different logical values for different bridge resistances. Consider gate C under FSIC 0011. Bridge resistance R_C, given by the crossing of Th_C and the solid characteristic V_a, is called *critical resistance* of gate C under FSIC 0011. For all $R_{sh} \in [0, R_C]$, gate C interprets logic-0, while for all other bridge resistances it interprets logic-1. Since logic-0 is the erroneous value, $[0, R_C]$ is called the (local) ADI at the (second) input of gate C. We write $[0, R_C]$ 0/1 to denote that the logical value on the line is 0 if R_{sh} is within the ADI and 1 otherwise.

The local ADI depends both on the logic threshold of the gate and the FSIC. For gate C and FSIC 0111, the local ADI would be $[0, R_C']$. For gate D and FSIC 0011, Th_D and $V_a(R_{sh})$ do not cross; there is no critical resistance and the local ADI is empty, i.e., the fault-free logical value is interpreted for all possible bridge resistances. Under vector 0111, a critical resistance (R_D') exists, and the local ADI is $[0, R_D']$. Critical resistances can be calculated using electrical equations (Renovell 1995) or looked up in a table pre-computed using an electrical-level simulator such as SPICE (Lee 2000).

4.2.2 ADI Propagation

Once all local ADIs have been calculated, they are propagated through the circuit (Line (7) of Procedure **RBF_FSIM**). This is illustrated in Fig. 4.2 for FSIC 0111. Consider the OR gate C. Its first input is 0 irrespective of the bridge resistance. As explained above, its second input interprets logic-0 if R_{sh} is within $[0, R_C{}']$ and logic-1 otherwise. Hence, its output value v is 1 whenever its second input interprets logic-1 and 0 whenever its second input interprets logic-0. In other words, the logic value at v is described by ADI $[0, R_C{}']$ 0/1, which is identical to the ADI on the second input of gate C. The ADI is propagated through gate C without modifications.

AND gate D's first input happens to have the controlling value of 0. Irrespective of the logic value interpreted by gate D's second input, the output value is 0. Hence, the ADI is eliminated during propagation through gate D. No fault effect is observed at gate D's output for any value of R_{sh}.

Inverter E's output f is 0 if its input is 1, i.e., if $R_{\text{sh}} \in \left[0, R'_E\right]$, and 1 otherwise. The propagation of input ADI $[0, R'_E]$ 1/0 through the inverter results in the inverted ADI $[0, R'_E]$ 0/1. (It could also have been equivalently written as $[R'_E, \infty]$ 1/0). Propagation through the inverting NAND gate F with non-controlling value 1 at its first input results in one more inversion of the interval, yielding the original ADI $[0, R'_E]$ at line w.

The XOR gate G has ADIs on both of his inputs. Gate G interprets logic-0 at input v and logic-1 at input w and produces 1 at the output z for $R_{\text{sh}} \in \left[0, R'_E\right]$ (remember that $R'_E < R_C{}'$ according to Fig. 4.3). For $R_{\text{sh}} \in \left[R'_E, R_C{}'\right]$, gate G interprets 0 at both inputs and produces 0 at z. For $R_{\text{sh}} \in [R_C{}', \infty]$, gate G interprets the fault-free values of logic-1 at v and logic-0 at w; the value at z is 0. In summary, the resulting ADI at z is $[R'_E, R_C{}']$ 0/1. A new interval which did not show up earlier is obtained by propagation through gate G. In general, it is possible that non-continuous sets of intervals are created during propagation. For instance, it would be possible to represent the obtained interval as $\left[0, R'_E\right] \cup [R_C{}', \infty]$ 1/0.

The circuit in Fig. 4.2 has two outputs: the output of gate D (to which no fault effect has been propagated) and line z. Since the fault-free value at z is 1, the resistive bridging fault is detected at z in interval $[R_E{}', R_C{}']$. This is the ADI A in Line (9) of Procedure **RBF_FSIM**. This interval will be merged with the C-ADI of the bridging fault between lines a and b calculated so far.

The practical implementation of the propagation process relies on a set of procedures for interval manipulation (complement, merging, intersection etc.) and a look-up-table which identifies the right operation from the type of the gate and the ADIs at its inputs. The efficiency of the approach is enhanced if all ADIs are *normalized*. An ADI of a line is called normalized if it contains all bridge resistances for which the logical value on the line is 1. All ADIs of shape [...] 0/1 are replaced by the equivalent ADIs of shape [...] 1/0. For instance, we observed earlier that we can write the ADI of line f as $[0, R_E{}']$ 0/1 or as $[R_E{}', \infty]$ 1/0. Only the second version is normalized. If all ADIs are normalized, we can omit "1/0" and simply write $[R_E{}', \infty]$. Values which are independent of bridge resistance can also be written as normalized ADIs: ∅ for logic-0 (because the logical value on the line is 1 for no bridge resistance) and $[0, \infty]$ for logic-1 (because the logical value on the line is 1 for all bridge resistances).

Now, we illustrate the propagation through gate D when the ADIs at the gate's inputs are normalized: ∅ at the first and $[R_C{}', \infty]$ at the second input. The propagation algorithm will consult the look-up table and determine that the ADI at the output of an AND2 gate is obtained by intersecting the ADIs at its inputs. In this case, the result will be ∅ or logic-0. Propagation through gate G consists of looking up the ADI construction rule for an XOR2 gate and application of that rule to the normalized intervals ($[R_C{}', \infty]$ at v and $[0, R_E{}']$ at w). The rule to construct output ADI A from input ADIs A_1 and A_2 is

$$A = \left(A_1 \cap \bar{A_2}\right) \cup \left(\bar{A_1} \cap A_2\right).$$

Its application results in $A = ([R_C{}', \infty] \cap [R_E{}', \infty]) \cup ([0, R_C{}'] \cap [0, R_E{}']) = [0, R_E{}'] \cup [R_C{}', \infty]$, which is the normalized version of $[R_E{}', R_C{}']$ 0/1.

4.2.3 *Fault Coverage Calculation*

To calculate $P - FC$ of one fault, the integral of function ρ over its C-ADI must be computed. This is done by approximating the integral by the weighted sum of ρ values for a large number of discrete bridge resistances. In our implementation[0] we consider all integer R_{sh} values for discretization. As mentioned above, *P-FC* values for individual faults are averaged to obtain the *P-FC* value for the circuit.

Calculation of *E-FC* requires the upper bound $R_{\max}$ for G-ADI. $R_{\max}$ is defined as the largest possible critical resistance. It is obtained by applying all possible FSICs, determining all critical resistances and selecting the maximal critical resistance as $R_{\max}$. A bridge resistance larger than $R_{\max}$ is guaranteed to induce intermediate voltage levels which will always be interpreted as fault-free logical values by all subsequent gates. Hence, $[0, R_{\max}]$ contains G-ADI (is an over-approximation).

A resistance in $[0, R_{\max}]$ may not be included in G-ADI because a defect with that resistance may require specific activation and propagation conditions which cause a conflict that cannot be resolved. An activation condition is the FSIC needed to detect the bridging defect. For instance, consider Fig. 4.3 again. A defect with resistance slightly below R_E can only be detected if FSIC 0011 is applied to the bridged gates; it would not be detected under FSIC 0111. If the circuit shown in Fig. 4.2 is part of a larger circuit, FSIC 0011 might not be justifiable at the fault site by any input vector. Then, the defect is untestable and is excluded from G-ADI, yet it is still included in $[0, R_{\max}]$. On the other hand, we have seen that an ADI can be reduced or even eliminated during propagation. This is particularly the case if multiple ADIs are propagated through reconverging paths. G-ADI contains only bridge resistances for which propagation to an output is possible and does not conflict with the above-mentioned activation conditions while $[0, R_{max}]$ contains all resistances which could theoretically result in an effect at an output.

To calculate *G-FC*, G-ADI information must be provided as an input. For both *E-FC* and *G-FC*, the integral in the denominator is computed using the approximation by weighted sum of ρ. To obtain *O-FC*, a check is performed whether the C-ADI is empty.

4.2.4 Experimental Results

An interval-based resistive bridging fault simulation based on the algorithms presented in this section has been implemented (Engelke 2006b). Table 4.1 shows results for selected circuits from the ISCAS (Int'l Symp. on Circuits and Systems) benchmark suite. We applied 1,000 random test patterns to 10,000 randomly selected non-feedback two-node bridging faults in each circuit. We derived the density function ρ from published data based on measurements (Rodríguez-Montañés 1992). All four fault coverage metrics introduced here are reported for combinational ISCAS-85 circuits and combinational cores of sequential ISCAS-89 circuits (indicated by prefix cs). The final row contains average results for all 42 ISCAS circuits.

As mentioned above, *G-FC* is the accurate metric, although its calculation is complex. Hence, the usefulness of other fault coverage definitions should be judged based on their ability to approximate *G-FC* at low computational cost. It turns out that *P-FC* yields results which are overly pessimistic, underestimating *G-FC* by more than 15% on average. On the other hand, *E-FC* and *O-FC* often provide a tight under- and over-approximation, respectively, cs00953 being an outlier. *E-FC* and *O-FC* define a "corridor" with an average width of some 2.5% in which *G-FC* is confined. For some circuits, the accurate value of *G-FC* is closer to *E-FC* (cs13207, cs15850), for some it is closer to *O-FC* (c5315, cs35932), and for some it is just in the middle of these values (c7552, cs38584).

Table 4.1 Fault coverages for 1,000 random test patterns and 10,000 random faults

Circuit	*P-FC*	*E-FC*	*G-FC*	*O-FC*
c5315	81.73	99.59	99.90	99.94
C7552	80.37	98.60	99.02	99.52
cs00953	82.13	92.00	97.19	98.33
cs13207	75.23	95.61	95.82	97.63
cs15850	76.46	96.37	96.69	98.04
cs35932	77.47	96.47	98.52	98.52
cs38417	79.79	95.58	97.72	99.22
cs38584	77.09	90.73	91.57	92.55
Average (42 ISCAS circuits)	80.67	95.08	96.98	97.59

4.2.5 Summary

Interval-based resistive bridging fault simulation is a relatively straightforward method to compute the coverage of resistive bridging faults in the circuit by a test set. It is based on an accurate local electrical analysis (described in Chapter 2) which yields intervals of bridge resistances called ADIs, and the propagation of the ADIs to the outputs. During propagation, ADIs may change their shape: they can be eliminated, inverted, intersected, or even get "holes" to become a disjoint set of intervals. This algorithm can be applied to moderately sized circuits of a few tens or hundred thousand gates. Experiments suggest that, out of the four alternative fault coverage metrics, *P-FC* is least useful. *E-FC* and *O-FC* provide reasonably tight bounds for the exact metric *G-FC* which, in general, requires information produced by resistive bridging fault ATPG (described later in this chapter).

4.3 High-Performance Fault Simulation

Interval-based resistive bridging fault simulation is computationally intensive compared to stuck-at fault simulation. A main reason for this is the complexity to store and process the resistance intervals. In contrast, a variety of successful speed-up techniques for stuck-at fault simulation relies on the efficient representation of logical values which show up during simulation. In this section, we present an approach which enables some of these techniques in context of RBF simulation. The approach is based on restricting an RBF to a small resistance range called *section* (Shinogi 2001). An RBF restricted to a section has properties similar to a multiple stuck-at fault. We demonstrate significant speed-ups for academic benchmark circuits of moderate size and applicability of the approach to industrial multi-million gate designs without any loss of accuracy.

4.3.1 Sectioning

Given an RBF, let $0 =: R_0 < R_1 < \ldots < R_m$ be the sorted list of all its critical resistances. Note that R_m corresponds to $R_{\max}$ defined in Section 4.1 of this chapter. A *section* is a resistance interval $[R_{i-1}, R_i]$ bounded by two critical resistances and containing no further critical resistance. For all defects with resistance from the same section, a gate driven by a bridged line will interpret the same value. (If a gate interprets logic-0 for one defect resistance and logic-1 for a different defect resistance, there must be a critical resistance between these resistances, so these resistances cannot be from the same section.) Hence, there exists the *detection status* of an RBF restricted to a section: either all defects with resistance from the section are detected by a test pattern, or no such defect is detected.

For a fixed FSIC and a fixed section, the behavior of the defective circuit can be represented by a multiple stuck-at fault (i.e., a number of stuck-at faults simultaneously present in the circuit). Consider again the circuit from Fig. 4.2, FSIC 0111 and section $[0, R_D']$. Gates C and D interpret the erroneous logical value of 0, while gate E interprets the erroneous logical value of 1. Recall that this holds for any defect with $R_{sh} \in [0, R_D']$. This behavior is represented by a triple stuck-at fault: stuck-at-0 at lines c and d and stuck-at-1 at line e. We denote this multiple stuck-at fault by $\{c/0, d/0, e/1\}$.

In sections $[R_D', R_C]$ and $[R_C, R_E']$, the equivalent multiple-stuck-at fault under FSIC 0111 is $\{c/0, e/1\}$. It is important that these sections are treated separately even though the critical resistance R_C has been calculated under a different FSIC (0011). In section $[R_E', R_C']$, the equivalent fault is actually the single stuck-at fault $\{c/0\}$. In section $[R_C', R_E]$ there is no equivalent fault: the circuit behaves as in the defect-free case.

The equivalent multiple-stuck-at fault does depend on the FSIC. Under FSIC 0011, the equivalent fault matches its counterpart under FSIC 0111 for section $[R_D', R_C]$: $\{c/0, e/1\}$. However, in section $[0, R_D']$ the equivalent fault is $\{c/0, e/1\}$ (and not $\{c/0, d/0, e/1\}$ as under FSIC 0111), and in section $[R_C, R_E']$ the equivalent fault is $\{e/1\}$ and not $\{c/0, e/1\}$. This implies that there is generally no such thing as a multiple stuck-at fault or a set of multiple stuck-at faults equivalent to an RBF. The logical behavior of the defective circuit is dependent from both the defect resistance (or section it belongs to) and the FSIC.

4.3.2 Sectioning-Based Simulation

The boundaries of any ADI which shows up in the interval-based simulation are critical resistances. This is because only critical resistances are possible as the right boundaries R_i of local ADIs $[0, R_i]$ when they are created at the fault site and all transformations of an ADI during propagation (complementation, intersection and merging) can only introduce a boundary of an existing ADI as a boundary of a new ADI. As a consequence, each ADI can be represented as a union of sections.

Table 4.2 contains the normalized ADIs calculated by interval-based RBF simulation (explained in detail in the previous section) and the logical values assumed in five considered sections. Note that the resistances which exceed the maximal critical resistance R_{max} (range $[R_E', \infty]$ in the example) are not considered because defects with these resistances are known to be undetectable. It is obvious that the information on the logical values is sufficient to reconstruct the ADI by merging all sections where the logical value of 1 is assumed. For example, the ADI on line w is obtained as $[0, R_D'] \cup [R_D', R_C] \cup [R_C, R_E'] = [0, R_E']$, which is the correct ADI determined by the interval-based simulation. In particular, the accurate ADI is computed for the circuit output z.

Sectioning-based RBF simulation determines the sections and performs, for each section, the simulation for an RBF restricted to that section. In the end, all sections

Table 4.2 Interval-based vs. sectioning-based simulation of circuit from Fig. 4.2

Circuit line	Fault-free value	ADI (normalized)	Value assumed in section				
			$[0, R_D]$	$[R'_D, R_C]$	$[R_C, R_E]$	$[R'_E, R_C]$	$[R'_C, R_E]$
c	1	$[R'_C, \infty]$	0	0	0	0	1
d	1	$[R'_D, \infty]$	0	1	1	1	1
e	0	$[0, R_E]$	1	1	1	0	0
f	1	$[R'_E, \infty]$	0	0	0	1	1
v	1	$[R'_C, \infty]$	0	0	0	0	1
w	0	$[0, R'_E]$	1	1	1	0	0
z	1	$[0, R'_E] \cup [R'_C, \infty]$	1	1	1	0	1

belonging to the same RBF are collected and the ADI is constructed. This ADI is equal to the interval which would have been determined by interval-based simulation. C-ADI is obtained by aggregating the ADI at the outputs for multiple test patterns.

As we have seen before, an RBF restricted to a section is equivalent to a multiple stuck-at fault if the FSIC is fixed (in case of sectioning-based simulation it is implied by the simulated test pattern). Hence, interval propagation is essentially replaced by a number of multiple stuck-at fault simulations. This allows the use of efficient speed-up techniques for (multiple) stuck-at faults. Sectioning-based simulation replaces Lines (6) and (7) of procedure **RBF_FSIM**, leaving other parts of the procedure largely unmodified.

4.3.3 SUPERB: Simulator Utilizing Parallel Evaluation of Resistive Bridges

Known performance enhancements of stuck-at simulation include *parallel-pattern single-fault processing* (PPSFP), *single-pattern, parallel-fault processing* (SPPFP), *deductive simulation* and *concurrent simulation* (Abramovici 1990). PPSFP and SPPFP are widely used in practice. On a K-bit computer, up to K patterns (PPSFP) or faults (SPPFP) are simulated in parallel, resulting in speed-ups of slightly below K. SUPERB connects sectioning-based RBF simulation with a 64-bit parallel multiple-stuck-at fault simulation engine which supports both PPSFP and SPPFP.

SUPERB calculates a *hash table* for each section of each RBF from the fault list as a pre-processing step. The hash table contains equivalent multiple stuck-at faults for each FSIC. For instance, the hash table for section $[0, R_D{}']$ of circuit from Fig. 4.2 has two entries: (0011 $\rightarrow$ $\{c/0, e/1\}$) and (0111 $\rightarrow$ $\{c/0, d/0, e/1\}$). Whenever the RBF restricted to section $[0, R_D{}']$ is simulated, the FSICs are evaluated and the equivalent multiple stuck-at fault is looked up in the hash table. For instance, if the FSIC is 0011, the equivalent fault is stuck-at 0 at line c and (simultaneously) stuck-at 1 at line e.

When SUPERB is used in the PPSFP (parallel-pattern) mode, one multiple stuck-at fault f (representing a section) is fault-simulated under 64 test patterns $t_1, \ldots, t_{64}$ simultaneously. Every signal line j is assigned a 64-bit string B_j represented using a machine word. The *ith* position of B_j stands for the logic value of signal line j under test pattern t_i when fault f is injected. The circuit is processed in topological order, i.e., from inputs to outputs. If signal line j is a primary input, its *ith* position is set to the corresponding value of test pattern t_i. If signal line j is an internal line it must be driven by some logic gate. We first assume that the inputs of that gate are not affected by the fault being simulated under any of the 64 test patterns. B_j is then obtained by applying the bit-wise logic function of the gate to the bit-strings of its inputs. For example, suppose that j is the output of a NOR3 gate with inputs k, l and m. Their bit-strings B_k, B_l and B_{m} have been calculated already. B_j is obtained as

$$B_j = \neg(B_k \vee B_l \vee B_m),$$

where $\neg$ is the bit-wise NOT and $\vee$ is the bit-wise OR operation.

The fault injection is performed by defining two *64-bit masks* for each signal line j: AND mask A_j and OR mask O_j. The *ith* position of A_j is set to 0 if a stuck-at-0 is injected at signal line j under test vector t_i. Otherwise (if a stuck-at-1 fault or no fault is injected), it is set to 1. Symmetrically, the *ith* position of O_j is set to 1 if a stuck-at-1 is injected at signal line j under test vector t_i and to 0 otherwise. A bit-wise AND operation with A_j and a bit-wise OR operation with O_j is integrated into the calculation of the bit-strings corresponding to the internal signals. The computation for the NOR3 gate mentioned above becomes

$$B_j = \neg\left((B_k \wedge A_k \vee O_k) \vee (B_l \wedge A_k \vee O_k) \vee (B_m \wedge A_k \vee O_k)\right).$$

The overall flow of SUPERB in the PPSFP mode for an RBF restricted to a section is as follows. After good-simulation of 64 test patterns, AND and OR masks are generated for all inputs of the gates driven by a bridged line. This information is extracted from the hash table corresponding to the section considered. For each of the 64 test patterns, the FSIC of the gates driving the bridged lines is determined from the good-simulation and the equivalent multiple stuck-at fault is looked up in the hash table. The *ith* position of A_j is set to 0 if the equivalent multiple stuck-at fault from the hash table contains a stuck-at-0 fault $j/0$; the *ith* position of O_j is set to 1 if it contains $j/1$. After that, simulation takes place in topological order, as outlined above.

In SPPFP (parallel-fault) mode, SUPERB simulates one test pattern for 64 multiple stuck-at faults (i.e., sections). The sections can but don't have to belong to one RBF. AND and OR masks are created at all lines involved in at least one simulated RBF. The FSICs of the gates driving the bridged lines are determined by good-simulation. The masks are filled by looking up in up to 64 hash tables, using the FSIC as the key. The subsequent simulation process is identical to the PPSFP case.

4.3.4 Experimental Results

Figure 4.4 compares the run times of SUPERB in PPSFP mode, SUPERB in SPPFP mode and the interval-based simulator, respectively. The fault list consists of randomly selected RBFs; their number equals the number of gates in a design multiplied by ten (this value was chosen to be close to typical numbers of realistic faults obtained by layout analysis). Apart from this modification, the experimental setup corresponds to that in the previous section. All experiments have been performed on the same 2.8 GHz Opteron Linux machine with 16 GB RAM. SUPERB in PPSFP mode is approximately ten times faster than SUPERB in SPPFP mode and approximately 800 times faster than the interval-based simulator. SUPERB also outperforms earlier resistive bridging fault simulators[0,0] by several orders of magnitude.

Table 4.3 reports the application of SUPERB to simulate large industrial circuits provided by NXP under 10,000 test patterns. For four largest circuits, the *E-FC* computed by SUPERB and its run time in PPSFP mode is given. In addition, the outcome of stuck-at fault simulation using the same simulation engine is reported. The final row contains average results for 18 NXP circuits. It can be seen that SUPERB can process multi-million gate designs in reasonable time (the largest time is approximately 8 h for the 2.5-million gate circuit p2921k). Preprocessing, i.e., hash

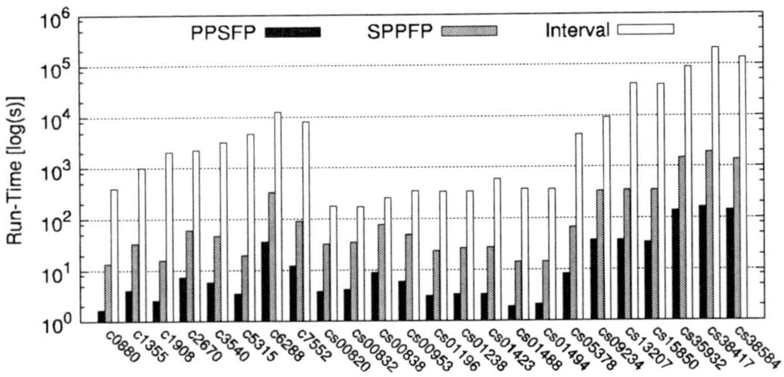

Fig. 4.4 Performance of SUPERB compared to the interval-based simulator (*logscale*)

Table 4.3 SUPERB results for combinational cores of industrial circuits provided by NXP

Circuit	Gates	RBFs	*E-FC*	Time	s-a faults	s-a FC	Time (s)
p388k	506,034	5,060,340	98.87	2,265.90	881,417	96.06	71.84
p951k	1,147,491	11,474,910	99.01	4,628.91	1,557,914	95.32	127.63
p1522k	1,193,824	11,938,240	93.26	15,874.83	1,697,662	80.91	287.23
p2927k	2,539,052	25,390,520	96.57	27,852.22	3,527,607	88.56	1,100.29
Average (18 NXP circuits)			94.29	6,580.10		85.90	412.63

table construction, consumes below 2 s for ISCAS circuits and up to three minutes for NXP circuits. The RBF coverage tends to exceed the stuck-at fault coverage of the same test set. The average RBF simulation time is some 19 times larger than stuck-at simulation time. This is competitive because the number of stuck-at faults is approximately five times smaller than the number of RBFs. Note that the number of sections and thus the number of the simulated equivalent multiple stuck-at faults is larger because an RBF has multiple sections (we observed the average number of sections per RBF being slightly above 3).

4.3.5 Summary

Sectioning-based resistive bridging fault simulation produces the same results as the interval-based simulation from the last section, yet the computation is accelerated by several orders of magnitude. Moreover, any improvements in the (multiple) stuck-at simulation engine are leveraged immediately. The main reason for this gain in efficiency is the mapping of a continuous problem (detectability of a fault as a function of its resistance) to discrete objects, i.e., sections, which can be manipulated by efficient discrete algorithms.

4.4 Automatic Test Pattern Generation

We have previously seen that, for a given RBF f, the circuit behavior on the logical level is identical for all defect resistances R_{sh} belonging to the same section $[R_{i-1}, R_i]$. This implies that a test pattern which detects the fault for one resistance from the section covers the entire section. We first propose procedure **gen_test** which finds a test pattern for an RBF restricted to a section. This procedure is called iteratively to cover all sections for all faults. RBF simulation is used to identify faults and sections covered by the patterns generated so far. Furthermore, **gen_test** can prove that an RBF restricted to a section is undetectable. Identification of all undetectable sections yields the global analogue detectability interval G-ADI which is required to calculate the accurate fault coverage *G-FC*.

4.4.1 Test Generation for a Section

Procedure **gen_test** takes a circuit CKT with n inputs and p outputs, a resistive bridging fault f and a section $S := [R_{i-1}, R_i]$ of fault f as inputs, and produces a test pattern which detects all resistive bridging defects described by f having resistances within section S, i.e., between R_{i-1} and R_i. The procedure is based on constructing a Boolean satisfiability instance and calling a SAT solver to obtain the pattern (SAT-based ATPG (Larrabee 1989)). Let C_i: $\mathbf{B}^n \rightarrow \mathbf{B}$ be the Boolean function of the CKT's i^{th} output in absence of any fault. For each output i, we define function $C_{f,S,i}$: $\mathbf{B}^n \rightarrow \mathbf{B}$ which describes the Boolean behavior of CKT in presence of RBF f restricted to S. Information necessary to define $C_{f,S,i}$ is contained in, e.g., hash tables discussed in the previous section.

Once $C_{f,S,i}$ has been defined, an assignment to Boolean variables $x_1, \ldots, x_n$ satisfying the formula

$$(C_1(x_1, \ldots, x_n) \oplus C_{f,S,1}(x_1, \ldots, x_n)) \vee \ldots \vee (C_p(x_1, \ldots, x_n) \oplus C_{f,S,p}(x_1, \ldots, x_n))$$

is sought. This is done by constructing the conjunctive normal form (CNF) of the formula and passing it to a SAT solver. If the SAT solver finds a satisfying assignment to $x_1, \ldots, x_n$, there must be at least one circuit output j for which $C_j(x_1, \ldots, x_n) \oplus C_{f,S,pj}(x_1, \ldots, x_n) = 1$or, equivalently, $C_j(x_1, \ldots, x_n) \neq C_{f,S,pj}(x_1, \ldots, x_n)$. This means that the assignment found induces different values on at least one circuit output in presence and in absence of the fault, i.e., it detects the fault.

The SAT solver may also report that there is no satisfying assignment. This is the formal proof that fault f restricted to section S is undetectable. Recall that this means that none of the defects with resistance from section S is detectable by any test pattern.

4.4.2 ATPG Algorithm

Figure 4.5 outlines the overall ATPG algorithm. The algorithm keeps two ADIs for each fault f in the fault list; $G(f)$ and L_f. $G(f)$ is the range of bridge resistances proved undetected so far. Whenever the call of procedure **gen_test** fails, the corresponding section is included in $G(f)$ in Line (11). L_f contains resistances left to detect. Resistances in L_f have neither been covered by test patterns generated so far nor proven undetectable. A fault with an empty L_f is dropped from the fault list in Lines (13) and (20). Test patterns are generated in Line (9) and fault-simulated in Line (18) until all faults are dropped.

The first fault in the fault list and the highest section of that fault undetected yet are targeted first in Line (8). The highest section is taken because high-resistance defects tend to be more difficult to detect than low-resistance defects, resulting in many specific constraints. Hence, it is more likely that a test pattern generated for a higher section will also cover lower sections of the same RBF than vice versa. However, it cannot be ruled out that an RBF requires multiple vectors to cover the entire range of resistances (Engelke 2006a). Procedure **RBF_ATPG** can resolve such instances: if not all sections of an RBF have been covered, the highest remaining section is targeted next.

The fault simulation procedure called in Line (18) could be either interval-based or sectioning-based. If interval-based simulation is used, procedure **RBF_ATPG**

```
Procedure RBF_ATPG
Input: Circuit CKT, technology parameters, fault list F = {f1, f2, ...}
Output: G(fi) for all fi ∈ F, Test set T = {t1, t2, ...} covering all faults
(1)  T := ∅;
(2)  for each f ∈ F begin
(3)     compute Rmax; G(f) := ∅; Lf := [0, Rmax];
(4)  end for
(5)  while (F not empty) begin
(6)     f := first fault of F;
(7)     compute critical resistances for f: 0 =: R0 < R1 < ... < Rm = Rmax;
(8)     int j := max{k | [Rk-1, Rk] not included in Lf};
(9)     test_pattern t := gen_test(CKT, f, [Rj-1, Rj]);
(10)    if (no t found) begin
(11)       G(f) := G(f) ∪ [Rj-1, Rj];
(12)       Lf := Lf \ [Rj-1, Rj];
(13)       if (Lf == ∅) F := F \ {f};
(14)    end if
(15)    else begin
(16)       T := T ∪ {t};
(17)       for each f' ∈ F begin
(18)          ADI D := RBF_FSIM(CKT, f', t);
(19)          Lf' := Lf' \ D;
(20)          if (Lf' == ∅) F := F \ {f'};
(21)       end for
(22)    end else
(23) end while
(24) return G(fi) for all fi ∈ F and T;
end RBF_ATPG
```

Fig. 4.5 Automatic test pattern generation algorithm for resistive bridging faults

avoids unnecessary generation of sectioning information by producing the list of critical resistances only for RBFs that are targeted explicitly in Line (7). No sectioning will be performed for a fault covered by a test pattern generated for a different fault in Line (20). If sectioning-based fault simulation is performed, critical resistances will be computed ahead of time for all faults and their repeated calculation in Line (7) can be omitted.

The algorithm terminates when the last fault has been dropped, i.e., ADIs L_f are empty for all faults. The ADIs $G(f)$ are equal to G-ADI in the end: they consist exclusively of sections for which no test pattern could be generated.

4.4.3 Experimental Results

Procedure **RBF_ATPG** has been implemented and applied to 10,000 faults in ISCAS circuits. Table 4.4 summarizes the results for the largest circuits. The number of RBFs undetectable for any R_{sh} values, the number of generated test patterns, the number of sections identified as undetectable and the run time on a 2 GHz Linux machine with 2 GB RAM are reported.

Table 4.4 Resistive bridging fault ATPG results (no compaction)

	Procedure RBF_ATPG				Stuck-at test sets		
Circuit	Undetect. faults	Test patterns	Undetect. sections	Time (s)	Stuck-at patterns	*G-FC* of s-a patterns	Top-up patterns
c5315	6	384	480	641.63	127	99.37	144
c7552	10	357	704	1,579.12	184	99.47	171
cs15850	8	1,060	501	4,684.61	197	99.07	218
cs35932	148	516	3,213	101,045.79	56	98.75	129
cs38417	1	1,178	1,678	52,233.31	194	98.80	320
cs38584	93	1,822	1,147	89,227.03	209	97.72	487

The tool can fully classify moderate-size circuits. The run times are relatively high. This is partly because the tool is not highly optimized for speed. For instance, it employs interval-based simulation and not the faster sectioning-based simulation (SUPERB was not available at the time the tool was developed) and does not use state-of-the art speed-up techniques for SAT-based ATPG. On the other hand, the number of undetectable sections is quite large. Each undetectable section translates to an unsatisfiable SAT instance which often requires long SAT solving time. On the other hand, some of these sections might be very small, so their impact on the fault coverage is negligible. It would be possible to start the SAT solver with a time limit and treat the sections which could neither be classified as testable or untestable as coverage loss.

The rightmost three columns of Table 4.4 report the performance of stuck-at test sets generated by a commercial tool in detecting resistive bridging faults. Their size, coverage (*G-FC*) and the number of test patterns which procedure **RBF_ATPG** generated to cover RBFs undetected by the stuck-at test sets to achieve *G-FC* of 100% (top-up patterns) are reported. It can be seen that stuck-at test sets do not cover all RBF. The smaller size of stuck-at test sets compared to RBF test sets is somewhat misleading, because no static or dynamic compaction of any kind is included in **RBF_ATPG** while the commercial tool employs sophisticated techniques to optimize the test set size.

We performed an investigation of the average number of faults covered by a test pattern (Engelke 2006a). It turned out that this number is higher for our tool than for academic stuck-at tools (with compaction switched off) and resistive bridging fault test generators published before Cusey (1997), Sar-Dessai (1999).

4.4.4 Summary

Resistive bridging fault ATPG can cover all possible bridge resistances by utilizing the sectioning technique. Previously published approaches (Cusey 1997; Sar-Dessai 1999) could not guarantee detection of all possible defects. In its present shape, our implementation can handle moderate-size circuits. Incorporating known optimization techniques for fault simulation and SAT-based ATPG would probably allow handling of industrial-size circuits. The relatively high pattern count could probably be reduced by implementing compaction procedures. On the other hand, many of the patterns have only minimal unique detection capability. If fault coverage slightly below 100% is acceptable, a large number of patterns could be excluded.

4.5 Extensions

This section discusses the extensions to the model required to handle faults in sequential circuits and feedback bridging faults, the dynamic effects of resistive faults, and test application under non-nominal conditions such as supply voltage and temperature.

4.5.1 Sequential Circuits

Even in case of simple fault models such as the stuck-at model, testing a sequential circuit poses significant challenges (Pomeranz 1993). Employing design-for-testability techniques such as scan chains eliminates most of the difficulties and enables the application of algorithms developed for combinational circuits. Resistive bridging fault simulation of a non-scan sequential circuit must consider the possibility that a fault effect, represented as an ADI, can be propagated to a flip-flop and fed back to the circuit in the next time frame, potentially showing up at the site of the bridging fault.

For instance, assume that the second input of gate A of circuit in Fig. 4.2 is driven by a flip-flop. Suppose that the RBF is simulated under a sequence of two test patterns, where the first pattern generates an ADI $[0,\ R']$ 1/0 on the line feeding that flip-flop for some resistance value R'. This means that the FSIC will be 0111 when the bridge resistance is between 0 and R', and 0011 otherwise. As we have seen in Fig. 4.3, FSICs 0111 and 0011 result in different voltage characteristics and thus different local ADIs.

From the simulation point of view, analysis for both FSICs must be performed. The local ADI computed for FSIC 0111 is valid for $R_{sh} \in [0,\ R']$, and the local ADI for FSIC 0011 is valid for all other values of R_{sh}. Thus, the ultimate local ADI must be composed of the respective local ADIs restricted to their ranges of validity. This phenomenon is known under the name "*multiple strength problem* (Engelke 2006b)" because more than one driving strength of the gates preceding the bridge must be considered.

While calculating C-ADI taking the multiple strength problem into account is feasible, the definition of G-ADI in sequential circuits is troublesome. For this reason, *E-FC* is used as the fault coverage metric for such circuits.

4.5.2 Feedback Faults

A bridging fault may involve two circuit lines with a sensitized path between these lines, e.g., lines f and z in Fig. 4.2. If the number of inverting gates on that path is odd, the circuit may oscillate for some bridge resistances. Suppose that the value on line v in Fig. 4.2 is 0 and that the fault-free values on lines f, w and z are 1, 0 and 0, respectively. For a given R_{sh} value, the bridge between lines f and z will impose an intermediate voltage $V_f(R_{\text{sh}})$ on line f. For some bridge resistance, $V_f(R_{\text{sh}})$ could fall below the threshold of NAND gate F and will be interpreted as logic-0. As a consequence, lines w and z will change their value to logic-1. This, in turn, will bring the voltage on line f back to V_{DD}, which will be interpreted as logic-1 by gate F. Lines w and z will thus oscillate between logic-1 and logic-0 with high frequency.

In general, a test pattern applied to a circuit having a feedback bridging defect with a given resistance could result in one of three possible circuit behaviors: either impose a faulty value on at least one circuit output; or lead to oscillation observable at an output; or have no effect. In the first case, the defect is detected; in the last case the defect is not detected. Whether the defect is detected if it implies oscillation depends on the characteristics of the automatic test equipment used. It is possible to calculate the resistance intervals for which oscillation takes place, similar to ADIs (Polian 2005). If the automatic test equipment detects oscillation, these intervals can be added to C-ADI and thus taken into account when calculating fault coverage.

Accurate calculation of resistance intervals for which the circuit exhibits oscillation is highly non-trivial. There are a variety of counter-intuitive situations in which oscillation could take place, including feedback loops not sensitized in the fault-free circuit (*disabled loops*). As a remedy, it is possible to pessimistically assume oscillation for all resistance ranges which cannot be resolved accurately [Polian 2005].

4.5.3 Dynamic Effects

This chapter concentrated on static effects of resistive bridging defects. All intermediate voltages are calculated in equilibrium, i.e., under assumption that the circuit is given sufficient time to stabilize. A resistive bridging defect typically slows down the switching speed of the gates driven by the bridged lines. A defect may not result in an intermediate voltage erroneously interpreted by a succeeding gate and thus be excluded from C-ADI. Yet the same defect could delay a transition at the succeeding gate. If the defect-induced extra delay prevents the circuit from completing the calculation of the output values within the clock cycle, the circuit will fail.

The dynamic effects of resistive bridging faults belong to the class of *delay faults*. While delay faults are broadly covered in Chapter 3, some simulation methods concentrate on delay faults induced by resistive bridging defects (Li 2003; Wang 2004). Similar to simulation of static RBF effects described above, the simulation of dynamic effects consists of two components: accurate analysis on the fault site and gate-level simulation. The fault-site analysis establishes the relationship between the defect resistance and the additional delay induced by the defect. It must take into account capacitive couplings between the bridged lines (crosstalk). The gate-level simulation determines the ranges of defect-induced delays for which the circuit will fail. Combining this information, one could derive C-ADI as the range of bridge resistances for which the circuit timing is violated.

4.5.4 Non-nominal Conditions

The detection capabilities of a test set with respect to some classes of defects are enhanced if test application is performed under non-nominal conditions such as lowered power supply voltage (V_{DD}) (Hao 1993) or ambient temperature (T) (Needham 1998). Resistive bridging faults constitute one such defect class (Liao 1996; Renovell 1996). Testing under non-nominal conditions is also effective in identifying *flaws*, i.e., defects which are present in the circuit yet are "too weak" to cause a failure. The flaws may deteriorate over time due to various aging mechanisms and lead to circuit failures during its life time. Detecting flaws is the main reason for performing costly stress tests such as burn-in (Pecht 1998). A further interest in dependence of defect detection capability from voltage and temperature derives from the increased popularity of circuits operating at multiple V_{DD} levels to reduce their power consumption (Khursheed 2008).

Using the framework introduced earlier in this chapter, one could define C-ADI and G-ADI under both nominal and non-nominal conditions. Both V_{DD} and T can be taken into account when critical resistances are calculated. Performing fault simulation and ATPG introduced above, C-ADI and G-ADI under nominal conditions are determined (we refer to them as C^{nom} and G^{nom}, respectively). Repeating the same procedures using critical resistances calculated using lower V_{DD} and/or T yields C-ADI and G-ADI under non-nominal conditions, called C^{nn} and G^{nn}. Note that $C^{\mathrm{nom}} \subseteq G^{\mathrm{nom}}$ and $C^{\mathrm{nn}} \subseteq G^{\mathrm{nn}}$ hold. C^{nom} is often (though not always) included in C^{nn}. Flaws are defects which cannot be detected under nominal conditions, i.e., defects with resistance $R_{\mathrm{sh}} \in [0, \infty] \setminus G^{\mathrm{nom}}$. We refer to defects with $R_{\mathrm{sh}} \in G^{\mathrm{nom}}$ as *hard defects*.

The detection capability under non-nominal conditions is measured using three fault coverage metrics (Engelk 2008) shown in Fig. 4.6 (all definitions are again with respect to one fault f, which is omitted for brevity). The *non-nominal fault coverage* FC^{nn} corresponds to the probability that non-nominal testing will detect a hard defect. The *combined fault coverage* FC^{comb} assumes two test applications: one under nominal and one under non-nominal conditions. A defect is considered detected if it has been detected during at least one of the test applications (i.e., it is included in either C^{nom} or C^{nn}). FC^{nn} and FC^{comb} both explicitly do not count flaw detections by restricting the integral in the numerator to G^{nom}. *Flaw coverage* FC^{flaw} calculates the probability to detect a flaw, i.e., the likelihood that a defect in ($[0, \infty] \setminus G^{\mathrm{nom}}$) is covered by C^{nn}. The figure also shows Venn diagrams illustrating

$$FC^{nn} = 100\% \cdot \frac{\int_{C^{nn} \cap G^{nom}} \rho(r)dr}{\int_{G^{nom}} \rho(r)dr}.$$

$$FC^{comb} = 100\% \cdot \frac{\int_{(C^{nom} \cup C^{nn}) \cap G^{nom}} \rho(r)dr}{\int_{G^{nom}} \rho(r)dr}.$$

$$FC^{flaw} = 100\% \cdot \frac{\int_{([0,\infty] \setminus G^{nom}) \cap C^{nn}} \rho(r)dr}{\int_{([0,\infty] \setminus G^{nom})} \rho(r)dr}.$$

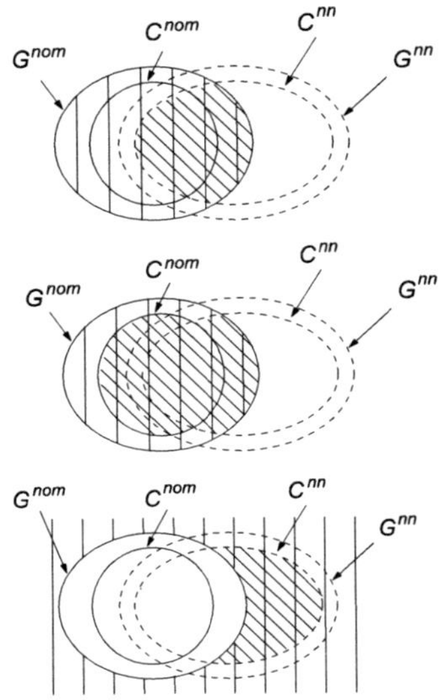

Fig. 4.6 Definitions and Venn diagrams of non-nominal coverage, combined fault coverage and flaw coverage

the fault coverage definitions. Diagonal lines and vertical lines refer to the numerator and the denominator of the formulae, respectively.

The experimental results (Engelk 2008) suggest that low-voltage testing does increase the coverage of hard defects even if performance degradation introduced by lowering V_{DD} is compensated by excluding some patterns from the test set. The coverage increase by low-temperature testing is limited. Given that low-voltage testing is associated with less equipment cost than low-temperature testing, it appears to be more efficient to detect hard defects. The detection of flaws is maximized when voltage and temperature are lowered simultaneously. It must be kept in mind that this conclusion may not be valid for defect classes other than resistive bridging faults.

4.6 Summary

Resistive faults are an important defect class in nanoscale CMOS. Traditional test methods based on the stuck-at fault model detect a significant fraction of resistive faults by incidence, yet this may not be sufficient to ensure an adequate coverage of such faults. Targeting resistive faults has been considered prohibitively complex in the past, mainly due to difficulties associated with modeling an infinite number of resistances a defect could have.

The work presented in this chapter demonstrates the feasibility of handling resistive faults directly. For an important sub-class of resistive faults, the resistive bridging faults, scalable fault simulation and ATPG methods are presented. Their efficiency is based on representing non-trivial electrical behavior by discrete objects which can be handled by fast algorithms. This allows to leverage speed-up techniques developed in the past for stuck-at faults while not compromising accuracy. SPICE-level precision becomes available for moderately-sized academic benchmark and even multi-million gate industrial circuits.

The utility of algorithms based on the resistive bridging fault model is not restricted to traditional roles of fault simulation and ATPG. We have mentioned above that they can help making informed choices when selecting the right strategy for testing under non-nominal conditions. A further application of the resistive bridging fault framework has helped to design a built-in self test (BIST) solution with sustainable non-target defect coverage (Tang 2006). The framework is generally useful to validate the performance of any test method optimized to detect stuck-at faults for other defect classes.

Although the results reported in this chapter are extensive, there remain a number of research challenges. One such challenge is the creation of adequate electrical models for both resistive bridging and resistive open faults in future technologies. One can generally assume that dynamic effects will play a dominant role in defect behavior. Complex interactions with other circuit elements, e.g., capacitive-coupled aggressor lines, may require more elaborate electrical modeling. Fault simulation and ATPG will probably somewhat resemble methods used today for delay faults. See Chapter 3 for an introduction to delay faults.

A further open question is the impact of statistical process variations on the quality of the obtained data. For the model presented in this chapter, the impact of process variations is expected to be limited, due to the following reason. Process variations will lead to different technology parameter and thus different critical resistances in different manufactured instances of the same circuit. As a consequence, C-ADI and G-ADI of a fault may differ throughout the manufactured circuit population and also may deviate from the intervals predicted without considering process variations. However, fault simulation and ATPG are concerned with the fault coverage, i.e., the *ratio* between C-ADI and G-ADI (weighted by ρ) rather than the exact boundaries of the intervals. Since C-ADI and G-ADI are computed from critical resistances, a larger C-ADI will typically be matched by a larger G-ADI, and the cumulative effect on the fault coverage will be reduced. This argumentation may not be valid for dynamic effects of resistive faults where fault detection generally depends on variations of all circuit components, not only of logic gates at the bridge site. Novel approaches to fault simulation and ATPG, possibly incorporating statistical information, may become mandatory in the future (Roy 2006).

Acknowledgments We are thankful to Dr. Piet Engelke of University of Freiburg for his contributions. The work was partially funded by the German Research Council (grant Be 1176/14-1).

References

Abramovici M, Breuer MA, Friedman AD (1990) Digital systems testing and testable design. Computer science press

Cheung H, Gupta SK (2007) Accurate modeling and fault simulation of Byzantine resistive bridges. Intäl conf. computer design, pp 347–353

Cusey JP, Patel JH (1997) BART: a bridging fault test generator for sequential circuits. Int'l Test Conf., pp 838–847

Engelke P, Polian I, Renovell M, Becker B (2006a) Automatic test pattern generation for resistive bridging faults. J Electron Test Theory Appl 22(1):61–69

Engelke P, Polian I, Renovell M, Becker B (2006b) Simulating resistive bridging and stuck-at faults. IEEE Trans Comput-Aided Des Integrat Circuits Sys 25(10):2181–2192

Engelke P, Braitling B, Polian I, Renovell M, Becker B (2007) SUPERB: simulator utilizing parallel evaluation of resistive bridges. Proceedings IEEE Asian test symposium, pp 433–438

Engelke P, Polian I, Renovell M, Kundu S, Seshadri B, Becker B (2008) On detection of resistive bridging defects by low-temperature and low-voltage testing. IEEE Trans Comput-Aided Des Integrat Circuits Sys 27(2):327–338

Ferguson FJ, Shen J (1988) Extraction and simulation of realistic CMOS faults using inductive fault analysis. In Proceedings of the international test conference, pp 475–484

Hao H, McCluskey EJ (1993) Very-low-voltage testing for weak CMOS logic ICs. In Proceedings of the international test conference, pp 275–284

Khare J, Maly W (1996) From contamination to defects, faults and yield loss. Kluwer Academic Publisher

Khursheed S, Ingelsson U, Rosinger P, Al-Hashimi BM, Harrod P (2008) Bridging fault test method with adaptive power management awareness. IEEE Trans Comput-Aided Des Integrat Circuits Sys 27(6):1117–1127

Larrabee T (1989) Efficient generation of test patterns using Boolean difference. In international test conference, pp 795–801

Lee C, Walker DMH (2000) PROBE: A PPSFP simulator for resistive bridging faults. In Proceedings of the VLSI test symposium, pp 105–110

Li Z, Lu X, Qiu W, Shi W, Walker DMH (2003) A circuit level fault model for resistive bridges. ACM Trans Des Auto Electron Sys 8(4):546–559

Liao Y, Walker DMH (1996) Fault coverage analysis for physically based CMOS bridging faults at different power supply voltages. In Proceedings of the international test conference, pp 767–775

Needham W, Prunty C, Yeoh EH (1998) High volume microprocessor test escapes, an analysis of defects our tests are missing. In Proceedings of the international test conference, pp 25–34

Pecht MG, Radojic R, Rao G (1998) Managing silicon chip reliability. CRC, Boca Raton, FL

Polian I, Engelke P, Renovell M, Becker B (2005) Modeling feedback bridging faults with non-zero resistance. J Electron Test Theory Appl 21(1):57–69

Pomeranz I, Reddy SM (1993) Classification of faults in synchronous sequential circuits. IEEE Trans Comput 42(9)

Renovell M, Huc P, Bertrand Y (1995) The concept of resistance interval: a new parametric model for resistive bridging fault. VLSI test symposium, pp 184–189

Renovell M, Huc P, Bertrand Y (1996) Bridging fault coverage improvement by power supply control. In Proceedings of the VLSI test symposium, pp 338–343

Rodríguez-Montañés R, Bruls EMJG, Figueras J (1992) Bridging defects resistance measurements in a CMOS process. In Proceedings of the international test conference, pp 892–899

Roy K, Mak TM, Cheng KT (2006) Test consideration for nanometer-scale CMOS circuits. IEEE Des Test Comp 23(2):128–136

Sar-Dessai V, Walker DMH (1999) Resistive bridge fault modeling, simulation and test generation. In Proceedings of the international test conference, pp 596–605

Shinogi T, Kanbayashi T, Yoshikawa T, Tsuruoka S, Hayashi T (2001) Faulty resistance sectioning technique for resistive bridging fault ATPG systems. In Proceedings of the Asian test symposium, pp 76–81

Tang Y, Wunderlich H-J, Engelke P, Polian I, Becker B, Schlöffel J, Hapke F, Wittke M (2006) X-masking during logic BIST and its impact on defect coverage. IEEE Trans VLSI Sys 14(2): 193–202

Wang L, Gupta SK, Breuer MA (2004) Modeling and simulation for crosstalk aggravated by weak-bridge defects between on-chip interconnects. In Proceedings of the Asian test symposium, pp 440–447

Chapter 5
Generalized Fault Modeling for Logic Diagnosis

Hans-Joachim Wunderlich and Stefan Holst

Abstract To cope with the numerous defect mechanisms in nanoelectronic technology, more and more complex fault models have been introduced. Each model comes with its own properties and algorithms for test generation and logic diagnosis. In diagnosis, however, the defect mechanisms of a failing device are not known in advance, and algorithms that assume a specific fault model may fail. Therefore, diagnosis techniques have been proposed that relax fault assumptions or even work without any fault model. In this chapter, we establish a generalized fault modeling technique and notation. Based on this notation, we describe and classify existing models and investigate the properties of a fault model independent diagnosis technique.

Keywords Logic diagnosis · Fault models

5.1 Introduction

Diagnosis is essential in modern chip production to increase yield, and debug constitutes a major part in the pre-silicon development process. Locating the structural problems by observing erroneous behavior is essential for debug and preparing physical analysis of prototypes and field-returns as well as for data-mining production fail data.

Traditionally, design, verification and diagnosis of microelectronic circuits have been viewed as separate tasks with individual challenges and techniques. However, in recent years more and more attention has been paid to the interaction of individual design steps in verification, diagnosis of prototypes, and field return analysis. These tasks support quality control and improvement during the complete lifecycle of the system by tackling faults occurring during design, manufacturing and operation.

H.-J. Wunderlich (✉) and S. Holst
Institut für Technische Informatik, Universität Stuttgart, Pfaffenwaldring 47, D-70569 Stuttgart, Germany
e-mail: wu@informatik.uni-stuttgart.de

H.-J. Wunderlich (ed.), *Models in Hardware Testing: Lecture Notes of the Forum in Honor of Christian Landrault*, Frontiers in Electronic Testing 43,
DOI 10.1007/978-90-481-3282-9_5,

Debug is the time-consuming task of identifying faulty modules and structures within the design. While some methods of formal verification are constructive and able to find the cause of malfunctions, simulation and emulation usually require additional efforts for fault location. As design complexity increases, verification is turning into a critical bottleneck in the development process. Estimates today are that more than 70% of the total design time is on verification (Chen 2003; Klein and Piekarz 2005). Despite the efforts spent by academia and industry on developing functional verification tools, logical and functional flaws remain the main cause of today's design respins.

Diagnosis is the process of locating faults in a physical chip at various levels down to real physical defects. Numerous parasitic and timing effects may show up in the first silicon (Roy et al. 2006), identifying them is part of silicon debug. With growing circuit complexity and shrinking geometries, the actual behavior of the silicon is hard to model (Krstic et al. 2003; Henderson and Soden 1997; Lavo et al. 1998), and cannot always be predicted and simulated (McPherson 2006).

In *volume diagnosis*, test data of a large number of failing chips are recorded and analyzed to find yield-limiting systematic defects and design issues. Diagnostic data from a single chip is not sufficient since systematic problems need to be differentiated from sporadic random defects. The extracted knowledge is used to support yield ramping and yield learning in advanced process technologies by improving design for manufacturability (Hora et al. 2002).

Precision diagnosis is performed on a small selected set of chips like first silicon or representatives for systematic defects determined by volume diagnosis to find the exact defect mechanisms in the individual chips. The constraints on computing time are reduced but high diagnostic resolution has to be provided to guide the physical inspection accurately.

Diagnosis is more related to defects and debug is closer to design errors, i. e. errors of the designer. However, both diagnosis and debug share many common objectives like achieving high diagnostic resolutions (Riley et al. 2006; Arnaout et al. 2006). Especially fault model independent approaches are suitable for both of these tasks.

For recent process technologies, defect mechanisms are increasingly complex, and continuous efforts are made to model these defects by using sophisticated fault models. Many debug and diagnosis algorithms are designed for specific fault models and reach more and more their limits. Choosing a fault model like the stuck-at fault model defines the set of all faults to be considered. With this set at hand, the algorithms select some candidates, whose behavior matches the observed responses. This works well, if the behavior of the defects closely resembles the behavior of the modeled faults. However in the current nanometer technology, there are numerous interactions between physical features in real silicon. Many of them have already been described with specialized fault models in the previous chapters; some of them are not well understood yet or too complex for useful modeling. This situation poses the most severe challenge for recent diagnosis algorithms. There is no single fault model anymore, that can describe all the possible defects, and there is no well defined set of faults anymore to choose from.

Instead of searching in a predefined set of fault candidates, diagnosis algorithms have to locate defective internal signals directly and identify the nature of the defect. To describe these defects, a generalized fault modeling calculus is needed. The requirements for this calculus are threefold. It must be able to express all the possible defective behaviors to a large extent, and the description of a defect must provide enough detail for the targeted failure analysis. Furthermore, the description of a defect should be as simple as possible in order to allow a ranking of the suspects corresponding to the complexity of the explanation.

Compared to the classical fault models, the calculus presented below does not impose any restrictions on the nature of a defect. Instead, the diagnosis method may restrict the set of defects to consider a certain class by additional constraints. Very common assumptions are for instance the deterministic behavior of a defect or that there is only one defect location in the device.

First, we will discuss the major application fields of logic diagnosis to describe the available input data and the formal requirements for representing diagnosis results. Section 5.3 will discuss the conditional line flip (CLF) calculus, provide examples and compare it to other general fault modeling approaches. Section 5.4 presents an application of the CLF calculus to logic diagnosis in order to locate defective regions in a circuit without any fault assumptions.

5.2 Applications

5.2.1 Prototyping and Characterization

Prototypes of a new design or a new technology may not work properly due to unknown effects. Systematic and random variations in modern technologies require the layout as well as the production process being fine-tuned to achieve acceptable yield. The failing prototypes must be thoroughly analyzed for determining the optimization tasks.

A typical case of using logic diagnosis in this environment is the preparation for physical failure analysis (PFA). PFA with its de-processing and imaging techniques is costly, time intensive and in most cases destructive. Before PFA, the erroneous behavior is analyzed to derive possible locations of a defect. As this analysis is done only for rather few cases, extensive fail data can be collected for logic diagnosis.

The available time permits the bypassing of any on-chip test compression logic and the application of large diagnostic pattern sets. Logic diagnosis, diagnostic pattern generation and pattern application can even be coupled to diagnose a defect adaptively. Based on the suspicious region determined by diagnosing responses to a standard pattern set, specially generated diagnostic patterns are generated and applied to improve diagnostic resolution.

The performance of a diagnosis algorithm in this application is measured by the area that has to be examined physically until the defect is found. This area is reduced by an efficient ranking of possible defect locations and by investigating the functional behavior of the defect itself.

Consider for instance a defect which sometimes drives an internal signal line to a faulty value. A stuck-at fault on this line can explain all erroneous responses of this circuit. However, the stuck-at fault might predict additional errors not observed in the device under test. These kinds of mispredictions can be used to record the set of test patterns for which the defect was inactive. This functional information can be correlated to the values of the neighboring line for each test pattern. If a strong correlation with another signal line can be found, the actual defect can be located close to both signals, the faulty one and the aggressing one. It is reasonable to investigate the areas first, where both signal lines are close together. Also, the line segments or vias of open defects can be determined precisely, if the states of the coupling signals are taken into account (Liu et al. 2007), and bridges can be characterized more precisely, if the Byzantine effect is considered (Khursheed et al. 2008).

Therefore, a candidate for PFA should be described with both the internal signal lines from where the erroneous output originates and the functional behavior or activation conditions of the victims. Combined with the layout of the design, this leads to accurate spatial coordinates for examination.

5.2.2 Yield Management

Yield ramp is part of the step from the prototyping phase into volume production and opens the next major application field of logic diagnosis. The goal in this phase is the optimization of the manufacturing process to achieve higher yield, to reduce the fabrication cost per functioning chip and to increase the product quality. The obtained diagnosis results are combined by using data-mining techniques to extract layout structures or features that fail in a significant number of chips and thus impact the yield. Based on this information, layout and process parameters are optimized.

The requirements for logic diagnosis on production fail data are different. While there is a huge amount of data to be analyzed for each production line, the fail data for a single device is rather limited. The limitation in response data has multiple reasons. The most obvious one is the test response compaction used to reduce the bandwidth needed between tester and device, and diagnosis has to be performed on compacted response data. Moreover, as tester memory is limited and the test time used on failing devices is usually kept as short as possible, only the first erroneous signatures are recorded. Another limitation stems from the used production test patterns. These pattern sets are kept small to reduce test time. Hence, each pattern excites many faults at the same time making it hard for diagnosis algorithms to distinguish between possible candidates (Chen et al. 2006).

Despite all these limitations of the input data for diagnosis, the algorithms must provide good predictions with a limited amount of computing power. With thousands of fail data sets coming in each minute during production, any increase in the analysis time of one data set has a great impact on the overall performance requirements.

The outcome of production diagnosis is used to infer yield limiters rather than performing physical failure analysis. Again, a diagnosis algorithm should provide a set of suspect signals for each failing device and both structural and functional data has to be generated for each candidate.

5.3 Fault Model Independence

Diagnostic fault models describe the defect candidates formally, and have structural and functional aspects. The structural aspects provide information on the locations of the defects in the circuit, and the functional aspects relate to the erroneous behavior. The conditional line flip (CLF) calculus is able to reflect these basic properties, and enables defect classification.

5.3.1 The Conditional Line Flip (CLF) Calculus

The CLF calculus is based on some observations described below. For the sake of simplicity, we will discuss the observations for the defects on gate level. The same arguments also hold for any gate-internal defects.

Each detectable defect disturbs the electrical behavior of one or more lines close to the defect site. The lines, which are influenced by the defect in this way, are called victims. A victim, like any signal line, has a driving gate and one or more receiving gates. In the general case, both drivers and receivers are influenced by the electrical conditions of a victim. This influence can also cause the neighborhood of this line to operate in a faulty way.

The receivers evaluate a degenerated voltage level at the victim line. Due to variations in the threshold voltages of the transistors in these gates, they may even come to different interpretations of the input voltage. While one receiver considers a certain voltage level being logic 1, another receiver of the same victim line reads logic 0. This is known as the Byzantine effect, and it is only one example of logically inconsistent operations caused by a defect (see Chapter 2).

Figure 5.1 shows an example of how a bridging defect influences two lines a and b. The logic behavior in this area is inconsistent, as gates x and y interpret the voltage level of line a differently. Outside the area of influence, the logic behavior is consistent again.

The proper operation of the victim's driving gate may be influenced too, by the defect. Here, a high current drain of the victim line may cause the driver to operate outside the specified limits and change its behavior.

The electrical influence of a defect can also propagate over multiple gates. If a receiving gate uses pass transistor logic, it might not produce a proper logic signal when connected to a degraded input signal. If a defect causes a short between two lines, it may compromise the power grid in the surrounding area. This is known as power droop and can cause multiple gates to fail (Tirumurti et al. 2004; Polian et al. 2006).

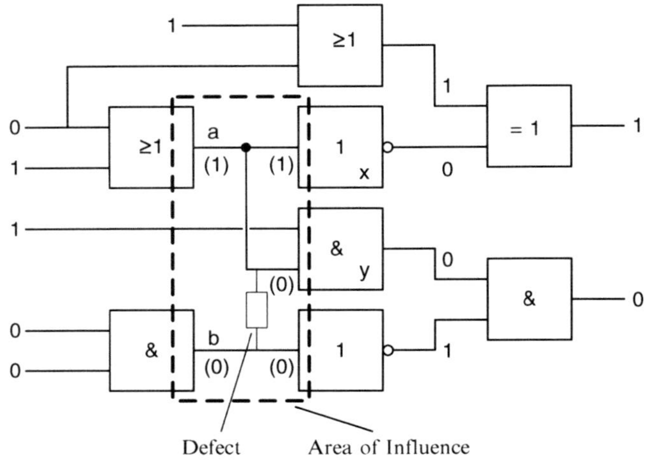

Fig. 5.1 Resistive bridge with Byzantine effect

Fig. 5.2 Defect site and surrounding logic

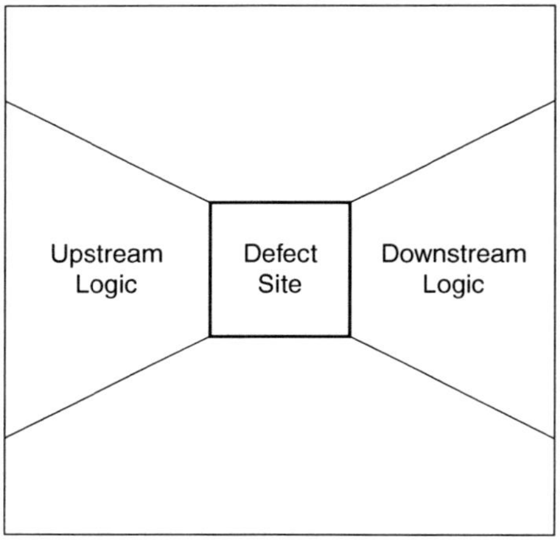

Partitioning the defective circuit into two parts helps to abstract from the electrical specifics. One part is the *defect site*. It contains all the gates and signals whose behavior cannot be explained or modeled at logic level. All signals with degraded electrical characteristics, all influenced, and all inoperable gates are contained in this defect site. The other part is the rest of the logic circuit, which implements the specified logic functionality and operates normally. The signals that connect these two parts are considered as ordinary logic signals.

Figure 5.2 shows the defect site and its surrounding logic circuit. The upstream logic is controlled by the inputs of the circuit and provides the defect site with logic values. The defect site performs some arbitrary function and provides logic values to the downstream logic. The downstream logic performs the specified functionality.

During test, patterns are applied at the inputs of the circuit and the logic values at the outputs are observed. As a consequence, logic diagnosis cannot observe the electrical conditions at the defect sites directly but has to rely on the logic signals leaving the defect sites towards the outputs. From the perspective of logic diagnosis, certain internal signals have logic values, which are inconsistent with the fault-free logic operation of the circuit. By finding for each test a small set of signal lines, which have to be flipped in order to explain the erroneous outputs, the defect site itself can be located. This set of flipped lines form the structural aspect of the diagnosis result.

With each new pattern, the electrical conditions at the defect sites change. The downstream logic receives different inputs and provides a different response. Even the same pattern can lead to different results each time, if the defect site behaves in a nondeterministic way and the line flips are not always active. The activation conditions are correlated to the electrical conditions within the defect site. They form the functional aspect of the diagnosis result.

The combination of the two aspects derived above lead to the notion of a conditional line flip (CLF). Unlike classic fault modeling approaches, which usually start with certain assumptions about the behavior of the anticipated defects, the construction of CLF does not pose any restrictions on the type of the defects. Therefore, practically any defective behavior, fault model and diagnosis result can be expressed in terms of CLFs to some extent.

A CLF is noted by the name of the victim line and an XOR-symbol followed by a condition clause:

$$line \oplus [condition]$$

The condition clause is evaluated using the original (fault-free) values of the lines affected by the described defect. A defect may be described by multiple CLFs on multiple signal lines or even on the same line, if the conditions are mutually exclusive.

In many approaches the conditions are assumed to be deterministic (Bhatti and Blanton 2006) or expressible as a Boolean function (Ubar 2003). In recent technologies, however, such assumptions become more and more restrictive, since defect behavior is often nondeterministic or timing related. If indeterminism is taken into account for the conditions, all technology related faults and design related errors can be expressed by multiple CLFs.

There are two important differences between the CLF calculus and classic fault models. First, there is no finite set of all possible faults in CLF, since the conditions can be arbitrary functions over time. Many classical fault models are designed to provide a finite set of target faults for ATPG. Second, there is no 1-to-1 correspondence between defects and faults. The behavior of a single defect can be described in many ways, and a single CLF can describe the behavior of many different defects. Classic fault models on the other hand are motivated by specific defect mechanisms, where one fault corresponds to a specific defect.

Besides describing the behavior of a defect site, the CLF calculus can also be used to describe faults. The faults may be of an arbitrary fault model, and by noting the faults in CLF, the properties and assumptions of the underlying fault model can be assessed.

Take for instance the single stuck-at fault model. A stuck-at fault ties a victim line to a fixed logic value. The excitation of a fault only depends on the original value on the victim line. If the line is driven to the logic value opposite to the value of the fault, an erroneous value is passed to the downstream logic. If the line is driven to the value of the fault, no effect can be observed. A stuck-at fault is therefore a CLF, which has a Boolean condition that only depends on the victim line:

$$a \oplus [a] : \text{stuck-at } 0$$

$$a \oplus [\bar{a}] : \text{stuck-at } 1$$

This notation reflects the basic properties of stuck-at faults. A fault only affects one line and can be expressed with a single CLF. The condition is a Boolean function; therefore the faults are deterministic and only depend on the current values in the combinational circuit.

Many defects also have temporal aspects. They may depend on the behavior of aggressing lines over time. Such conditions are noted in CLFs by allowing the condition to depend on past values of lines. These past values of a line are noted with a subscript giving the difference to the current time. For instance, the last value of line a is denoted as a_{-1}, and the value before is a_{-2}.

As seen in Chapter 1, stuck-open faults receive new attention in recent technologies (see also Wadsack 1978; Soden et al. 1989; Li and McCluskey 2005; Rodríguez-Montañés et al. 2007; Flottes Landrault et al. 1991). Gates with a stuck-open fault retain their previous output value for a certain input pattern. If, for instance, the p-transistor of the input a in a NAND gate is disconnected, the output $y = a$ NAND b will be disconnected, too, and hold its charge for the input pattern (a,b) = (0,1). This fault is described by the CLF

$$y \oplus (\bar{a} b \bar{y}_{-1})$$

Another example is a crosstalk fault, where transitions on an aggressor line a cause glitches on an influenced (victim) line b. A glitch is only produced, if line a changes from the value equal to b towards the opposite value (see Fig. 5.3). For now, we consider the glitch itself to be long enough to fully propagate through the downstream logic to the outputs. The CLF notation of such a fault is:

$$b \oplus [(a_{-1} \oplus a) \cdot (a \oplus b)]$$

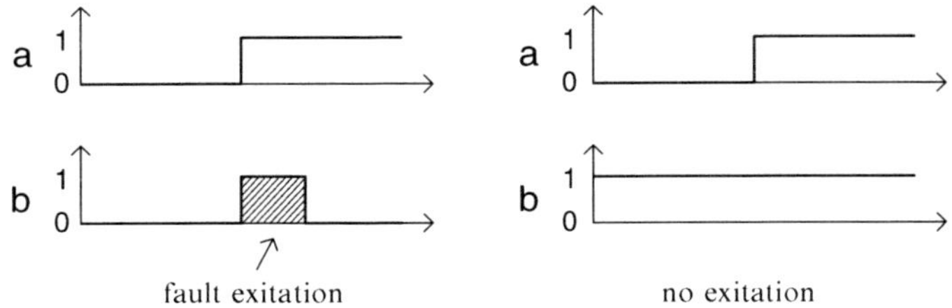

Fig. 5.3 Crosstalk fault

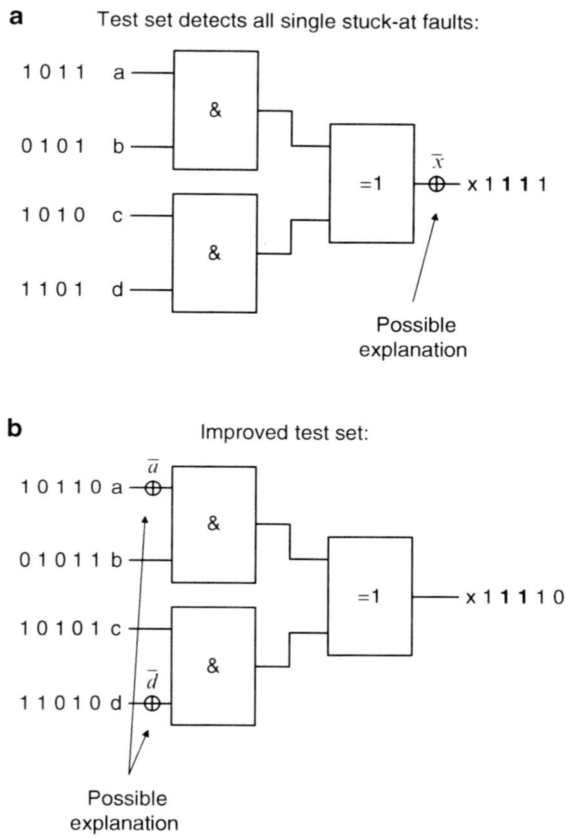

Fig. 5.4 Example of aliasing in diagnosis. The response to a test set in (**a**) is explained by a single stuck-at fault. The defective behavior is actually more complex because the additional test in (**b**) produces a 0 at the output

The first part of the condition is true, if there is an event on line a, and the second part is true, if the final value of a is different from the current value of line b.

At the first glance, the explanations for observed responses with the minimum number of CLFs are the most reasonable ones, however, there is the risk of aliasing as demonstrated in Fig. 5.4. Thus, not only the number of CLFs, but also the complexity of their conditions should be considered.

In most cases, the goal for production test generation is to achieve high stuck-at fault coverage. It is likely, that standard ATPG would generate the four patterns shown in case (a). This test set provides complete single stuck-at fault coverage and leads to two fails. The most reasonable explanation of this behavior is a stuck-at 1 at the output x. However, if one additional pattern is added to the test set like in case (b), the circuit produces a 0. This response cannot be explained anymore by a stuck-at fault at the output. In fact, there exists no single stuck-at fault, which would produce such a response. One possible explanation involves two stuck-at faults at lines a and d.

5.3.1.1 Other General Fault Models

The idea of generalizing fault modeling to describe complex defects is not new. However, the main motivation of the previous works was more related to test generation than to diagnosis. For efficient test generation, the initial values of internal signals, the preconditions and the fault effects have to be given explicitly in a formal way. Therefore, these notations are more restrictive in their formulation of conditions than CLFs. We will take a quick look at three modeling approaches and discuss their relation to the CLF calculus.

Pattern faults (Keller 1996) distinguish between static and dynamic faults. Static faults have a condition in the form of a set of required signal values. If the condition is met, the fault is active and its impact is described as a set of value changes on internal signals. The following example shows the description of a static OR-bridge:

```
STATIC {
   REQ {
      net a 1
   }
   PROP {
      net b 0/1
   }
}
```

Signal b changes from 0 to 1 if the aggressor signal a is 1. Two conditions have to be met in order to detect this fault. Signal a has to be 1, and signal b has to be 0. In CLF notation, this fault is equivalent to $b \oplus [a\bar{b}]$.

In general, a pattern fault may require multiple signals to carry a specific value. This corresponds to a conjunction of these signals in the condition of a CLF. If the condition of CLF is a Boolean formula with only one minterm, the fault can be expressed in the pattern fault model. The fault $a \oplus [b\bar{c}]$ for instance can be expressed as:

```
STATIC {
   REQ {
      net b 1
      net c 0
   }
   PROP {
      net a 0/1
      net a 1/0
   }
}
```

In contrast to the CLF calculus, the propagation description has two terms.

Dynamic pattern faults contain an additional block describing an initial condition for a set of signals. This initial condition has to be met first, and then the signals must change to match the values given in the REQ section. The signal values given in the initial condition correspond to the indexed (x_{-1}) values in CLF notation. A dynamic pattern fault corresponds to a CLF with one minterm in the condition. In addition, the minterm may contain both current and indexed previous signal values. An example of a dynamic pattern fault is described below where a transition on signal a causes a faulty value on signal c:

```
DYNAMIC {
  INIT {
    net a 0
    net b 0
  }
  REQ {
    net a 1
    net b 0
  }
  PROP {
    net c 1/0
  }
}
```

In CLF, this fault corresponds to $c \oplus [\bar{a}_{-1}\bar{b}_{-1} \cdot a\bar{b} \cdot c]$. The previous values of the signals a and b have to be 0, the current value of signal a has to be 1, signal b must stay at 0 and signal c must be 1. If the condition of a CLF is not Boolean, it has no representation in the pattern fault notation.

A similar notation is used in Kundu et al. (2006) which also targets test generation. The fault effect can be described as slow-to-rise or slow-to-fall signal with a certain delay. This way, ATPG can be advised to sensitize a path of sufficient length from the fault site to an observation point to observe the fault effect. This explicit definition of the temporal behavior of the fault impact has no direct representation in CLF as it cannot be directly observed in logic diagnosis.

Another very general fault modeling technique with a wide application field uses fault tuples (Blanton et al. 2006). A single fault tuple covers either a condition in the form of a required signal value or a fault impact in the form of a new value for a victim signal. For example, the condition fault tuple $(a, 0, i)^c$ requires the signal a to carry the value 0 at time i, and the excitation fault tuple $(b, 0, i)^e$ describes a stuck-at 0 on line b at time i. The product of fault tuples combines conditions and excitations, so that the described fault impact is only present, if all condition fault tuples are satisfied. For instance, the product of the two tuples above models a bridge, where signal a AND-dominates signal b. Multiple products can be combined with the OR-operation to model more complex faults.

This modeling technique is very similar to pattern faults or the notation in Kundu et al. (2006). Again, any CLF with a Boolean function can be noted with fault tuples, more complex conditions cannot be expressed.

5.3.1.2 A Taxonomy of Static Bridging Faults

As already described in the second chapter, bridges are an important fault class. They usually involve two signal lines which interact in a certain manner. Depending on the type of bridge and the current values of the signal lines, one or both signals may change their logic value. The types of bridges are described by two CLFs at most.

Static bridges provide a good example of how the CLF calculus can be used to express a class of fault models. There are many different fault models available for static bridges (e.g. wired-logic, dominant-driver). Rousset et al. (2007) presents a taxonomy for the most common models. Common to all these fault models is the fact that they do not model timing related behavior. The conditions can therefore be expressed using Boolean functions which depend on the current values of the involved signals.

Another basic property of static bridge fault models is the fact that errors only occur, if the two involved signal lines carry different values. This necessary precondition is described by an XOR-term in the conditions. If this precondition is true, the actual behavior of the two signals is determined by two Boolean functions f_a and f_b. The function f_a depends only on signal b, because the value of signal a is already determined by the precondition. Similarly, function f_b depends only on signal a. This leads to the following generalized CLF formulation of an arbitrary bridge between two signal lines a and b:

$$a \oplus [f_a(b) \cdot (a \oplus b)],$$
$$b \oplus [f_b(a) \cdot (a \oplus b)]$$

There are exactly four basic expressions for f_a and f_b, respectively. An expression may be constant 0, constant 1 or may use the positive or the inverted value of the other signal in the bridge:

$$f_a(b) \in \{0, 1, \bar{b}, b\}, \quad f_b(a) \in \{0, 1, \bar{a}, a\}$$

Any more complex Boolean formula can be simplified by using the precondition and Boolean identities. The formulas given above therefore model every possible static bridge configuration. There are $4^2 = 16$ possible configurations that are derived by choosing one of the four possible expressions for f_a and f_b. From these 16 configurations, there are six, that are actually derived from other bridges by interchanging the roles of the signals a and b. This leads to ten unique bridge types including the fault free case (Table 5.1).

Table 5.1 The ten possible static bridge types

$f_a(b)$	$f_b(a)$	Bridge type
0	0	Fault free
0	1	a dominates b
0	$\bar{a}$	a AND-dominates b
0	a	a OR-dominates b
1	1	a and b swap values (4-way bridge)
1	$\bar{a}$	b dominates a & a AND-dominates b
1	a	b dominates a & a OR-dominates b
$\bar{b}$	$\bar{a}$	wired-AND
$\bar{b}$	a	b AND-dominates a & a OR-dominates b
b	a	wired-OR

All common bridge fault models are present in this table. There are also three more exotic bridges described which are not widely used. These are combinations of different dominations from a to b and from b to a.

5.4 Logic Diagnosis

In this section, we apply the CLF calculus to logic diagnosis. The method presented below identifies possible faulty regions in a combinational circuit based on its input/output behavior and independent of a fault model. The approach combines a flexible and powerful effect-cause pattern analysis algorithm with high-resolution ATPG.

5.4.1 Effect Cause and Cause Effect

The classic diagnosis algorithms follow two different paradigms: Effect-cause analysis looks at the failing outputs and starts reasoning using the logic structure of the circuits (Abramovici and Breuer 1980; Waicukauski and Lindbloom 1989). One example of effect-cause analysis is the 'Single Location At a Time' (SLAT) technique introduced in Bartenstein et al. (2001). A diagnostic test pattern has the SLAT property, if there is at least one observable stuck-at fault which produces a response on that pattern identical with the response of the device under diagnosis (DUD). In SLAT diagnosis, the explaining stuck-at faults for all available SLAT patterns are combined to form possible explanations for the erroneous behavior of the DUD as a whole. Cause-effect analysis is based on a fault model. For each fault of the model, fault simulation is performed, and the behavior is matched with the outcome of the DUD.

Standard debug and diagnosis algorithms usually work in two passes. First, a fast effect-cause analysis is performed to constrain the circuits region where possible culprits may be located. Second, for each of the possible fault sites, a cause-effect simulation is performed for identifying those faults, which match the real observed behavior (Desineni et al. 2006; Amyeen et al. 2006). The resolution of a test set corresponds to the number of faults which cannot be distinguished any further (Veneris et al. 2004; Bartenstein 2000; Bhatti and Blanton 2006). The main drawback of the cause-effect paradigms is the dependency on a fault model.

5.4.2 Fault Dictionaries vs. Adaptive Diagnosis

Cause-effect diagnosis can be speeded up, if for each fault and each failing pattern the erroneous output is determined by simulation and then stored in a dictionary (Pomeranz and Reddy 1992). Even after an effect-cause pass, the size of such a dictionary may explode, and significant research effort has been spent for reducing the size of fault dictionaries (Boppana et al. 1996; Chess and Larrabee 1999; Liu et al. 2008). During debug and during diagnosis of first silicon, there exists an efficient alternative to precomputed fault dictionaries in so-called adaptive diagnosis (Gong and Chakravarty 1995).

Here, we use faulty and fault free responses of the device under diagnosis (DUD) in order to guide the automatic generation of new patterns for increasing the resolution. A pattern analysis step extracts information from responses of the DUD and accumulates them in a knowledge base. This knowledge in turn guides an automatic test pattern generator (ATPG) to generate relevant patterns for achieving high diagnostic resolution. Such a diagnostic ATPG does not rely on a precomputed fault dictionary, and significant memory savings are obtained. The loop ends, when an acceptable diagnostic resolution is reached (Fig. 5.5). The definition of the exact abort criterion depends on the number and confidence levels of fault candidates. In the subsequent sections we present the 'Partially Overlapping Impact couNTER' (POINTER) approach (Holst and Wunderlich 2009).

5.4.3 Pattern Analysis

In this section, we present a method to analyze the behavior of the DUD for a given test set and a measure to quantify how well it is reflected by a certain CLF. The SLAT paradigm will be just the special case of a perfect match for one pattern. Let *FM(f)* be a fault machine, i.e. the circuit with stuck-at fault f injected. For each test pattern $t \in T$, we define the evidence

$$e(f,t) = (\Delta\sigma_t, \Delta\iota_t, \Delta\tau_t)$$

as tuple of natural numbers $\Delta\sigma_t, \Delta\iota_t, \Delta\tau_t \in \mathbb{N}$ (see Fig. 5.6) where:

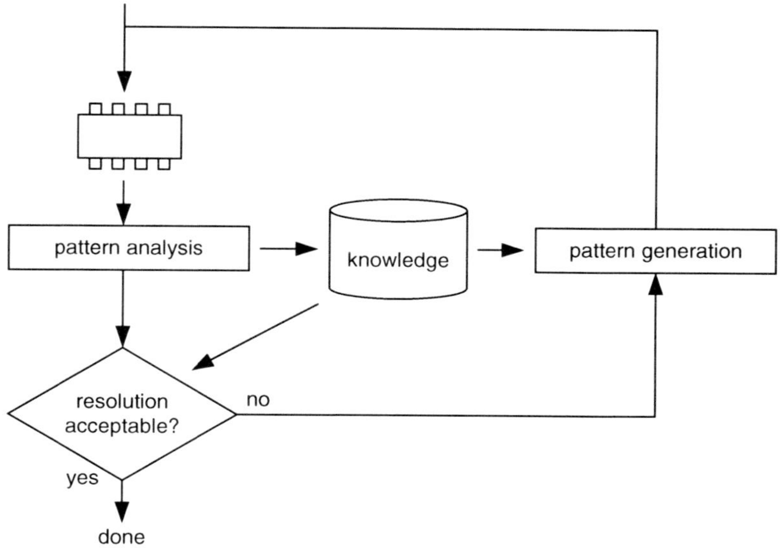

Fig. 5.5 Adaptive diagnosis flow

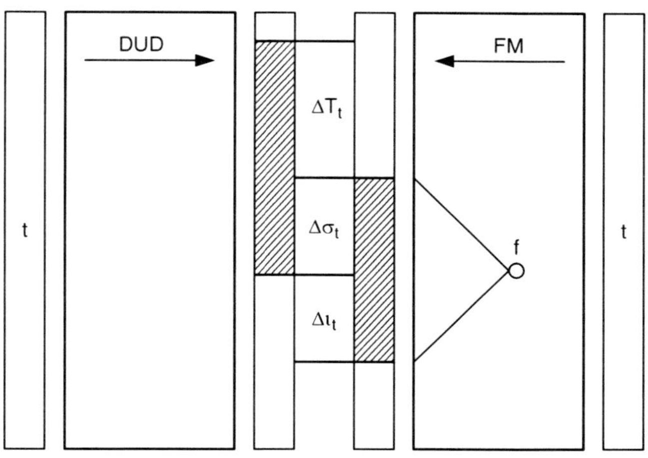

Fig. 5.6 Definition of evidence

$\Delta\sigma_t$ is the number of failing outputs where both the DUD and the fault machine FM match. It can be interpreted as the number of predictions by assuming fault f as the culprit.

$\Delta\iota_t$ is the number of outputs which fail in FM but are correct in DUD. This is the number of mispredictions by assuming fault f.

$\Delta\tau_t$ is the number of outputs which fail in DUD but are correct in FM. These are error outputs which cannot be explained by fault f.

For a SLAT test pattern t, the evidence will provide maximum $\Delta\sigma_t$ and $\Delta\iota_t = \Delta\tau_t = 0$ as this fault explains all the errors and there is no single stuck-at fault with a higher number of predictions.

The evidence of a fault f and a test set T is

$$\begin{aligned} e(f,T) &= (\sigma_T, \iota_T, \tau_T), \text{ with} \\ \sigma_T &= \sum_{t \in T} \Delta\sigma_t, \\ \iota_T &= \sum_{t \in T} \Delta\iota_t, \text{ and} \\ \tau_T &= \sum_{t \in T} \Delta\tau_t \end{aligned}$$

Again, if the real culprit is indeed the stuck-at fault f, we get $\iota_T = \tau_T = 0$ and σ_T will be maximum.

While processing pattern after pattern, $t_1, \ldots, t_i$, the knowledge base is constructed by the evidences $e(f, T_i)$, $T_i = \{t_1, \ldots, t_i\}$ for all the stuck-at faults f. If a fault is not observable under a certain pattern, no value change takes place and this fault is not considered within this iteration. If the DUD gives the correct output under a pattern t, only ι_T is increased for faults which are observable under this pattern and hence lead to a misprediction. In this way, candidates can be excluded using passing patterns, too. The maximum achievable diagnostic resolution is bounded by the size of the equivalence classes of the faults in the knowledge base.

If the fault in the DUD is not always active due to nondeterministic behavior or some unknown activation mechanism, the measure still provides consistent evidences.

For instance, let f' be a slow to rise transition fault. For some patterns t, fault f' will appear as a stuck-at 0 fault, for others it is not observable. In this case, $e(f,t) = (\Delta\sigma_t, \Delta\iota_t, \Delta\tau_t)$ provides $\Delta\sigma_t \geq \Delta\tilde{\sigma}_t$ for all the other evidences $e(\tilde{f}, t) = (\Delta\tilde{\sigma}_t, \Delta\tilde{\iota}_t, \Delta\tilde{\tau}_t)$. As a consequence, we have $\sigma_T \geq \tilde{\sigma}_T$ for all evidences $e(\tilde{f}, T)$ and the evidence $e(f, T)$ still contributes information for locating the fault. However, the value ι_T will not be zero anymore and can be used for ranking fault candidates.

Now we define $\Delta\gamma_t = \min\{\Delta\sigma_t, \Delta\iota_t\}$ and $\gamma_T = \sum_{t \in T} \Delta\gamma_t$.

Under the single fault assumption, let f be a stuck-at fault which models at least a part of the DUD behavior for some patterns under some conditions. If the conditions are satisfied for a pattern $t \in T$, the set of failing outputs of *FM(f)* corresponds to the fails of the DUD and there is no misprediction ($\Delta\iota_t = 0$). Otherwise, the failing outputs of *FM(f)* and DUD are disjoint ($\Delta\sigma_t = 0$). Hence, all $\Delta\gamma_t$ and also γ_T are zero for fault f. If there is a pattern t with $\Delta\gamma_t > 0$ like in Fig. 5.6, the corresponding conditional stuck-at is not a single fault candidate. When assuming multiple faults, we observe that mutual fault masking is rather rare, and ranking the stuck-at fault according to the size of γ_T provides a good heuristic.

Table 5.2 Fault models and evidence forms

Classic model	ι_T	τ_T	γ_T
Single stuck-at	0	0	0
Stuck-at, more fault sites present	0	>0	0
Single conditional stuck-at	>0	0	0
Cond. stuck-at, more fault sites present	>0	>0	0
Delay fault, i.e. long paths fail	>0	0	>0

This fault model independent pattern analysis approach is able to identify circuit parts containing arbitrary faulty behavior. However, if the behavior of the DUD can be explained using some classic fault models, certain evidence forms are observed. Table 5.2 shows suspect evidences for some classic models.

If ι_T, τ_T and γ_T are all zero, a single stuck-at fault explains the DUD behavior completely. If τ_T and γ_T are zero, a faulty value on a single signal line under some patterns $T' \subset T$ provides complete explanation. With $\iota_T = \gamma_T = 0$, such a stuck-at fault explains a subset of all fails, but some other faulty behavior is present in the DUD. These other fault sites are independent from the stuck-at fault at hand, i.e. for each pattern an output is either influenced by the stuck-at fault only or by some other fault sites. With only $\gamma_T = 0$, a faulty value on the corresponding single signal line explains a part of DUD behavior and more fault sites are present again. If only τ_T is zero, the suspect fails are a superset of DUD fails.

If all suspects show positive values in all components ι_T, τ_T, γ_T, the responses were caused by multiple interacting fault sites, and all simplistic fault models would fail to explain the DUD behavior.

For further analysis, the evidences in the knowledge base are ordered to create a ranking with the most suspicious fault sites at the beginning (lowest rank). Firstly, evidences are sorted by increasing γ_T, i.e.

$$\gamma_T^a > \gamma_T^b \Rightarrow \text{rank}(e(f^a, T)) > \text{rank}(e(f^b, T))$$

moving single conditional stuck-at faults in front. Evidences with identical γ_T are sorted by decreasing σ_T moving candidates in front, which explain most failures:

$$\sigma_T^a > \sigma_T^b \Rightarrow \text{rank}(e(f^a, T)) < \text{rank}(e(f^b, T)).$$

Finally evidences with identical γ_T and σ_T are ordered by increasing ι_T values:

$$\iota_T^a > \iota_T^b \Rightarrow \text{rank}(e(f^a, T)) > \text{rank}(e(f^b, T)).$$

For a brief example of the pattern analysis approach, consider the circuit in Fig. 5.7. It contains two gates and four exemplary stuck-at faults for fault simulation.

The exhaustive test set and the response from the DUD are shown in the first two columns of Table 5.3. The erroneous bits are shown in bold, the DUD has failed on output x in the third pattern.

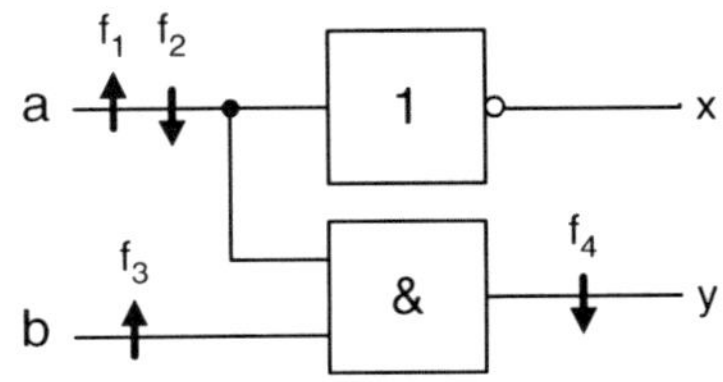

Fig. 5.7 Circuit model for fault simulation

Table 5.3 Syndrome and result from stuck-at fault simulation

Pattern ab	Syndrome xy	f_1 xy	f_2 xy	f_3 xy	f_4 xy
00	10	**00**	10	10	10
01	10	**01**	10	10	10
10	**10**	00	**10**	**01**	00
11	01	01	**10**	01	**00**

Table 5.4 Evidences and rank of the four faults

Fault	σ_T	ι_T	τ_T	γ_T	Rank
f_1	0	3	1	0	4
f_2	1	2	0	0	1
f_3	0	1	1	0	2 or 3
f_4	0	1	1	0	3 or 2

Now, the four faults are simulated for the given pattern set and their signatures are shown in the remaining columns in Table 5.3. The fault f_1 is observable in three response bits, but it fails to explain the erroneous bit in the syndrome. This leads for this fault to an evidence of $e(f_1, T) = (\sigma_T, \iota_T, \tau_T, \gamma_T) = (0, 3, 1, 0)$. The evidence is derived for the other stuck-at faults as well; Table 5.4 shows the result.

All evidences show $\gamma_T = 0$, so the ranking procedure continues with σ_T. Only f_2 has positive σ_T, so this fault is ranked above all other faults. The other faults are ranked by increasing ι_T. The top-ranked evidence f_2 shows positive σ_T and positive ι_T. Therefore, none of the simulated faults can explain the syndrome completely, but f_2 explains a subset of all fails. This leads to a CLF of the form $a \oplus [a \cdot \text{cond}]$ with some arbitrary condition.

5.4.4 Volume Diagnosis and Pattern Generation

If the resolution provided by the evidences of a test pattern set T is not sufficient during adaptive diagnosis or design debug, we have the option to use the evidences for guiding further diagnostic ATPG. In volume diagnosis, the pattern set is fixed, and we have to extract as much diagnostic information as possible from rather limited information. Usually, only the first i failing patterns are recorded, and in addition, all the passing patterns up to this point can be used for diagnosis.

The number of suspects reported by logic diagnosis must be limited in order to be used for volume analysis. If the number of suspects exceeds a parameter k, significance for certain flaws is hardly obtained and further analysis may be too expensive. If diagnosis successfully identified the culprit, the rank describes the position of the corresponding evidence within the ordered list.

For each fault f with $e(f, T) = (\sigma_T, \iota_T, \tau_T)$ we have $\sigma_T + \iota_T > 0$, if T detects f. Otherwise, f may be undetected due to redundancy, or T must be improved to detect f.

Even if there are no suspects with $\sigma_T > 0$, the possible fault sites are ranked by ι_T. This way, multiple faults on redundant lines can be pointed out. For the special case of $\iota_T = 0$, at least a subset of DUD failures can be explained with an unconditional stuck-at fault.

The faults with $e(f, T) = (\sigma_T, \iota_T, \tau_T)$ and $\sigma_T > 0$ are the suspects, and by simple iteration over the ranking, pairs of suspects f^a, f^b are identified with equal evidences $e(f^a, T) = e(f^b, T)$. To improve the ranking, fault distinguishing patterns have to be generated (Veneris et al. 2004; Bartenstein 2000) and applied to the DUD. To reduce the number of suspects and the region under consideration further, diagnostic pattern generation algorithms have to be employed which exploit layout data (Desineni et al. 2006).

5.5 Evaluation

5.5.1 *Single Line Defects*

The fault machine for a stuck-at fault f at a line a will mispredict, if the condition of the CLF $a \oplus [cond]$ is not active while the CLF is actually modeling the defective behavior of line a. We split the conditions into $cond = cond_0 \vee cond_1$ with $cond_0 = \bar{a} \wedge cond$ and $cond_1 = a \wedge cond$. Now, $a \oplus cond_0$ models a conditional stuck-at 1 fault and $a \oplus [cond_1]$ models a conditional stuck-at 0 fault.

The unconditional stuck-at 0 fault at line a explains all the errors introduced by $a \oplus [cond_1]$, and there is no unconditional fault which can explain more errors. The same argument holds for the stuck-at 1 fault at line a and $a \oplus [cond_0]$. As a consequence, assuming faults at line a will explain all the errors, and there is no line where assumed unconditional faults could explain more errors. However, there may be several of those lines explaining all the errors, and the ranking explained in the section above prefers those with a minimum number of mispredictions.

In **?** (**?**) the calculus described above is applied to large industrial circuits up to 1 million of gates, and analysis of stuck-at faults was used for validating the method. For a representative sample of stuck-at faults, the ranked lists of evidences are generated, and for all the fault candidates f with $e(f, T) = (\sigma_T, 0, 0)$ and a maximum number σ_T of predictions, additional distinguishing patterns are generated as far as possible.

Even for the largest circuits, an average rank better than 1.2 was obtained, and the real culprit was most often on top of the list. Only in cases where distinguishing patterns could not be generated and the faults seemed to be equivalent, multiple trials were required.

If volume diagnosis is performed, the test set cannot be enhanced and only a limited number of failing patterns is observed. By storing eight failing pattern outputs at maximum, the method described above puts the real culprit in average at rank 1.5 within the candidate fault list. This value is highly sufficient for deciding about further adaptive diagnosis in a second step.

The conditions for single stuck-at faults are rather simple, and diagnosis of more complex single line faults is more challenging. An example which fits for both logic debug and complex CMOS cells is the analysis of gates of a wrong type. For instance, the exchange of an $a = b$ OR c by an $a = b$ AND c is described by the CLF $a \oplus [b \oplus c]$. Experiments are reported about randomly changing the gate type while the rank of the real culprit is still better than 1.5 on average.

Similar results are known, if timing has to be considered in the activating condition of the CLF. An example is crosstalk fault as described above where the rank of the real culprits still remained at the top level.

5.5.2 Multiple Line Defects

If multiple lines are faulty, the corresponding fault effects may mask each other. As a consequence, predictions and mispredictions on an actual CLF may be affected in the presence of other active CLFs. Yet, it is known that test sets for single stuck-at faults are able to detect a large part of multiple stuck-at faults. The same reasoning does also hold for CLFs, however, it is not any more true that the (unconditional) stuck-at fault at one of the defect lines always explains the highest number of errors.

The reasonings described above form just a heuristic and still works in a rather efficient way as evidenced by the results reported. The 4-way bridges discussed above affect two lines, and just by looking only into 8 failing output patterns the algorithm described above points to the defect region with an average rank of 2.

5.6 Summary

Faults in circuits, those implemented in modern technology, show a more and more complex behavior. Diagnosis algorithms cannot assume any more a simplified fault model but have to do both locating the flaws in the structure and layout and extracting the faulty behavior at these lines. The chapter introduced a method to model faulty behavior of defective lines sufficiently precise for debug and diagnosis.

The method can be used for implementing an effect-cause analysis and allows identifying faults sites under all technology dependent fault models like delay faults, opens, bridges or even more complex functional faults.

References

Abramovici M, Breuer MA (1980) Fault diagnosis based on effect-cause analysis: an introduction. In Proceedings 17th design automation conference (DAC) 1980, pp 69–76, doi:10.1145/800139.804514

Amyeen ME, Nayak D, Venkataraman S (Oct 2006) Improving precision using mixed-level fault diagnosis. In Proceedings 37th IEEE international test conference (ITC) 2006, pp 22.3, doi:10.1109/TEST.2006.297661

Arnaout T, Bartsch G, Wunderlich H-J (Jan 2006) Some common aspects of design validation, debug and diagnosis. In 3rd IEEE international workshop on electronic design, test and applications (DELTA) 2006, pp 3–10, doi:10.1109/DELTA.2006.79

Bartenstein T (2000) Fault distinguishing pattern generation. In Proceedings 31st IEEE international test conference (ITC) 2000, pp 820–828, doi:10.1109/ TEST.2000.894285

Bartenstein T, Heaberlin D, Huisman LM, Sliwinski D (2001) Diagnosing combinational logic designs using the single location at-a-time (SLAT) paradigm. In Proceedings 32nd IEEE international test conference (ITC) 2001, pp 287–296, doi:10.1109/TEST.2001.966644

Bhatti NK, Blanton RD (Oct 2006) Diagnostic test generation for arbitrary faults. In Proceedings 37th IEEE international test conference (ITC) 2006, pp 19.2, doi:10.1109/TEST.2006.297647

Blanton RD, Dwarakanath KN, Desineni R (2006) Defect modeling using fault tuples. IEEE Trans CAD Integrat Circuits Sys 25(11):2450–2464, doi:10.1109/TCAD.2006.870836

Boppana V, Hartanto I, Fuchs WK (1996) Full fault dictionary storage based on labeled tree encoding. In Proceedings 14th IEEE VLSI test symposium (VTS) 1996, pp 174–179, doi:10.1109/VTEST.1996.510854

Chen KC (2003) Assertion-based verification for SoC designs. In Proceedings 5th International conference on ASIC 1:12–15

Chen G, Reddy SM, Pomeranz I, Rajski J (2006) A test pattern ordering algorithm for diagnosis with truncated fail data. In Proceedings 43rd design automation conference (DAC) 2006, pp 399–404, doi:10.1145/1146909.1147015

Chess B, Larrabee T (Mar 1999) Creating small fault dictionaries. IEEE Trans Comput-Aided Des Integrat Circuits Sys 18(3):346–356, doi:10.1109/43.748164

Desineni R, Poku O, Blanton RD (Oct 2006) A logic diagnosis methodology for improved localization and extraction of accurate defect behavior. In Proceedings 37th IEEE international test conference (ITC) 2006, pp 12.3, doi:10.1109/TEST.2006.297627

Flottes M-L, Landrault C, Pravossoudovitch S (1991) Fault modeling and fault equivalence in CMOS technology. J Electron Test, vol 2, no 3, pp 229–241, doi:10.1007/BF00135440

Gong Y, Chakravarty S (1995) On adaptive diagnostic test generation. In Proceedings IEEE international conference on computer-aided design (ICCAD) 1995, p 181, doi:10.1109/ICCAD.1995.480010

Henderson CL, Soden JM (1997) Signature analysis for IC diagnosis and failure analysis. In Proceedings 28th IEEE international test conference (ITC) 1997, pp 310–318, doi:10.1109/TEST.1997.639632

Holst S, Wunderlich H-J (May 2007) Adaptive debug and diagnosis without fault dictionaries. In Proceedings 12th European test symposium (ETS) 2007, pp 7–12, doi:10.1109/ETS.2007.9

Holst S, Wunderlich H-J (2009) Adaptive debug and diagnosis without fault dictionaries. In J Electron Test, vol 25, no 4–5, pp 259–268, doi:10.1007/s10836-009-5109-3

Hora C, Segers R, Eichenberger S, Lousberg M (2002) An effective diagnosis method to support yield improvement. In Proceedings 33rd IEEE international test conference (ITC) 2002, pp 260–269, doi:10.1109/TEST.2002.1041768

Keller BL (Aug 1996) Hierarchical pattern faults for describing logic circuit failure mechanisms, US Patent 5,546,408

Khursheed S, Rosinger P, Al-Hashimi BM, Reddy SM, Harrod P (2008) Bridge defect diagnosis for multiple-voltage design. In Proceedings 13th European Test Symposium (ETS) 2008, pp 99–104, doi:10.1109/ETS.2008.14

Klein R, Piekarz T (2005) Accelerating functional simulation for processor based designs. Proceedings International Workshop on System-on-Chip for Real-Time Applications 2005, pp 323–328, doi:10.1109/IWSOC.2005.34

Krstic A, Wang L-C, Cheng K-T, Liou J-J, Abadir MS (2003) Delay defect diagnosis based upon statistical timing models – the first step. In Proceedings 6th Design, Automation and Test in Europe (DATE) 2003, pp 10,328–10,335

Kundu S, Sengupta S, Goswami D (Apr 2006) Generalized fault model for defects and circuit marginalities, US Patent 7,036,063

Lavo DB, Chess B, Larrabee T, Hartanto I (1998) Probabilistic mixed-model fault diagnosis. In Proceedings 29th IEEE international test conference (ITC) 1998, pp 1084–1093, doi:10.1109/TEST.1998.743308

Li C-MJ, McCluskey EJ (2005) Diagnosis of resistive-open and stuck-open defects in digital CMOS ICs. IEEE Trans CAD Integrat Circuits Sys 24(11):1748–1759, doi:10.1109/TCAD.2005.852457

Liu C, Zou W, Reddy SM, Cheng W-T, Sharma M, Tang H (Oct 2007) Interconnect open defect diagnosis with minimal physical information. In Proceedings 38th International Test Conference (ITC) 2007, pp 7.3, doi:10.1109/TEST.2007.4437580

Liu C, Cheng W-T, Tang H, Reddy SM, Zou W, Sharma M (Nov 2008) Hyperactive faults dictionary to increase diagnosis throughput. In Proceedings 17th Asian test symposium (ATS) 2008, pp 173–178, doi:10.1109/ATS.2008.16

McPherson JW (2006) Reliability challenges for 45 nm and beyond. In Proceedings 43rd Design Automation Conference (DAC) 2006, pp 176–181, doi:10.1145/1146909.1146959

Polian I, Czutro A, Kundu S, Becker B (Oct 2006) Power droop testing. In Proceedings international conference on computer design (ICCD) 2006, pp 243–250, doi:10.1109/ICCD.2006.4380824

Pomeranz I, Reddy SM (1992) On the generation of small dictionaries for fault location. In Proceedings IEEE/ACM International Conference on Computer-Aided Design (ICCAD) 1992, pp 272–279, doi:10.1109/ICCAD.1992.279361

Riley M, Chelstrom N, Genden M, Sawamura S (Oct 2006) Debug of the CELL processor: moving the lab into silicon. In Proceedings 37th IEEE international test conference (ITC) 2006, pp 26.1, doi:10.1109/TEST.2006.297671

Rodríguez-Montañés R, Arumí, D, Figueras J, Eichenberger S, Hora C, Kruseman B (2007) Impact of gate tunnelling leakage on CMOS circuits with full open defects. Electron Lett 43(21): 1140–1141, 11. doi:10.1049/el:20072117

Rousset A, Bosio A, Girard P, Landrault C, Pravossoudovitch S, Virazel A (Oct 2007) Fast bridging fault diagnosis using logic information. In Proceedings 16th Asian Test Symposium (ATS) 2007, pp 33–38, doi:10.1109/ATS.2007.75

Roy K, Mak TM, Cheng K-T (2006) Test consideration for nanometer-scale CMOS circuits. IEEE Des Test Comput 23(2):128–136, doi:10.1109/MDT.2006.52

Soden JM, Treece RK, Taylor MR, Hawkins CF (Aug 1989) CMOS IC stuck-open-fault electrical effects and design considerations. In Proceedings 20th international test conference (ITC) 1989, pp 423–430, doi:10.1109/TEST.1989.82325

Tirumurti C, Kundu S, Sur-Kolay S, Chang Y-S (2004) A modeling approach for addressing power supply switching noise related failures of integrated circuit. In Proceedings 7th Design, Automation and Test in Europe (DATE) 2004, pp 1078–1083, doi:10.1109/DATE.2004.1269036

Ubar R (2003) Design error diagnosis with resynthesis in combinational circuits. J Electron Test Theory Appl 19:73–82, doi:10.1023/A:1021948013402

Veneris AG, Chang R, Abadir MS, Amiri M (2004) Fault equivalence and diagnostic test generation using ATPG. In Proceedings IEEE international symposium on circuits and systems (ISCAS) 2004, pp 221–224

Wadsack R (1978) Fault modeling and logic simulation of CMOS and MOS integrated circuits. Bell Sys Techn J 57:1449–1488

Waicukauski JA, Lindbloom E (Aug 1989) Failure diagnosis of structured VLSI. IEEE Des Test Comput 6(4):49–60, doi:10.1109/54.32421

Chapter 6
Models in Memory Testing
From Functional Testing to Defect-Based Testing

Stefano Di Carlo and Paolo Prinetto

Abstract Semiconductor memories have been always used to push silicon technology at its limit. This makes these devices extremely sensible to physical defects and environmental influences that may severely compromise their correct behavior. Efficient and detailed testing procedures for memory devices are therefore mandatory. As physical examination of memory designs is too complex, working with models capable of precisely representing memory behaviors, architectures, and fault mechanisms while keeping the overall complexity under control is mandatory to guarantee high quality memory products and to reduce the overall test cost. This is even more important as we are fully entering the Very Deep Sub Micron era. This chapter provides an overview of models and notations currently used in memory testing practice highlighting challenging problems waiting for solutions.

Keywords Memory testing · Memory modeling · Fault models · March test

6.1 Introduction

Since 1945 when the ENIAC, the first computer system with its memory of mercury and nickel wire delay lines went into service, through the relatively expensive core memory used in about 95% of computers by 1976, memory has played a vital role in the history of computing.

With the advent of semiconductor memories for commercial applications (the Intel™ 1103 shown in Fig. 6.1 was the first 1 Kbit dynamic RAM commercial chip), for the first time a significant amount of information could be stored on a single chip. This represented the basis for modern computer systems.

S. Di Carlo (✉) and P. Prinetto
Politecnico di Torino, Control and Computer Engineering Department, Corso duca degli Abruzzi 24, 10129, Torino, Italy
e-mail: stefano.dicarlo@polito.it

H.-J. Wunderlich (ed.), *Models in Hardware Testing: Lecture Notes of the Forum in Honor of Christian Landrault*, Frontiers in Electronic Testing 43,
DOI 10.1007/978-90-481-3282-9_6,

Fig. 6.1 Intel 1103, first DRAM commercial chip with 1024 bits

Nowadays, the role of memory devices in the semiconductor industry is even clearer. Applications such as computer graphics, digital signal processing, and rapid retrieval of huge volumes of data, demand an exponentially increasing amount of memory. A constantly growing percentage of Integrated Circuits (ICs) area is thus dedicated to implement memory structures. According to the International Technology Roadmap for Semiconductors (ITRS) (ITRS 2007), a leading authority in the field of semiconductors, memories occupied 20% of the area of an IC in 1999, 52% in 2002, and are forecasted to occupy up to 90% of the area by the year 2011.

Due to this considerable usage of memories in ICs, any improvement in the design and fabrication process of these devices has a considerable impact on the overall ICs characteristics. Reducing the energy consumption, increasing the reliability and, above all, reducing the cost of memories directly reflect on the systems they are integrated in.

This continuous research for improvement has historically pushed the memory technology at its limit, making these devices extremely sensible to physical defects and environmental influences that may severely compromise their correct behavior. Efficient and detailed testing of memory components is therefore mandatory. A large portion of the price of a memory derives today from the high cost of memory testing, which has to satisfy very high quality constraints, ranging from 50 failing parts per million (ppm) for computer systems to less than 10 ppm for mission-critical applications (such as those in the automotive industry).

As physical examination of memory designs is too complex, working with models capable of precisely representing memory behaviors, architectures, and fault mechanisms while keeping the overall testing problem complexity under control is mandatory to guarantee high quality memory products and to reduce the test cost. This is even more important as we fully enter the *very deep sub-micron* (VDSM) era.

This chapter provides an overview of models and notations currently used in the memory testing practice, and concludes by highlighting challenging and still open problems.

6.2 Models for Memory Testing: A Multidimensional Space

Tens of models have been proposed in the literature to support the different aspects and phases of the memory life-cycle. From design to validation and verification, from manufacturing to testing, and from diagnosis to repair, most of the proposed models fulfill specific needs and target well defined goals. Such a proliferation of different "*custom*" models is not surprising at all. Memory representation and modeling is a typical multidimensional space, where, depending on the specific goal, peculiar information items need to be modeled and characterized. Figure 6.2 shows some of the most significant dimensions of this space, not necessarily orthogonal to each other.

Among the others, it is worth mentioning:

- *Abstraction level*: identifies the desired degree of details included in a memory model. Typical values for this dimension are system level, register transfer (RT) level, logic level, device level, and layout level. They will be deeply analyzed in the next section.
- *Representation domain*: for each abstraction level, this orthogonal dimension allows us to focus on different sets of aspects of interest. Typical values for this dimension include: behavioral domain, structural domain, physical domain, and geometrical domain.

The behavioral domain focuses on the behavior of the system, only, without any reference to its internal organization. The structural domain focuses on the structure (i.e., the topology) of the system, in terms of connection of blocks. Such a description is usually technology independent. The physical domain introduces the physical properties of the basic components used in the structural domain, and finally the geometrical domain adds information about geometrical entities to the design.

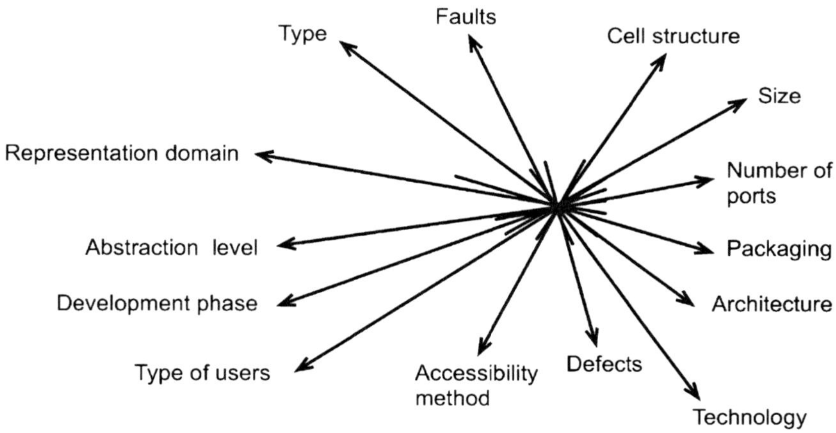

Fig. 6.2 The memory modeling space

- *Type*: several types of semiconductor memories have been historically defined, the most representative ones being: random-access memories (RAMs), read-only memories (ROMs), and content-addressable memories (CAMs).

  RAMs are memories whose cells can be randomly accessed to perform write and/or read operations, while ROMs are memories whose cells can be read indefinitely but written just a limited number of times.

  Read-only memories can be further characterized according to the number of possible write operations and to the way in which these can be performed. ROMs usually identify memory devices that can be written by the manufacturer, only once. Programmable ROMs (PROMs) can be programmed by the user just once, while Erasable PROMs (EPROMs) can be programmed by the user several times ($<10^2$), resorting to ultraviolet light. Electrically EPROMs (EEPROMs) can be reprogrammed several times ($<10^5$) by the user using high voltage signals. Finally, Flash memories can be easily written several times ($<10^5$) without the need of high voltage signals, even if they can only written page-by-page, and not on a word-by-word basis.

  CAMs represent a completely different type of memory devices that differs from RAMs and ROMs in the way data are accessed. While both in RAMs and ROMs the user supplies a memory address and the device either returns the data stored at that address (read operation) or modifies the content of the addressed location (write operation), CAMs are designed in such a way that the user supplies a data word and the device searches for the data in the memory to check whether it is stored or not. If the data word is found, the CAM returns a list of one or more storage addresses where the word was found and, in some implementations, the data word and/or additional associated information items, as well.
- *Cell structure*: identifies a set of characteristics of each storing element (cell) of the memory. Typical values for this dimension, related to RAM memories, allow us to distinguish between static RAMs (SRAMs) and dynamic RAMs (DRAMs).

  In an SRAM a bistable electrical component is used to store a binary value. When a cell is set into one of its two possible states ('0' or '1'), it stays in it until the device is powered off. DRAMs cells are instead constructed of simple capacitive elements that store electrical charges to represent a given logical level, and are characterized by gradual loss of the stored information, due to a physical phenomenon known as transistor leakage current (Roy et al. 2003). As a consequence, DRAM cells need to be periodically "recharged" ("refreshed", "rewritten") not to loose the stored value. Figure 6.3 shows an example of device level structural model of: (a) an SRAM cell and (b) a DRAM cell.
- *Number of access ports*: defines the number of users that can access the memory concurrently. While a single-port RAM allows one access at a time, multi-port RAMs, being equipped with several I/O ports, allow multiple read or write operations to occur at the same time, or nearly at the same time.
- *Type of users*: identifies the target user of the memory. It is a key point to define an appropriate memory model. Different users typically resort to different models

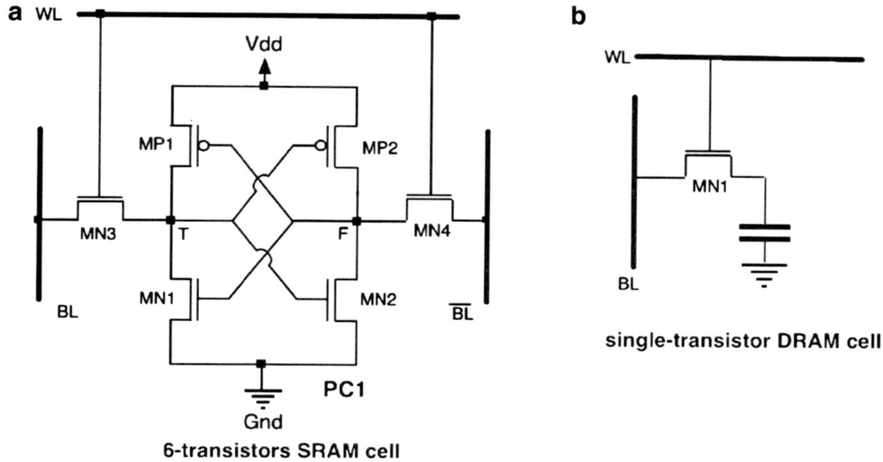

Fig. 6.3 (**a**) SRAM, and (**b**) DRAM device level structural model

due to both their peculiar goals and the amount of available details. Several types of users can be considered in the supply chain of semiconductor memories: the designer, the test engineer, the device manufacturer, the system integrator, and the end-users.

- *Faults and defects*: defines the classes of failure mechanisms as well as physical defects that may occur in a given memory. They will be deeply analyzed in the next sections.

6.3 Models for Structures, Behaviors, and Architectures

Dealing with the multidimensional space presented in the previous section is always a complex and hard to manage task. Appropriate sub-spaces, or views, are therefore used to reduce the complexity of the modeling process. Van de Goor (1991) proposes a very interesting sub-space based on different nested levels (Fig. 6.4). Going from the external levels to the internal ones, the amount of information about the actual implementation of the memory decreases while the information about the way the system is expected to work increases. This sub-space has the main disadvantage of mixing, at the same time, abstraction levels and representation domains.

Here we prefer to introduce a different sub-space that explicitly considers the abstraction level and the representation domain (see Section 6.2) as two orthogonal dimensions. Figure 6.5 shows this sub-space where for the sake of simplicity only the most representative representation domains are included. Based on this model, in the sequel of this section we shall analyze the matrix proposed in Fig. 6.5, outlining

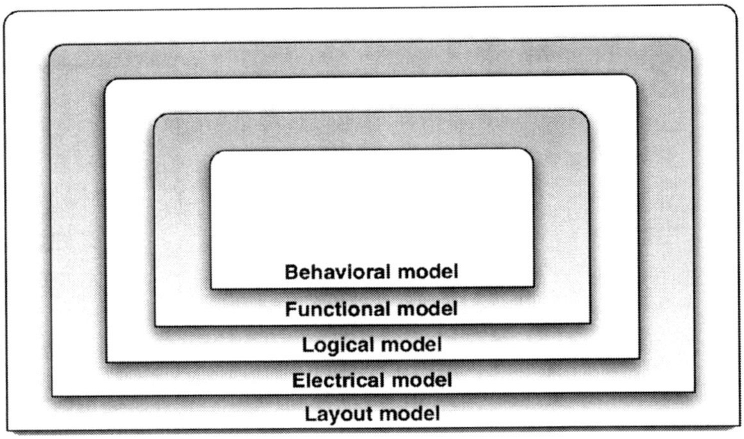

Fig. 6.4 Different memory modeling levels proposed by Van de Goor

Fig. 6.5 Abstraction level vs. representation domain modeling sub-space

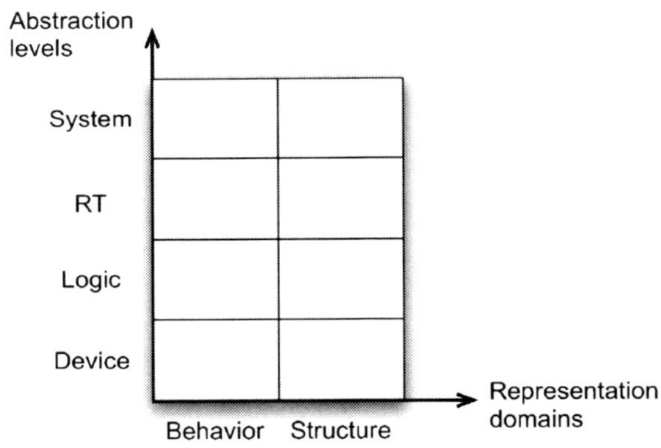

a set of well established models used to represent memory structures, behaviors, and architectures. In particular we shall focus on those models usually used in memory testing.

The system level is the highest representation level and is usually very close to the specifications. When dealing with memories, behavioral descriptions at system level describe the memory behavior in terms of Input/Output (I/O) relationships. Timing diagrams can be used to graphically represent the minimum and maximum timing constraints and the I/O signal relationships for the various memory operations. On the other hand, a system level structural model focuses on the definition of the I/O signals of the memory. Figure 6.6 shows an example of (a) a single-port and (b) a two-ports system level structural model.

Moving to the RT abstraction level, behavioral modeling is usually accomplished resorting to state transition graphs (STG). STG based descriptions for memories, first introduced in Brzozowski et al. (1992), have been used for long time to

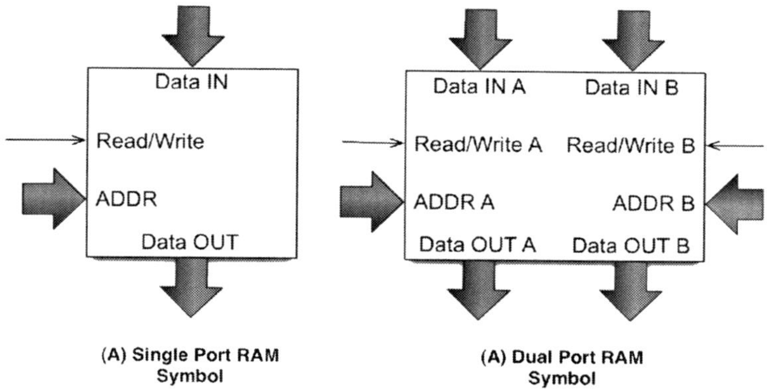

Fig. 6.6 System level structural model

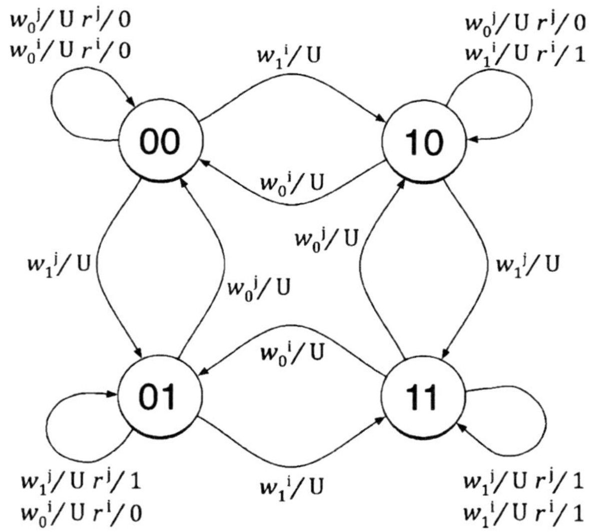

Fig. 6.7 RT-Level behavioral memory model

efficiently model the behavior of faulty memories (Zarrineh et al. 1998; Niggemeyer et al. 2000), and to support the automatic generation of memory test algorithms (see Section 6.5.1). Figure 6.7 shows an extension of the model of Brzozowski et al. (1992) proposed in Benso et al. (2008). It is based on the use of edge labeled directed graphs, and allows modeling both the internal memory state and the memory output resulting from read operations. Figure 6.7 shows the model of a memory composed of 2 one-bit cells identified by i and j, respectively. Nodes represent memory states while arcs represent transitions among states. Arcs are labeled with triggering conditions and corresponding memory outputs. Due to the explosion in

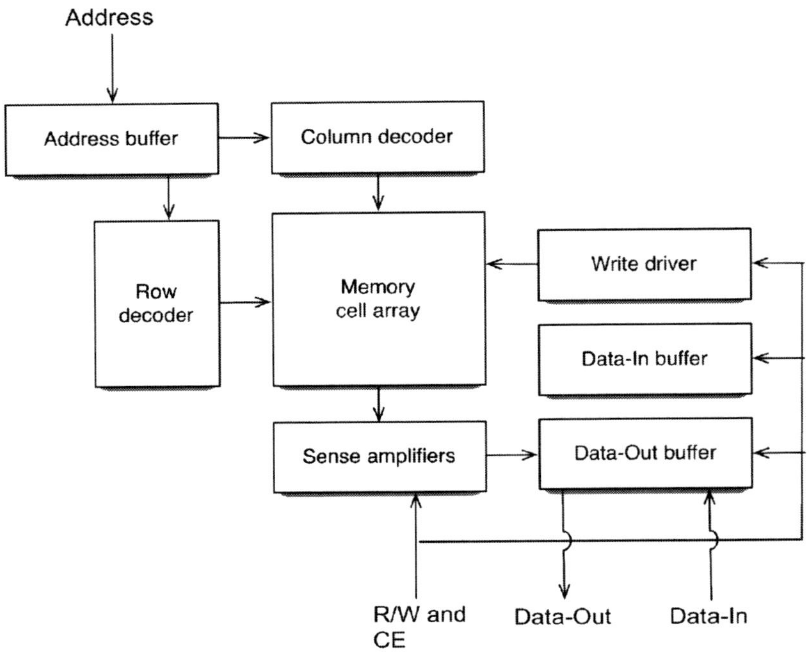

Fig. 6.8 Detailed RT-Level memory structural model

terms of complexity (number of states equal to $2^{\# memory-cells}$), very small portions of memories are usually represented and the memory regularity is then exploited to properly extend the model.

RT-level structural descriptions (Fig. 6.5) play a significant role in testing-oriented memory modeling. Memories are here modeled as a proper interconnection of interacting sub-systems, each characterized by a specific function, and cooperating to achieve the target external memory behavior. Figure 6.8 shows the detailed RT-level structural model of an SRAM proposed by Van de Goor in Van de Goor (1991). The memory cell array is the kernel of the memory and contains the set (or "array", or "matrix") of memory cells, properly arranged in rows and columns. To access each cell, the cell address is first partitioned to obtain the row address and column address and then decoded by the row and column decoders. The outputs of these two decoders are usually referred to as word lines (WLs) and column selects (CSs), respectively. The address buffer, data-in buffer, and data-out buffer store the addresses and data, while the sense amplifiers allow identifying the data stored in the cells of the memory cell array.

When the memory model is used for test purpose, without exploiting the actual internal memory structure, the model of Fig. 6.8 is usually simplified as in Fig. 6.9 (Van de Goor 1991), where the address decoder and the read/write logic include all the logic required to perform address decoding and read/write operations, respectively.

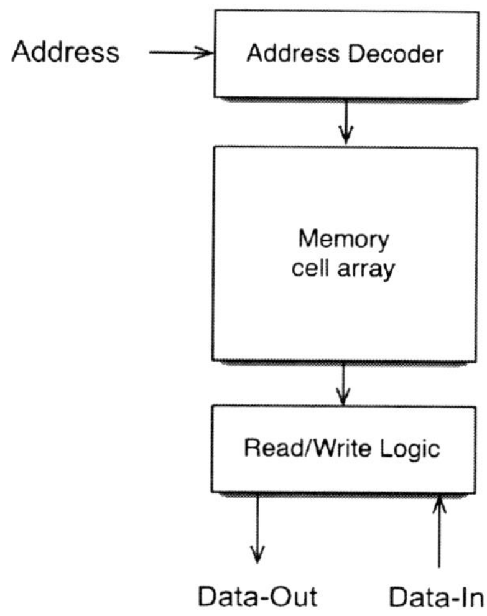

Fig. 6.9 Simplified RT-Level memory structural model

Moving down to the logic level, behavioral descriptions (Fig. 6.5) are usually accomplished resorting to Boolean expressions representing the Boolean function of each output, whereas structural descriptions at the same level consist of netlists, i.e., interconnections of gates and/or flip-flops and latches. Since memories are very seldom modeled resorting to their logic structures, the logic level is here of practical no interest.More interesting for memories is the device level. Device level behavioral models (Fig. 6.5) consist of descriptions of the electrical characteristics of the physical devices used to build the memory, i.e. voltage levels, currents, etc. with related timings. Structural descriptions at the same level consist of proper interconnections of the basic electrical components used to implement the memory, i.e., transistors, resistors, capacitors, and connectors. Examples of structural descriptions of a typical cells for SRAMs and DRAMs are in Fig. 6.3a and b respectively, whereas Fig. 6.10 shows the complete structure of a 3×3 cells SRAM.

6.4 Fault Models

Fault modeling is probably one of the most important elements of memory testing and fault analysis. As physical examination of memory is usually too complex, test engineers resort to functional fault models (FFM) to build efficient functional test algorithms. Functional test makes the proof of both test completeness and non-redundancy a logic problem easy to manage and automate.

Fault modeling for memory has a long history. Early work on FFMs for RAMs was done by Thatte and Abraham (Thatte et al. 1977) followed by Van de Goor in the early 1980s (Ven de Goor et al. 1990). These publications represent the basis for more than 20 years of work in defining different types of FFMs and corresponding test algorithms. This huge amount of work was motivated by the continuous

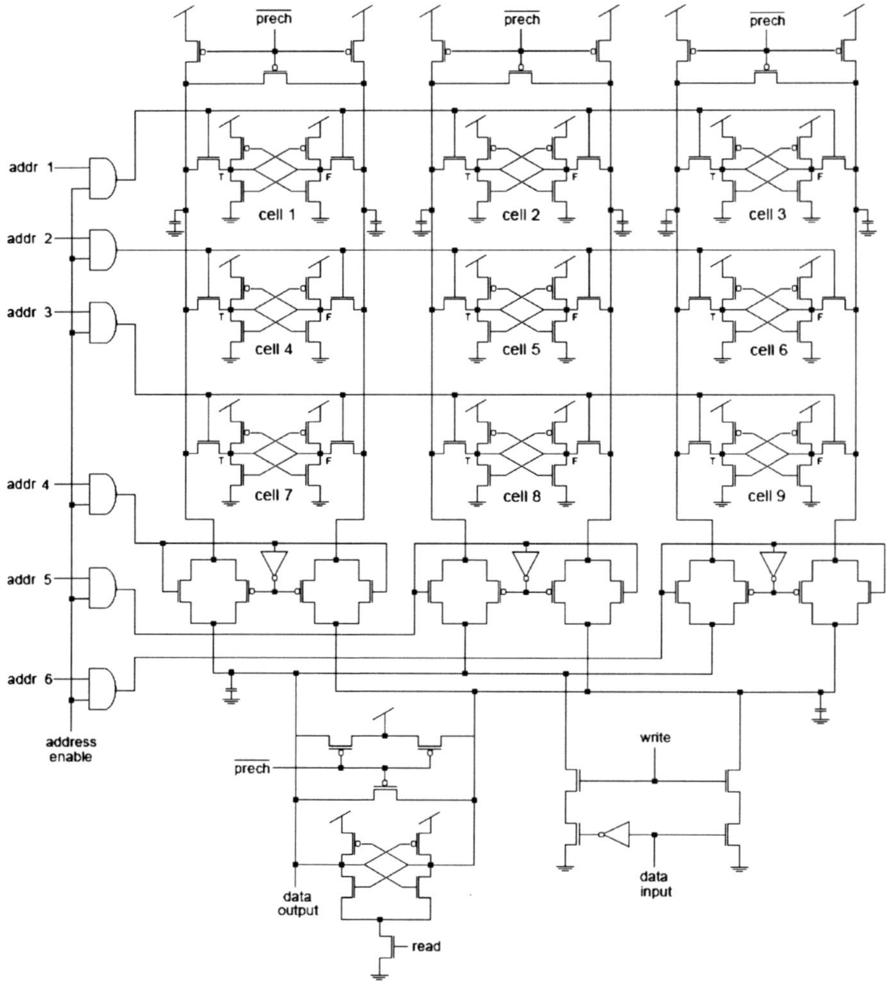

Fig. 6.10 Device level structural model of a 3 × 3 SRAM

improvement of silicon technology leading to an increased sensitivity to fabrication defects and environmental stress.

Given this premise, a complete description of all proposed FFMs and related fault mechanisms would be too long to fit a single book chapter. Moreover, the continuous technology improvement would make it outdated in a short time. For this reason this section will focus on the basic concepts and theories required to understand and to work with memory FFMs, introducing a reduced set of well established FFMs, only.

According to the simplified RT-Level structural model proposed in Fig. 6.9, FFMs can be classified according to the targeted functional block as: FFMs for the address decoder logic usually referred to as *address faults (AF)*, FFMs for the memory cell array usually referred to as *memory cell faults*, and FFMs for the read/write logic.

Address faults have been one of the first studied memory fault models. Thatte et al. (1977) and Nair (1979) provide a non-formal definition of four categories of address faults:

1. *Type 1 AFs*: with a certain address no cell is accessed.
2. *Type 2 AFs*: a certain cell is not accessible by any address.
3. *Type 3 AFs*: with a certain address, multiple cells are accessed simultaneously.
4. *Type 4 AF*: a certain cell is accessible with multiple addresses.

In Van de Goor (1991) and Van de Goor et al. (1990) a set of proofs show that, under well defined conditions, test algorithms designed for memory cell faults are able to detect address faults. The same happens for faults in the read/write logic. In Nair et al. (1978), Thatte et al. (1977), Van de Goor (1991), and Marinescu (1982), Van de Goor and others give a proof that faults in the read/write logic of the memory are equivalent to faults in the memory array.

For this reason this section will focus on memory cell faults, only.

6.4.1 FFMs Definition and Representation

Defining FFMs has been up to now a very ad-hoc process, usually performed by inserting electrical defects into an electrical model of a memory followed by an intensive campaign of simulations to understand the effect of the defects on the memory behavior. Inductive fault analysis (IFA) (Shen et al. 1985) is an additional common practice to analyze the chip layout, and to determine which FFMs correctly describe the actual physical defects that occur. The challenge in this process is the choice of the stimuli applied during simulation. To show the presence or absence of a new FFM the precise sequence of stimuli able to sensitize the fault needs to be identified and applied. This in turn requires knowing a priori the behavior of the fault, or precisely defining the space of potential faults to be later explored for the identification of new FFMs.

A memory *functional fault model* can be informally defined as the deviation of the observed memory behavior from the specified one, under a given sequence of performed memory operations.

Two basic elements are therefore required to define a generic FFM:

1. A sequence of memory operations able to sensitize the fault, usually referred to as *sensitizing operation sequence (SOS)*
2. A list of observed deviations of the memory behavior from the observed one, usually referred to as *faulty behavior (FB)*

A common and very effective way to represent this information is the *fault primitive* (FP) formalism first introduced in Van de Goor et al. (2000), and then refined in Benso et al. (2008) (Eq. 6.1).

$$FP =< SOS/FB > \tag{6.1}$$

The *SOS* represents the sensitizing operation sequence and includes the *initial data* C stored in a set of memory cells involved in the target fault, followed by an optional sequence of *memory operations* to be performed in order to sensitize the fault (Eq. 6.2). The cells initialized and accessed during the *SOS* are denoted as *f-cells.*

$$SOS = (C, op_1\, op_2\, \ldots\, op_m), \text{ with } m \geq 0 \tag{6.2}$$

where:

- $C = \left(s^{f-cell_1} s^{f-cell_2} \ldots s^{f-cell_k}\right)$. $s^{f-cell_i} \in \{0, 1, -\}$ denotes the content of the *f-cell* with address *f-cell*$_i$, and '$-$' denotes a don't care condition in which the initial value of the cell is not relevant;
- $op_i \in \{r^i_{[d]}, w^i_d\}$. w^i_d, where $d \in \{0, 1\}$, represents a write operation of the value d on the cell with address i while $r^i_{[d]}$ represents a read operation of the content of the cell with address i. If the optional value d is specified, it is interpreted as the expected value returned by the read operation and the operation itself is denoted as a *read and verify* operation.

The sequence of memory operations can be omitted when the faulty behavior is sensitized just by the *f-cells* being in a certain condition (e.g., State Coupling fault (Dekker et al. 1990)).

Examples of *SOSs* are:

- $SOS = 0^i$. It corresponds to an FB sensitized by the state of the *f-cell* with address i equal to 0.
- $SOS = -, w^i_1$. It corresponds to an FB sensitized by writing 1 into the *f-cell* with address i, regardless the current state of the cell.
- $SOS = 1^i, w^i_1 r^i$. It corresponds to an FB sensitized by a write operation of the value 1 on the faulty cell with address i immediately followed by a read operation of the content of the same cell. In this case the FB is sensitized only if the two operations are applied when the cell i initially contains the value 1.

FB (Eq. 6.1) identifies the behavior of the faulty memory with respect to the fault free one. In the early 1980s, and during the first part of the 1990s, the only element considered relevant for the definition of a faulty behavior was the logic state stored in the memory cells (Van de Goor 1991). Starting from the end of the 1990s, due to the increased memory complexity the output value of a read operation started to be relevant to correctly describe realistic and observed fault models (Van de Goor 1999). An FB can be expressed using the notation of Eq. 6.3.

$$F/R \tag{6.3}$$

where

$$F = \left(f^{f-cell_1} f^{f-cell_2} \ldots f^{f-cell_k}\right), \text{ with } f \in \{0, 1\} \tag{6.4}$$

represents the content of the *f-cells* after applying the operations contained in the *SOS*, while

$$R = \left(r^{f-cell_1} r^{f-cell_2} \ldots r^{f-cell_k}\right) \text{ with } r \in \{0, 1, -\} \tag{6.5}$$

represents a set of erroneous outputs of the memory obtained when reading the *f-cells*. This element of the FB is required to represent classes of faults where a read operation returns an erroneous value while the content of the memory cell is correct. The symbol '–' is used whenever a write operation, and not a read operation, is responsible for sensitizing the fault.

Definition: A memory FFM is a non-empty set of fault primitives.

6.4.2 Generic Space of Faults

As stated in Section 6.4.1, the definition of new memory FFMs requires a precise identification and exploration of the generic space of possible faults. As we define FFMs in terms of collections of FPs, this problem is actually reduced to the definition of the space of possible FPs.

FPs can be classified into different classes based on the characteristic of the related *SOS*. In particular, let $|f|$ be the number of *f-cells* involved in the *SOS*, and m (Eq. 6.2) the number of operations composing *SOS*.

Depending on $|f|$, FPs can be categorized into *single-cell* FPs when $|f| = 1$, and *n-coupling FPs* when $|f| = n$. In case of n-coupling FPs, the set of *f-cells* is split into *victims (v)* and *aggressors (a)*. In a FP the faulty behavior always manifests into the victim cells, while the aggressors contribute to the sensitization of the fault.

Depending on the number m of operations composing the *SOS*, an FP can be classified as *static* when $m \leq 1$, i.e., a single memory operation is sufficient to sensitize the target faulty behavior, or *dynamic* when $m > 1$, i.e., n memory operations are required to sensitize the target faulty behavior.

Figure 6.11 shows a graphical representation of the taxonomy of the FP space. The two classifications criteria introduce two different hierarchies in the FP space. Since FFMs have been defined as a non-empty set of FPs, we can inherit the same type of classification and hierarchy when considering FFMs. For example, a FFM composed of a collection of static FPs will be referred to as a static FFM, etc.

6.4.3 Static Fault Models

Static faults consist of groups of static FPs ($m \leq 1$) sensitized by at most a single memory operation. They were historically the first set of faults observed into memory devices. In the following the most important classes of static faults will be presented.

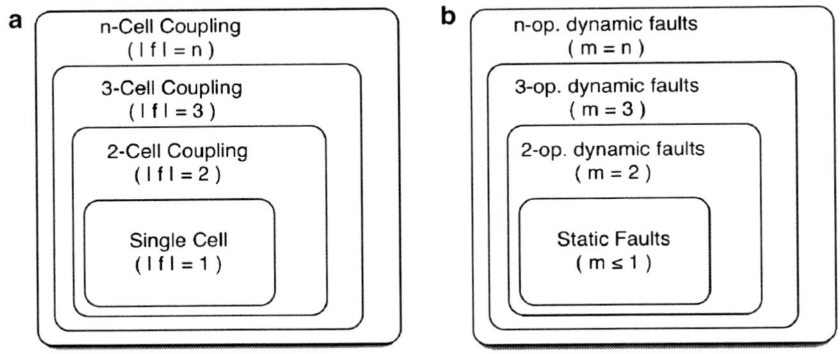

Fig. 6.11 Fault primitive classification based on (**a**) number of *f-cells*, and (**b**) number of sensitizing operations

Table 6.1 Static FPs space

#	FP	Fault model
1	$< 0/1/- >$	State-0 fault (SF_0)
2	$< 1/0/- >$	State-1 fault (SF_1)
3	$< 0, w_0/1/- >$	Write-0 destructive fault (WDF_0)
4	$< 0, w_1/0/- >$	Up transition fault (TF_1)
5	$< 1, w_0/1/- >$	Down transition fault (TF_0)
6	$< 1, w_1/0/- >$	Write-1 destructive fault (WDF_1)
7	$< 0, r_0/0/1 >$	Incorrect read-0 fault (IRF_0)
8	$< 0, r_0/1/0 >$	Deceptive read-0 destructive fault ($DRDF_0$)
9	$< 0, r_0/1/1 >$	Read-0 destructive fault (RDF_0)
10	$< 1, r_1/0/0 >$	Read-1 destructive fault (RDF_1)
11	$< 1, r_1/0/1 >$	Deceptive read-1 destructive fault ($DRDF_1$)
12	$< 1, r_1/1/0 >$	Incorrect read-1 fault (IRF_1)

6.4.3.1 Single-Cell Static Faults

In a single-cell static fault, a single generic memory cell i is responsible either for sensitizing, and manifesting the effect of the fault. As we work with a single cell, the notation presented in Section 6.4.1 can be simplified by omitting the information about the cell address.

The single-cell static fault condition, i.e., $|f| = 1$, and $m \leq 1$, leads to the restricted set of static *SOSs* in Eq. 6.6:

$$SOS \in \{'0';'1';'0, w_0';'0, w_1';'1, w_0';'1, w_1';'0, r_0','1, r_1'\} \qquad (6.6)$$

that allows the definition of the space of static FPs reported in Table 6.1 where the right most column reports the common name of the fault model associated with the given FP.

This set of FPs can be grouped to define a set of six well established and characterized FFMs:

1. *State fault* (SF_x): the logic value of the target cell flips in correspondence of a given initialization value, even if no operation is performed. The state fault should be understood in the static sense, i.e., the cell should flip in the short time period after initialization and before accessing the cell. This fault is special in the sense that no operation is needed to sensitize it and, therefore, it only depends on the initial value stored in the cell. Two types of state faults exist: $SF_0 = \{FP_1\}$, and $SF_1 = \{FP_2\}$.
2. *Transition fault* (TF_x): the target cell fails to undergo an up ($0 \rightarrow 1$) or a down ($1 \rightarrow 0$) transition. Two types of transition faults exist: $TF_0 = \{FP_5\}$, and $TF_1 = \{FP_4\}$.
3. *Read destructive fault* (RDF_x): a read operation performed on the target cell changes the content of the cell and returns an incorrect value on the memory output. Two types of read destructive faults exist: $RDF_0 = \{FP_9\}$, and $RDF_1 = \{FP_{10}\}$.
4. *Write destructive fault* (WDF_x): a non-transition write operation performed on the target cell, i.e., $0, w_0$, or $1, w_1$, causes the cell to flip. It is similar to the TF. In both cases a write operation fails to work properly. Two types of write destructive faults exist: $WDF_0 = \{FP_3\}$, and $WDF_1 = \{FP_6\}$.
5. *Incorrect read fault* (IRF_d): a read operation performed on the target cell returns the incorrect logic value while keeping the correct cell content. Two types of incorrect read faults exists: $IRF_0 = \{FP_7\}$, and $IRF_1 = \{FP_{12}\}$.
6. *Deceptive read destructive fault* ($DRDF_d$): a read operation performed on the target cell returns the correct value while changing the content of the cell [Adams et al., 1996]. Two types of deceptive read destructive faults exists: $DRDF_0 = \{FP_8\}$, and $DRDF_1 = \{FP_{11}\}$.

The proposed set of FFMs is able to completely cover the set of FPs proposed in Table 6.1, and therefore any test able to cover these fault models is able to detect any single-cell static faulty behaviors.

Additional fault models have been defined in the literature; nevertheless, using the proposed classification, they result in a combination of the six proposed FFMs. For example the well known *stuck-at fault model* (Van de Goor 1991), i.e., a cell is stuck at a given value for all performed operations, can be modeled as follows:

- $SAF_0 = SF_1 \cup TF_1 \cup WDF_1$, denoting the stuck-at-0.
- $SAF_1 = SF_0 \cup TF_0 \cup WDF_0$, denoting the stuck-at-1.

In this case, SFA_0 is defined as a set of three FPs. Each fault primitive in this set is able to sensitize this fault, i.e., any test that covers at least one of the fault primitives in this set is able to cover the fault model.

6.4.3.2 2-Coupling Static Faults

2-coupling static FFMs are faults described by FPs involving two *f-cells* ($|f| = 2$) and sensitized by the application of at most a single memory operation ($m \leq 1$). In this condition, one of the two *f-cells* (usually denoted by the generic address v) is the victim cell where the effect of the faulty behavior manifests, while the second cell (usually denoted by the generic address a) is the aggressor cell, responsible with the victim for producing the faulty behavior. With this distinction three classes of *SOSs* can be generated:

1. No cell accessed: the state of the cells sensitizes the fault.
2. Only the aggressor cell is accessed.
3. Only the victim cell is accessed: the aggressor contributes to the fault simply with its initial state.

Starting with this classification it is possible to enumerate the space of 2-coupling FPs of Table 6.2 composed of 36 different FPs. Only those combinations of operations that actually represent a faulty behavior have been considered.

As for the single-cell static FFMs, this set of FPs can be grouped to define a set of seven well established and characterized FFMs:

1. *State Coupling Fault (CFst)*: the victim cell is forced into a given logic state when the aggressor cell is in a given state, without performing any operation. As for the state fault, this FFM is special, as no operation is required to sensitize the fault. Four types of state coupling faults exist, defined as $CF_{st(xy)} = \{< x^a y^v / \bar{y}^v / - >\}$, where $x, y \in \{0, 1\}$. This covers FP_1, FP_2, FP_3, and FP_4.

Table 6.2 2-coupling FP space

#	FP	#	FP
1	$< 0^a 0^v / 1^v / - >$	19	$< 0^a 0^v, w_1^v / 0^v / - >$
2	$< 0^a 1^v / 0^v / - >$	20	$< 1^a 0^v, w_1^v / 0^v / - >$
3	$< 1^a 0^v / 1^v / - >$	21	$< 0^a 1^v, w_0^v / 1^v / - >$
4	$< 1^a 1^v / 0^v / - >$	22	$< 1^a 1^v, w_0^v / 1^v / - >$
5	$< 0^a 0^v, w_0^a / 1^v / - >$	23	$< 0^a 1^v, w_1^v / 0^v / - >$
6	$< 0^a 1^v, w_0^a / 0^v / - >$	24	$< 1^a 1^v, w_1^v / 0^v / - >$
7	$< 0^a 0^v, w_1^a / 1^v / - >$	25	$< 0^a 0^v, r_0^v / 0^v / 1^v >$
8	$< 0^a 1^v, w_1^a / 0^v / - >$	26	$< 1^a 0^v, r_0^v / 0^v / 1^v >$
9	$< 1^a 0^v, w_0^a / 1^v / - >$	27	$< 0^a 0^v, r_0^v / 1^v / 0^v >$
10	$< 1^a 1^v, w_0^a / 0^v / - >$	28	$< 1^a 0^v, r_0^v / 1^v / 0^v >$
11	$< 1^a 0^v, w_1^a / 1^v / - >$	29	$< 0^a 0^v, r_0^v / 1^v / 1^v >$
12	$< 1^a 1^v, w_1^a / 0^v / - >$	30	$< 1^a 0^v, r_0^v / 1^v / 1^v >$
13	$< 0^a 0^v, r_0^a / 1^v / - >$	31	$< 0^a 1^v, r_1^v / 0^v / 0^v >$
14	$< 0^a 1^v, r_0^a / 0^v / - >$	32	$< 1^a 1^v, r_1^v / 0^v / 0^v >$
15	$< 1^a 0^v, r_1^a / 1^v / - >$	33	$< 0^a 1^v, r_1^v / 0^v / 1^v >$
16	$< 1^a 1^v, r_1^a / 0^v / - >$	34	$< 1^a 1^v, r_1^v / 0^v / 1^v >$
17	$< 0^a 0^v, w_0^v / 1^v / - >$	35	$< 0^a 1^v, r_1^v / 1^v / 0^v >$
18	$< 1^a 0^v, w_0^v / 1^v / - >$	36	$< 1^a 1^v, r_1^v / 1^v / 0^v >$

2. *Disturb coupling fault* (CF_{ds}): an operation (write or read) performed on the aggressor cell forces the victim cell into a given logic state. Any operation performed on the aggressor is accepted as sensitizing operation (a read, a transition write, or a non-transition write). Twelve types of disturb coupling faults exist, defined as $CF_{ds(xz,w_y)} = \{< x^a z^v, w_y^a/\bar{z}^v/- >\}$, and $CF_{ds(xz,r_y)} = \{< x^a y^v, r_x^a/\bar{y}^v/- >\}$ where $x, y, z \in \{0, 1\}$. This covers FP_5, FP_6, FP_7, FP_8, FP_9, FP_{10}, FP_{11}, FP_{12}, FP_{13}, FP_{14}, FP_{15}, and FP_{16}.
3. *Transition coupling fault* (CF_{tr}): the state of the aggressor cell causes the failure of a transition write operation performed on the victim cell. This fault is sensitized by a write operation on the victim cell, while the aggressor is in a given state. Four types of transition coupling faults exist, defined as $CF_{tr(x0)} = \{< x^a 0, w_1^v/0^v/- >\}$, and $CF_{tr(x1)} = \{< x^a 1, w_0^v/1^v/- >\}$ where $x \in \{0, 1\}$. This covers FP_{19}, FP_{20},FP_{21}, FP_{22}.
4. *Write destructive coupling fault* (CF_{wd}): a non-transition write operation performed on the victim cell while the aggressor cell is in a given state results in a transition of the cell itself. Four types of write destructive coupling faults exist, defined as $CF_{wd(xy)} = \{< x^a y^v, w_y^v/\bar{y}^v/- >\}$, where $x, y \in \{0, 1\}$. This covers FP_{17}, FP_{18}, FP_{23}, FP_{24}.
5. *Read destructive coupling fault* (CF_{rd}): a read operation performed on the victim cell, while the aggressor cell is in a given state, destroys the data stored in the victim. Four types of read destructive coupling faults exist, defined as $CF_{rd(xy)} = \{< x^a y^v, r_y^v/\bar{y}^v/\bar{y}^v >\}$, where $x, y \in \{0, 1\}$. This covers FP_{29}, FP_{30}, FP_{31}, FP_{32}.
6. *Incorrect read coupling fault (CFir)*: a read operation performed on the victim cell returns the incorrect logic value, while the aggressor is in a given state. Four types of incorrect read coupling faults exist, defined as $CF_{ir(xy)} = \{< x^a y^v, r_y^v/y^v/\bar{y}^v >\}$, where $x, y \in \{0, 1\}$. This covers FP_{25}, FP_{35}, FP_{26}, FP_{36}.
7. *Deceptive read destructive coupling fault (CFdr)*: a read operation performed on the victim cell returns the correct logic value and changes the contents of the victim while the aggressor is in a given logic state. Four types of deceptive read destructive coupling faults exist, defined as $CF_{dr(xy)} = \{< x^a y^v, r_y^v/\bar{y}^v/y^v >\}$, where $x, y \in \{0, 1\}$. This covers FP_{27}, FP_{33}, FP_{28}, FP_{34}.

The presented set of FFMs allows covering all FPs proposed in Table 6.2, and any test covering these FFMs is therefore able to cover all possible 2-coupling static faults. Other sets of fault models have been presented in the literature, such as:

- *Idempotent coupling fault* (CF_{id}): a transition write operation on the aggressor cell forces the victim in a given state: $CF_{id(xy,w_{\bar{x}})} = \{< x^a y^v, w_{\bar{x}}^a/\bar{y}^v/- >\}$, where $x, y \in \{0, 1\}$.
- *Inversion coupling fault* (CF_{in}): a transition write operation on the aggressor cell flips the content of the victim cell: $CF_{in(x,w_{\bar{x}})} = \{< x^a 0^v, w_{\bar{x}}^a/1^v/- >, < x^a 1^v, w_{\bar{x}}^a/0^v/- >\}$, where $x \in \{0, 1\}$.

- *Non-transition coupling fault* (CF_{nt}): a non-transition write operation performed on the the aggressor cell forces the victim cell in a given state: $CF_{nt(xy,w_x)} = \{< x^a y^v, w_x^a/\bar{y}^v/- >\}$, where $x, y \in \{0, 1\}$.

Nevertheless, all these FFMs are either subsets of the seven FFMs presented before or can be expressed as a combination of these basic FFMs.

6.4.4 Dynamic Fault Models

As operations are added to the *SOS* we enter into the dynamic fault space that results in a theoretically infinite number of potential FFMs. Equation 6.7 describes a relation between the number of possible FPs and the number m of operations in *SOS* for single-cell dynamic faults (Al-Ars 2005):

$$\#FP_{\text{single-cell}} = \begin{cases} 2 & m = 0 \\ 10 \cdot 3^{m-1} & m \geq 1 \end{cases} \tag{6.7}$$

The equation clearly shows an exponential relation between the number of FPs and the number of operations in *SOS*. This actually reduces the ability of exploring this huge space of faults for defining FFMs, due to limited availability of simulation time and computation power.

In order to cope with this problem, experiments on an extensive set of memory devices showed that the probability of dynamic fault decreases when m increases (Al-Ars et al. 2002). Based on this assumption, two-operations dynamic faults have been the most studied in the literature and will be considered in this chapter. As for static fault models, two-operations dynamic faults can be additionally clustered according to the number of *f-cells* ($|f|$) involved in the fault. We shall focus on: (i) single-cell two-operations dynamic faults ($|f| = 1$, $m = 2$), and (ii) 2-coupling two-operations dynamic faults ($|f| = 2$, $m = 2$). This leads to a space of 30 single-cell FPs, plus 192 2-coupling FPs.

This space is in some way already too huge to be explored. For this reason in Van de Goor et al. (2000), a limited set of these FPs has been simulated on realistic defective memory devices and the following established FFMs have been defined:

1. *Dynamic Read Disturb Fault (dRDF)*: a write operation immediately followed by a read operation on the same cell changes the logical value stored in the faulty memory cell and returns an incorrect output. Four types of dRDFs exist, defined as $dRDF_{(xy)} = \{< x, w_y r_y/\bar{y}/\bar{y} >\}$, where $x, y \in \{0, 1\}$.
2. *Dynamic Deceptive Read Disturb Fault (dDRDF)*: a write operation immediately followed by a read operation on the same cell changes the logical value stored in the faulty memory cell, but returns the expected output. Four types of dDRDFs exist, defined as $dDRDF_{(xy)} = \{< x, w_y r_y/\bar{y}/y >\}$, where $x, y \in \{0, 1\}$.
3. *Dynamic Incorrect Read Disturb Fault (dIRF)*: a write operation immediately followed by a read operation on the same cell does not change the logical value

stored in the faulty memory cell, but returns an incorrect output. Four types of dIRFs exist, defined as $IRF_{(xy)} = \{< x, w_y r_y / y / \bar{y} >\}$, where $x, y \in \{0, 1\}$.

4. *Dynamic Disturb Coupling Fault* (*dCFds*): a write operation followed immediately by a read operation performed on the aggressor cell causes the victim cell to flip. Eight types of dCFdss exist, defined as $dCFds_{(xyz)} = \{< x^a y^v, w_z^a r_z^a / \bar{y}^v / - >\}$, where $x, y, z \in \{0, 1\}$.
5. *Dynamic Read Disturb Coupling Fault* (*dCFrd*): a write operation immediately followed by a read operation on the victim cell when the aggressor cell is in a given state changes the logical value stored in the victim, and returns an incorrect output. Eight types of dynamic dCFrds exist, defined as $dCFrd_{(xyz)} = \{< x^a y^v, w_z^v r_z^v / \bar{z} / \bar{z} >\}$, where $x, y, z \in \{0, 1\}$.
6. *Dynamic Deceptive Read Disturb Coupling Fault* (*dCFdr*): a write operation immediately followed by a read operation on the victim cell when the aggressor cell is in a given state changes the logical value stored in the victim cell, but returns the expected output. Eight types of dCFdrs exist, defined as $dCFdr_{(xyz)} = \{< x^a y^v, w_z^v r_z^v / \bar{z} / z >\}$, where $x, y, z \in \{0, 1\}$.
7. *Dynamic Incorrect Read Disturb Coupling Fault* (*dCFir*): a write operation immediately followed by a read operation on the victim cell when the aggressor cell is in a given state does not affect the logical value stored in the victim but returns an incorrect output. Eight types of dCFirs, defined as $dCFir_{(xyz)} = \{< x^a y^v, w_z^v r_z^v / z / \bar{z} >\}$, where $x, y, z \in \{0, 1\}$.

It is clear that the set of FFMs defined here addresses a very restricted number of FPs with respect to the complete fault space. This makes dealing with dynamic faults a very complex task that can be solved only moving from higher abstraction levels to lower ones where the knowledge of the physical memory layout and structure, and of the set of realistic defects can be used to restrict the fault space (see Section 6.6)

6.4.5 n-Coupling Fault Models

In-coupling faults represent fault models where n different memory cells are involved in the fault mechanism ($f\text{-}cells = n$). They are usually referred to as *pattern sensitive faults*. In general the content of a cell i (or the ability of i to change its state) is influenced by the contents of all other memory cells, or by the operations performed on them. A pattern sensitive fault is the most general definition of n-coupling fault in which n is equal to the size of the memory.

In a more realistic situation, the so called *neighborhood pattern sensitive faults (NPSFs)* are usually considered, in which a reduced set of cells spatially located in adjacent positions are responsible for the fault mechanism. The neighborhood is the total number of cells in this set. Traditionally the victim cell is called in this context *base cell*, while the aggressor cells are called the *deleted neighborhood*.

In the PSF the neighborhood can be anywhere in the memory while in the NPSF the neighborhood must be in a single position surrounding the base cell. These type

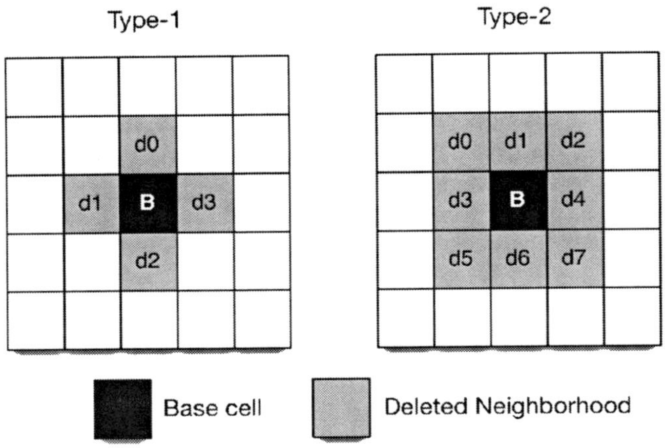

Fig. 6.12 Type-1 and Type-2 NPSF

of fault models are particularly indicated when dealing with high density DRAMs, due to the reduced memory cell capacitance.

In general two types of neighborhood patterns are considered: Type-1 including four deleted neighborhood cells, and Type-2 including eight deleted neighborhood cells (Suk et al. 1979). The type-2 model is more complex and allows to model diagonal coupling effects in the memory matrix. Figure 6.12 shows the two types of neighborhood.

Three types of NPSF have been considered in the literature:

1. Active NPSF (ANPSF) (Suk et al. 1980), also called dynamic NPSF (Saluja et al. 1985) where the base cell changes its value based on a change in the pattern of the deleted neighborhood. In particular, a cell of the deleted neighborhood has a transition while the rest of the neighborhood including the base cell has a given pattern. For example $< x_1^{d0} x_2^{d1} x_3^{d2} x_4^{d3} x_5^{B}, w_{\bar{x}_1}^{d0} / \bar{x}_5^{B} / - >$, where $x_i \in \{0, 1\}$, denotes a generic FP belonging to the ANPSF FFM.
2. Passive NPSF (Suk et al. 1980): a certain neighborhood pattern prevents the base cell to change.
3. Static NPSF (Saluja et al. 1985): the base cell is forced into a particular state when the deleted neighborhood contains a particular pattern. This differs from the ANPSF as no transition is required to excite the fault.

6.4.6 Multiple Faults

It may happen that the effects of two FFMs link together. If the faults share the same aggressor cell and/or the same victim cell, the FFMs are said to be *linked*. As an example let's consider the CF_{ds} denoted by the following two FPs: $FP_1 = < 0^a 0^v, w_1^a / 1^v / - >$, and $FP_2 = < 0^a 1^v, w_1^a / 0^v / - >$.

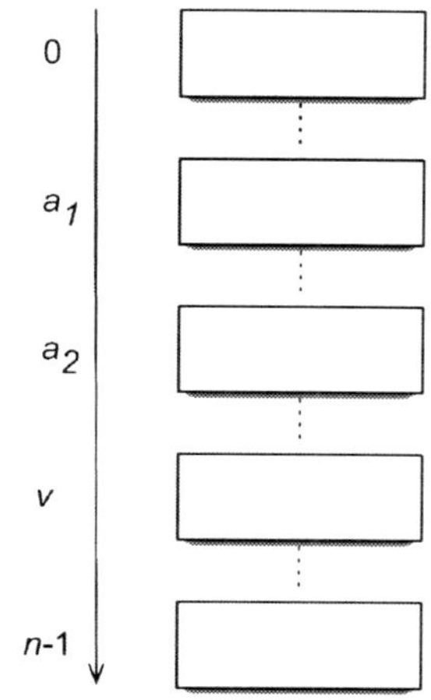

Fig. 6.13 Example of linked fault

Figure 6.13 shows a memory with n cells affected by FP_1 and FP_2 having different aggressor cells with addresses a_1 and a_2, the same victim cell with address v, and $a_1 < a_2 < v$. According to FP_1, starting with a_1 equal to 0 and by performing $w_1^{a_1}$, the victim cell v flips from 0 to 1; then, starting with a_2 equal to 0 and performing $w_1^{a_2}$, according to FP_2 the victim cell v changes its value again, from 1 to 0. The global result is that the fault effect is masked by the application of FP_2, since FP_2 has a faulty behavior opposite to FP_1.

Based on this example, two FPs, $FP_1 = <SOS_1/FB_1>$, and $FP_2 = <SOS_2/FB_2>$ are linked, and denoted by $FP_1 \rightarrow FP_2$, if both of the following conditions are satisfied:

- FP_1 masks FP_2, i.e., $FB_2 \rightarrow \overline{FB_1}$.
- SOS_2 is applied after SOS_1, on either the aggressor cell or the victim cell of FP_1.

To detect linked faults (LFs), it is necessary to detect in isolation at least one of the FPs that compose the fault (i.e., preventing the other FP to mask the fault) (Hamdioui et al. 2004).

Among the extended space of possible linked FFMs, based on several simulations on defective memory devices, the following established realistic linked FFMs have been defined (Hamdioui et al. 2004):

- *Single cell linked faults*: involve a single memory location where all FPs are sequentially applied. Table 6.3 reports the list of realistic single-cell linked faults.
- *2-coupling linked faults*: 2-coupling linked faults involve two distinct memory cells: one aggressor cell a, and one victim cell v. Two different situations may happen: (i) $a < v$, and (ii) $v < a$. Based on this distinction realistic 2-coupling linked faults can be clustered in three different classes: (i) linked faults based on a combination of 2-coupling FPs that share both the aggressor and the victim cell ($LF2_{aa}$), (ii) linked faults where FP_1 is a 2-coupling FP and FP_2 is a single-cell FP ($LF2_{av}$), and (iii) linked faults where FP_1 is a single-cell FP and FP_2 is a

Table 6.3 Single-cell linked faults

FFM	FPs
$TF_{\bar{x}} \rightarrow WDF_x$	$< x, w_{\bar{x}}/x/- > \rightarrow < x, w_x/\bar{x}/- >, x \in \{0, 1\}$
$WDF_x \rightarrow WDF_{\bar{x}}$	$< x, w_x/\bar{x}/- > \rightarrow < \bar{x}, w_{\bar{x}}/x/- > x \in \{0, 1\}$
$DRDF_x \rightarrow WDF_{\bar{x}}$	$< x, r_x/\bar{x}/x > \rightarrow < \bar{x}, w_{\bar{x}}/x/- >, x \in \{0, 1\}$
$TF_{\bar{x}} \rightarrow RDF_x$	$< x, w_{\bar{x}}/x/- > \rightarrow < x, r_x/\bar{x}/\bar{x} >, x \in \{0, 1\}$
$WDF_x \rightarrow RDF_{\bar{x}}$	$< x, w_x/\bar{x}/- > \rightarrow < \bar{x}, r_{\bar{x}}/x/x >, x \in \{0, 1\}$
$DRDF_x \rightarrow RDF_{\bar{x}}$	$< x, r_x/\bar{x}/x > \rightarrow < \bar{x}, r_{\bar{x}}/x/x >, x \in \{0, 1\}$

2-coupling FP ($LF2_{va}$). Table 6.4 reports the list of realistic 2-coupling linked faults where the following notation is used: $op \in \{r, w\}$, $x_2 = y_1$, $x_i = y_i$ if $op_i = r$.

- *3-coupling linked faults*: 3-coupling linked faults are composed of FPs sharing the same victim cell but having different aggressor cells (a_1 and a_2). Considering the possible mutual positions of a_1,a_2, and v, realistic fault models proposed in [Hamdioui et al. 2004] belong to the following two situations: (i) $a_1 < v < a_2$, and (ii) $a_2 < v < a_1$. Realistic 3-coupling linked faults can be represented by the same FPs used to represent 2-coupling linked faults.

6.4.7 Fault Models for Specific Technologies and Architectures

The space of fault models defined in the previous sections is far from representing a complete taxonomy of possible memory faults. It actually focuses on a set of very high level, technology independent faults that can be easily applied to any type of memory.

As we start exploring all the dimensions of the multidimensional space introduced in Section 6.2, several specific functional fault models can be defined, as for example:

- Fault models for multi-port memories (Hamdioui et al. 2001)
- Fault models for cache memories (Al-Ars et al. 2008)
- Fault models for DRAMs (Al-Ars 2005)

A detailed analysis of all these fault models is out of the scope of this chapter, and, if interested, the reader should refer to specific publications.

6.5 From Fault Models to Memory Testing

In order to inspect memory devices for possible faulty behaviors, all memory components are usually tested at the end of production and sometimes in the field. As already stated in Section 6.1, common practice for memory testing is to apply functional test patterns that try to cover FFMs.

Table 6.4 2-coupling linked faults

2-coupling linked faults L_{aa}	
$CF_{ds(x_1 0.op1_{y1})} \rightarrow CF_{ds(x_2 1.op2_{y2})}$	$CF_{wd(x0)} \rightarrow CF_{wd(x1)}$
$CF_{ds(x_1 1.op1_{y1})} \rightarrow CF_{ds(x_2 0.op2_{y2})}$	$CF_{wd(x1)} \rightarrow CF_{wd(x0)}$
$CF_{tr(x0)} \rightarrow CF_{ds(x1.op_y)}$	$CF_{dr(x0)} \rightarrow CF_{wd(x1)}$
$CF_{tr(x1)} \rightarrow CF_{ds(x0.op_y)}$	$CF_{dr(x1)} \rightarrow CF_{wd(x0)}$
$CF_{wd(x0)} \rightarrow CF_{ds(x1.op_y)}$	$CF_{ds(x0.op_y)} \rightarrow CF_{wd(y1)}$
$CF_{wd(x1)} \rightarrow CF_{ds(x0.op_y)}$	$CF_{ds(x1.op_y)} \rightarrow CF_{wd(y0)}$
$CF_{dr(x0)} \rightarrow CF_{ds(x1.op_y)}$	$CF_{tr(x0)} \rightarrow CF_{rd(x1)}$
$CF_{dr(x1)} \rightarrow CF_{ds(x0.op_y)}$	$CF_{tr(x1)} \rightarrow CF_{rd(x0)}$
$CF_{ds(x0.op_y)} \rightarrow CF_{wd(y1)}$	$CF_{wd(x0)} \rightarrow CF_{rd(x1)}$
$CF_{ds(x1.op_y)} \rightarrow CF_{wd(y0)}$	$CF_{wd(x1)} \rightarrow CF_{rd(x0)}$
$CF_{tr(x0)} \rightarrow CF_{wd(x1)}$	$CF_{dr(x0)} \rightarrow CF_{rd(x1)}$
$CF_{tr(x1)} \rightarrow CF_{wd(x0)}$	$CF_{dr(x1)} \rightarrow CF_{rd(x0)}$
2-coupling linked faults L_{av}	
$CF_{ds(x0.op_y)} \rightarrow WDF_1$	$CF_{ds(x0.op_y)} \rightarrow RDF_1$
$CF_{ds(x1.op_y)} \rightarrow WDF_0$	$CF_{ds(x1.op_y)} \rightarrow RDF_0$
$CF_{tr(x0)} \rightarrow WDF_1$	$CF_{tr(x0)} \rightarrow RDF_1$
$CF_{tr(x1)} \rightarrow WDF_0$	$CF_{tr(x1)} \rightarrow RDF_0$
$CF_{wd(x0)} \rightarrow WDF_1$	$CF_{wd(x0)} \rightarrow RDF_1$
$CF_{wd(x1)} \rightarrow WDF_0$	$CF_{wd(x1)} \rightarrow RDF_0$
$CF_{dr(x0)} \rightarrow WDF_1$	$CF_{dr(x0)} \rightarrow RDF_1$
$CF_{dr(x1)} \rightarrow WDF_0$	$CF_{dr(x1)} \rightarrow RDF_0$
2-coupling linked faults L_{va}	
$TF_0 \rightarrow CF_{ds(x1.op_y)}$	$DRDF_0 \rightarrow CF_{ds(x1.op_y)}$
$TTF_1 \rightarrow CF_{ds(x0.op_y)}$	$DRDF_1 \rightarrow CF_{ds(x0.op_y)}$
$TF_0 \rightarrow CF_{wd(x1)}$	$DRDF_0 \rightarrow CF_{wd(x1)}$
$TF_1 \rightarrow CF_{wd(x0)}$	$DRDF_1 \rightarrow CF_{wd(x0)}$
$TF_0 \rightarrow CF_{rd(x1)}$	$DRDF_0 \rightarrow CF_{rd(x1)}$
$TF_1 \rightarrow CF_{rd(x0)}$	$DRDF_1 \rightarrow CF_{rd(x0)}$
$WDF_0 \rightarrow CF_{ds(x1.op_y)}$	
$WDF_1 \rightarrow CF_{ds(x0.op_y)}$	
$WDF_0 \rightarrow CF_{wd(x1)}$	
$WDF_1 \rightarrow CF_{wd(x0)}$	
$WDF_0 \rightarrow CF_{rd(x1)}$	
$WDF_1 \rightarrow CF_{rd(x0)}$	

Memories are among of the most complex digital circuits. They involve many analog parts and the resulting circuitry is denser than any other type of digital device. No single pattern is therefore sufficient to test a memory for all types of real defects. Actually a suite of patterns is required to detect the real defects that may happen in the manufacturing environment (Dean et al. 1993).

Several testing approaches have been proposed in the literature to build functional memory test algorithms. One of the first proposed algorithms was the GALPAT (Van de Goor 1991). It is composed of the following steps:

1. Initialize all memory cells with '0'
2. For each cell i do:
   a) Complement the cell content
   b) For each cell $j \neq i$ read the content of j and the content of i
   c) Complement the content of i
3. Repeat step 2 starting with the memory initialized with '1'

The main drawback of this approach is that its complexity is $O(4n^2)$ where n is the number of memory cells.

Several improvements of this algorithm have been proposed:

- Galloping Diagonal Test: similar to GALPAT (Van de Goor 1991), but it moves diagonally checking both column and row decoders simultaneously. Its complexity is $O(n^{\frac{3}{2}})$.
- Walking Pattern: similar to GALPAT except that the test cell is read once and then all other cells are read. Its complexity is $O(2n^2)$.

All these tests have two common drawbacks: (i) the complexity is in general too high as it is not linear with the number of memory cells, and (ii) the fault coverage is in general low as they to not systematically try to address specific fault models. For these two reasons these tests have been abandoned and nowadays common practice is to resort to a well-known category of test algorithms known as *march tests*.

The idea of march tests is to construct a number of operation sequences and to perform each sequence on all memory cells, one after the other, before performing the next sequence in the test. A march test is therefore defined as a sequence of *march elements*, where a march element is a sequence of memory operations performed sequentially on all memory cells. In a march element, the way one proceeds from one cell to the next is specified by the address order, which can be increasing (denoted by $\Uparrow$) or decreasing (denoted by $\Downarrow$). The $\Uparrow$ address order has to be the exactly opposite of the $\Downarrow$ address order. For some march elements, the address order can be chosen arbitrarily as increasing or decreasing and denoted by the $\Updownarrow$ symbol. In a march element, it is possible to perform a write 0 (w_0), write 1 (w_1), read 0 (r_0), and read 1 (r_1) operation. The 0 and 1 after the read operations represent the expected values of the read. By arranging a number of march elements one after the other, a march test is constructed.

Among all published march tests, a very interesting march algorithm able to cover all static, dynamic, and linked FFMs proposed in the pervious sections of this chapter is the March AB (Bosio et al. 2008) reported in Eq. 6.8.

$$\begin{aligned} &\Updownarrow (w_1) \\ &\Downarrow (r_1 w_0 r_0 w_0 r_0) \\ &\Downarrow (r_0 w_1 r_1 w_1 r_1) \end{aligned} \tag{6.8}$$

$$\Uparrow (r_1 w_0 r_0 w_0 r_0)$$
$$\Uparrow (r_0 w_1 r_1 w_1 r_1)$$
$$\Updownarrow (r_0)$$

March tests are a preferred method for RAM testing either by means of external testers or through built in self test (BIST) solutions. Their linear complexity, regularity, and symmetry are the reason for this preference. However, tests for NPSFs (see Section 6.4.5) cannot be performed by march tests (Mazumder et al. 1996), since the base cell needs to be addressed differently from the cells in the deleted neighbor, thus requiring test algorithms with higher complexity difficult to implement in embedded test environments.

6.5.1 Generation of March Tests

The generation of a march test begins with the analysis of a set of target FPs used to identify so-called detection conditions providing the minimum requirements a march test has to achieve in order to detect the target faulty behaviors. Detection conditions can be then combined together to provide a complete march test.

As an example, starting with the following FP $< 0, w_1/0/- >$ modeling a TF_1 transition fault, it is easy to derive that any march test containing the following conditions: $\Updownarrow (\ldots w_0 \ldots)$ $\Updownarrow (\ldots w_1 \ldots)$ $\Updownarrow (\ldots r_1 \ldots)$, is able to detect the target faulty behavior. Multiple detection conditions needed to detect a number of different FPs have to be combined together to generate a single march test to fully test the memory for all targeted faulty behaviors.

The automatic generation of march test is a deeply studied and analyzed problem and several generation algorithms are available in the literature: Smit et al. (1994), Zarrineh et al. (1998), Wu et al. (2000), Zarrineh et al. (2001), Cheng et al. (2002), Benso et al. (2002), Al-Harabi et al. (2003), Niggemeyer et al. (2004), Benso et al. (2005, 2006a,b, 2008).

6.6 From Fault-Based to Defect-Based Memory Testing: Trends and Challenges

Functional tests and functional fault models proved to be very helpful in generating functional test algorithms independent of the target technology and able to guarantee high fault coverage and therefore high quality in memory products. Unfortunately, as technology continuously scales down, and we fully enter the VDSM era, the sensitivity of memories to physical defects is strongly increasing. This turns into the continuous identification and definition of new dynamic faulty behaviors (see Section 6.4.4) to model the effect of new memory defects.

As a consequence, the traditional test generation flow proposed in Section 6.5.1, where a list of FFMs described in terms of FPs defines a set of conditions able to detect the target faulty behaviors to be later combined into a resulting march test, is becoming a bottleneck. Due to the increased number of FPs to consider, the complexity of the resulting test algorithms is drastically increasing. Increased complexity means increased test time and therefore increased test cost (see Section 6.1). In several situations such a significant overhead is not justified with respect to the very marginal improvement in defect coverage they provide.

This makes it mandatory introducing a stronger link between functional test and physical defects, thus moving from *fault-based test approaches* to *defect-based test approaches*.

Defect-based testing typically aims at targeting the following questions:

> *What can go wrong with this design? How would the design's behavior change if this happen, and how can that be measured*? (Aitken et al. 2003)

Several publications already proved that, working with device level memory models, the set of realistic fault models for a specific memory architecture and technology can be drastically reduced. Moreover, resorting to the detailed information about memory architecture and technology, optimized test algorithms can be implemented, drastically reducing the overall test time and complexity while guaranteeing very high fault coverage (Dilillo et al. 2003, 2005a,b, 2006, 2007).

While defect-based test represents a key element to reduce test cost, it presents the main drawback that test algorithms should be deeply customized to the target memory technology and architecture. Defect-based testing for memory concentrates on defect analysis of key parts of the layout and the development of test patterns that will test for likely failures. This is completely in contrast with the architecture and technology independent form of traditional march tests. In order to be effectively applicable in an industrial scenario, defect-based memory testing requires a strong investment in automating all steps, from defect analysis and simulation, to realistic fault models extraction, and to test generation. Few publications addressed these problems so far Cheng et al. (2003), Al-Ars et al. (2005), and Di Carlo et al. (2008) with all the proposed solutions still far from being applicable in real scenarios. Such a big challenge will most likely be leading several researchers in the field of defect-based memory testing in next years.

6.7 Summary

We would like to conclude this chapter with a thought about the future of memory modeling and testing. The first era of memories lasted roughly 10 years, the second one 20 years. We are now around 30 years of semiconductor memories. What's next? Which technologies will allow us to store the hundreds of terabytes we are going to need tomorrow? How shall we model and test these monster devices?

Not having a so powerful crystal ball, we simply conclude wishing that the era of purely academic test algorithms is going to finish soon, to be quickly replaced by the era of new automated approaches to generate effective and efficient defect-based algorithms, capable of supporting memory testing, diagnosing, repairing, and, why not, on-the-flight real-time autonomic reconfigurations.

References

Adams RD, Cooley ES (1996) Analysis of deceptive destructive read memory fault model and recommended testing. In Proceedings of the IEEE North Atlantic test workshop

Aitken R (2003) Applying defect based test to embedded memories. In Proceedings of the IEEE international workshop on memory technology, design and testing, pp 72–77

Al-Ars Z, van de Goor AJ (2002) Approximating infinite dynamic behavior for DRAM cell defects. In Proceedings of the 20th IEEE VLSI test symposium, pp 401–406

Al-Ars Z (2005) DRAM fault analysis and test generation. PhD Thesis, TU Delft

Al-Ars Z, Hamdioui S, Mueller G, van de Goor A (2005) Framework for fault analysis and test generation in drams. In Proceedings of design automation and test in Europe, pp 1020–1021

Al-Ars Z, Hamdioui S, Gaydadjiev G, Vassiliadis S (2008) Test set development for cache memory in modern microprocessors. IEEE Trans VLSI Sys 16(6):725–732

Al-Harbi SM, Gupta SK (2003) Generating complete and optimal March tests for linked faults in memories. In Proceedings of the 21st IEEE VLSI test symposium, pp 254–261

Benso A, Di Carlo S, Di Natale G, Prinetto P (2002) An optimal algorithm for the automatic generation of March tests. In Proceedings of the design, automation and test in Europe conference and exhibition, pp 938–943

Benso A, Bosio A, Di Carlo S, Di Natale G, Prinetto P (2005) Automatic March tests generation for static and dynamic faults in SRAMs. In Proceedings of the 10th IEEE European test symposium, pp 22–25

Benso A, Bosio A, Di Carlo S, Di Natale G, Prinetto P (2006a) Automatic March tests generations for static linked faults in SRAMs. In Proceedings of the design, automation and test in Europe conference and exhibition, pp 1–6

Benso A, Bosio A, Di Carlo S, Di Natale G, Prinetto P (2006b) Automatic March tests generation for multi-port SRAMs. In Proceedings of the 3rd IEEE international workshop on electronic design, test and applications, pp 17–19

Benso A, Bosio A, Di Carlo S, Di Natale G, Prinetto P (Dec 2008) March test generation revealed. IEEE Trans Comput 57(12):1704–1713

Bosio A, Di Carlo S, Di Natale G, Prinetto P (May 2007) March AB, a state-of-the-art march test for realistic static linked faults and dynamic faults in SRAMs. IET Comput Dig Proc 2(3): 237–245

Brzozowski J, Jurgensen H (Aug 1992) A model for sequential machine testing and diagnosis. J Electron Test Theory Appl 3(3):219–234

Cheng K-L, Wang C-W, Lee J-N, Chou YF, Huang C-T, Wu C-W (Apr 2002) Fault simulation and test algorithm generation for random access memories. IEEE Trans Comput-Aided Des Integrat Circuits Sys 21(4):480–490

Cheng K-L, Wang C-W, Lee J-N, Chou Y-F, Huang C-T, Wu C-W (2003) Fame: a fault-pattern based memory failure analysis framework. In Proceedings of the international conference on computer aided design, pp 595–598

Dean CA, Zorian Y (1993) Do you practice safe test? What we found out about your habits. In Proceedings of the international test conference, pp 887–892

Dekker R, Beenker F, Thijssen L (Jun 1990) A realistic fault model and test algorithms for static random access memory. IEEE Transaction on Comput-Aided Des Integrat Circuits Sys 9(6):567–572

Di Carlo S, Prinetto P, Scionti A, Al-Ars Z (2008) Automating defects simulation and fault modeling for SRAMs. In Proceedings of the IEEE international high level design validation and test workshop, pp 169–176

Dilillo L, Girard P, Pravossoudovitch S, Virazel A, Borri S (2003) Comparison of open and resistive-open defect test conditions in SRAM address decoders. In Proceedings of the 12th IEEE Asian test symposium, pp 250–255

Dilillo L, Girard P, Pravossoudovitch S, Virazel A, Hage-Hassan MB (2005a) Resistive-open defect influence in SRAM pre-charge circuits: analysis and characterization. In Proceedings of the IEEE European test symposium, pp 116–121

Dilillo L, Girard P, Pravossoudovitch S, Virazel A, HageHassan M (2005b) Data retention fault in SRAM memories: analysis and detection procedures. In Proceedings of the 23rd IEEE VLSI test symposium, pp 183–188

Dilillo L, Girard P, Pravossoudovitch S, Virazel A, Bastian M (2006) March pre: an efficient test for resistive-open defects in the SRAM pre-charge circuit. In Proceedings of IEEE design and diagnostics of electronic circuits and systems, pp 254–259

Dilillo L, Al-Hashimi BM (2007) March CRF: an efficient test for complex read faults in SRAM memories. In Proceedings if IEEE design and diagnostics of electronic circuits and systems, pp 1–6

Hamdioui S, Van de Goor AJ, Eastwick D, Rodgers M (2001) Detecting unique faults in multi-port SRAMs. In Proceedings of the 10th IEEE Asian test symposium, pp 37–42

Hamdioui S, Al-Ars Z, van de Goor AJ, Rodgers M (May 2004) Linked faults in random access memories concept fault models test algorithms and industrial results. IEEE Trans Comput-Aid Des Integrat Circuits Sys 23(5):737–757

Hamdioui S, Al-Ars Z, van de Goor AJ (Dec 2006) Opens and delay faults in CMOS ram address decoders. IEEE Trans Comput 55(12):1630–1639

International technology roadmap for semiconductors [Online]. Available: http://www.itrs.net/

Marinescu M (1982) Simple and efficient algorithms for functional RAM testing. In Proceedings of the IEEE international test conference, pp 236–239

Mazumder P, Chakraborty K (1996) Testing and testable design of high-density random-access memories. Kluwer AcademicPublishers

Nair R, Thatte SM, Abraham JA (Jun 1978) Efficient algorithms for testing semiconductor random-access memories. IEEE Trans Comput C-27(6):572–576

Nair R (Mar 1979) Comments on an optimal algorithm for testing stuck-at faults in random access memories. IEEE Trans Comput 28(3):258–261

Niggemeyer D, Redeker M, Rudnick EM (2000) Diagnostic testing of embedded memories based on output tracing. In Proceedings of the IEEE international workshop memory technology, pp 113–118

Niggemeyer D, Rudnick EM (Sep 2004) Automatic generation of diagnostic memory tests based on fault decomposition and output tracing. IEEE Trans Comput 53(9):1134–1146

Roy K, Mukhopadhyay S, Mahmoodi-Meimand H (Feb 2003) Leakage current mechanisms and leakage reduction techniques in deep-submicrometer CMOS circuits. Proceedings of the IEEE 91(2):305–327

Saluja K, Kinoshita K (Mar 1985) Test pattern generation for API faults in RAM. IEEE Trans Comput C-34(3):284–287

Shen JP, Maly W, Ferguson FJ (Dec 1985) Inductive fault analysis of MOS integrated circuits. IEEE Des Test Comput 2(6):13–26

Smit B, van de Goor AJ (1994) The automatic generation of March tests. In Proceedings of the IEEE international workshop on memory technology, design and testing, pp 86–92

Suk DS, Reddy SM (1979) An algorithm to detect a class of pattern sensitive faults in semiconductor random access memories. In Proceedings of the international fault-tolerant computing symposium, pp 219–225

Suk DS, Reddy SM (Jun 1980) Test procedures for a class of pattern-sensitive faults in semiconductor random-access memories. IEEE Trans Comput C-29(6):419–429

Thatte SM, Abraham JA (Jun 1977) Testing of semiconductor random access memories. In Proceedings of the international fault-tolerant computing symposium, pp 81–87
van de Goor AJ, Verruijt CA (Mar 1990) An overview of deterministic functional RAM chip testing. ACM Comput Surv 22(1): 5–33
van de Goor AJ (1991) Testing semiconductor memories, theory and practice. Wiley, Chichester, UK
van de Goor AJ (1999) Industrial evaluation of DRAM test. In Proceedings of design, automation and test in Europe, pp 623–630
van de Goor AJ, Al-Ars Z (2000) Functional memory faults: a formal notation and taxonomy. In Proceedings IEEE VLSI test symposium, pp 281–289
Wu C-F, Huang C-T, Cheng K-L, Wu C-W (2000) Simulation-based test algorithm generation for random access memories. In Proceedings of the 18th IEEE VLSI test symposium, pp 291–296
Zarrineh K, Upadhyaya SJ, Chakravarty S (1998) A new framework for generating optimal march tests for memory arrays. In Proceedings of the IEEE international test conference, pp 73–82
Zarrineh K, Upadhyaya SJ, Chakravarty S (Dec 2001) Automatic generation and compaction of March tests for memory. IEEE Trans VLSI Sys 9(6):845–857

Chapter 7
Models for Power-Aware Testing

Patrick Girard and Hans-Joachim Wunderlich

Abstract Power consumption of circuits and systems receives more and more attention. In test mode, power consumption is even more critical than in system model and has severe impact on reliability, yield and test costs. This chapter describes the different types and sources of test power. Power-aware techniques for test pattern generation, design for test and test data compression are presented which allow efficient power constrained testing with minimized hardware cost and test application time.

Keywords Low power test · Design for test

7.1 Introduction

Before 2005, the trend stopped to exponentially increase system frequency while scaling down the geometrical dimensions. Instead, scaling is now mainly used for implementing highly parallel systems and increasing performance not by frequency but by parallelism. The main reason of this development is found in the increased power consumption which reaches economical and technical limits (Borkar et al. 2005).

Dynamic power consumption is increased due to the higher switching activity; the standard way to overcome this is reducing system voltage which may in turn lead to increased static power consumption due to higher leakage currents. Power consumption affects battery life time, heat dissipation, reliability, packaging, cooling and many other factors of quality and cost.

P. Girard (✉)
LIRMM/CNRS, 161rue Ada, 34392 Montpellier, France
e-mail: girard@lirmm.fr

H.-J. Wunderlich
Institut für Technische Informatik, Universität Stuttgart, Pfaffenwaldring 47,
D-70569 Stuttgart, Germany

H.-J. Wunderlich (ed.), *Models in Hardware Testing: Lecture Notes of the Forum in Honor of Christian Landrault*, Frontiers in Electronic Testing 43,
DOI 10.1007/978-90-481-3282-9_7,

The power issues are severe in design and system mode, but they have been seen earlier during design for test and in test mode (Nicolici and Al-Hashimi 2003; Girard 2002). Test has to exercise all devices of the circuit in short time, and, if countermeasures are not taken, switching activity will be 2 to 4 times as high as in the system mode (Sde-Paz and Salomon 2008). The increased current may have impact on the circuit's lifetime or may even damage it and the overstress may change the circuit's behaviour and result in yield loss. The classical workaround in industry consists in partitioning and scheduling the test (Zorian 1993), reducing the test frequency or even both. These measures will increase test time and incur additional costs, and the reduced test speed makes it difficult to detect delay faults as described in the previous chapters.

Power considerations during test are motivated by cost and reliability aspects. The next section will describe appropriate models for power estimation during functional mode and during test mode. While average power is related to heat dissipation, instantaneous and peak power introduce additional robustness problems. Modeling and estimating test power introduce also complexity issues as the exact computation of the power consumption during scan shifting is rather expensive.

Section 7.3 discusses in detail the impact of test power on reliability, yield and test costs. Automatic test pattern generation algorithms can take care of this to a large extent; methods for supporting external and built-in testing during ATPG are discussed in Section 7.4.

Section 7.5 presents power-aware design for test solutions mainly for scan based techniques. For systems-on-a-chip of today's size, test data compression and compaction are mandatory to limit test time and fulfil throughput requirements. Yet, these techniques may introduce additional switching activity, if special precautions are not taken as described in Section 7.6.

7.2 Models for Power Estimation

As power consumption is now considered as a constraint during test, power estimation is required to measure the saving in power and evaluate the effectiveness of a given test power reduction technique. Models are needed for test power estimation. In this section, we describe the models used to estimate the various components of power consumption during functional mode and test mode. We also discuss how test power can be estimated at the various levels of abstraction of the design process.

7.2.1 Functional Power Modeling

The main components of CMOS power consumption are from dynamic and static sources. *Dynamic power* is typically defined as the power consumed whenever the circuit is switching, while *static power* is the power consumed when the circuit is idle (Pedram and Rabaey 2002).

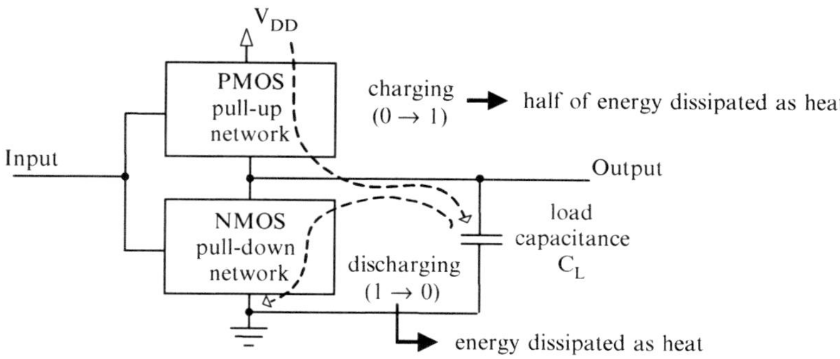

Fig. 7.1 Dynamic switching power

Dynamic power is divided into *dynamic short-circuit* power and *dynamic switching* power. Dynamic short-circuit power is due to the direct current path from V_{DD} to G_{ND} that occurs during output switching. The short-circuit current of a CMOS logic gate is proportional to the ratio between the input slew of the gate and the load capacitance at the output of the gate. The short-circuit power represents a small fraction of the total dynamic power and is often neglected.

Dynamic switching power is due to charging and discharging of the output load capacitance during switching. Let us consider the generic representation of a CMOS logic gate shown in Fig. 7.1. During output switching from 0 to 1, a charge $Q = C_L.V_{DD}$ is delivered to the load capacitance C_L. The power rail must supply this charge at voltage V_{DD}, so the energy supplied is $Q.V_{DD} = C_L.V_{DD}^2$. However, the energy E stored on a capacitance C_L charged to V_{DD} is only half of this, i.e., $E = 1/2.C_L.V_{DD}^2$. According to the energy conservation principle, the other half must be dissipated by the PMOS transistors in the pull-up network. Similarly, when the inputs change again causing the output to discharge (from 1 to 0), all the energy stored on the capacitance C_L is dissipated in the pull-down network, as no energy can enter the ground rail ($Q.V_{GND} = Q.0 = 0$). In both cases, the energy is dissipated as heat (Athas et al. 1994).

The dynamic switching power is consumed during the charge of the load capacitance C_L, when a current I flows between power and ground rails through the capacitance. The power consumed during the time interval $[0, T]$ is therefore: $P_{dyn} = V_{DD}.I = V_{DD}.Q.1/T$ where $Q = C_L.V_{DD}$. As several transitions may occur during the time interval $[0, T]$, the dynamic switching power consumption can be expressed as follows:

$$P_{dyn} = C_L.V_{DD}^2.N_{0\to 1}.1/T \qquad (7.1)$$

Where $N_{0\to 1}$ represents the number of rising transitions at the gate output during the time interval $[0, T]$. Without loss of generality, it can be assumed that the number of rising transitions is equal to half of the total number of N transitions at the gate

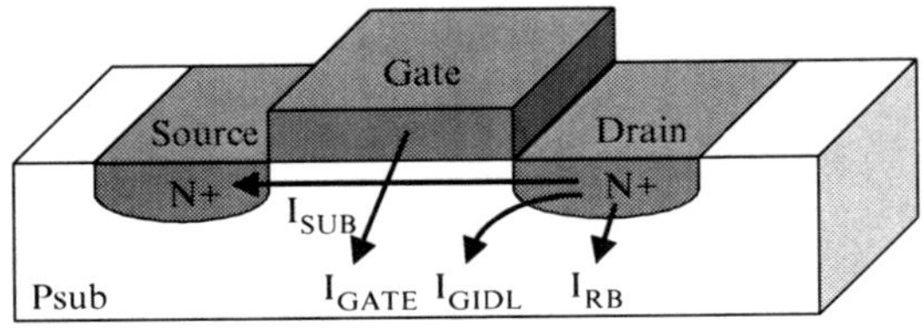

Fig. 7.2 Static leakage power

output. The dynamic switching power consumed by the logic gate during the time interval $[0, T]$ can finally be expressed as:

$$P_{dyn} = 1/2.C_L.V_{DD}^2.N.1/T \tag{7.2}$$

The above analysis shows that dynamic switching power consumption occurs during the charge of output capacitance, whereas power (or energy) dissipation occurs during the charge or discharge of each node. Considering that average power is given by the ratio between energy and time, it can be observed that the power dissipated by N rising or falling transitions during the time interval $[0, T]$ is given by $E/T = 1/2.C_L.V_{DD}^2.N.1/T$. This expression is equivalent to the above expression of the dynamic switching power consumption. It can be concluded that the terms "power consumption" and "power dissipation" can be used without distinction.

Static (or leakage) power is the power consumed when the circuit is idle and is due to four main components (see Fig. 7.2): the reverse-biased junction leakage current (I_{RB}), the gate induced drain leakage current (I_{GIDL}), the gate direct-tunneling leakage current (I_{GATE}) and the sub-threshold leakage current (I_{SUB}). The latter is the main contributor to static power dissipation and is proportional to the ratio between V_{DD} and the threshold voltage of transistors inside the gate (Roy et al. 2003).

7.2.2 Test Power Modeling

In order to explain the dynamic switching power dissipation during test, let us consider a circuit composed of N nodes and a test sequence of length L used to achieve a given fault coverage (Girard et al. 2007). The average energy consumed at node i per switching is $1/2.C_i.V_{DD}{}^2$ where C_i is the equivalent output capacitance at node i and V_{DD} the power supply voltage (Cirit 1987). A good approximation of the energy consumed at node i in a time interval t is $1/2.C_i.S_i.V_{DD}^2$ where S_i is the average number of transitions during this interval (also called *switching activity factor* at node i). Furthermore, nodes connected to more than one logic gate in the circuit are nodes with a higher output capacitance. Based on this fact, and in a first approximation, it can be stated that output capacitance C_i is proportional to the fanout at node i, denoted as F_i (Wang and Roy 1995). Therefore, an estimation of the energy E_i consumed at node i during the time interval t is given below, where C_0 is the minimum output capacitance of the circuit.

$$E_i = 1/2.S_i.F_i.C_0.V_{DD}^2 \tag{7.3}$$

According to this formulation, the energy consumed after application of a pair of successive input vectors ($V_{k\text{-}1}$, V_k) can be expressed by:

$$E_{Vk} = 1/2.C_0.V_{DD}^2.\sum_i S_i(k).F_i \tag{7.4}$$

Where i ranges across all the nodes of the circuit and $S_i(k)$ is the number of transitions provoked by V_k at node i. Now, the *total energy* consumed in the circuit after application of the complete test sequence of length L is given below, where k ranges across all the vectors of the test sequence.

$$E_{total} = 1/2.C_0.V_{DD}^2.\sum_k \sum_i S_i(k).F_i \tag{7.5}$$

By definition, power is given by the ratio between energy and time. The *instantaneous power* is generally calculated as the amount of power required during a small instant of time t_{small} such as the portion of a clock cycle immediately following the system clock rising or falling edge. Consequently, the instantaneous power dissipated in the circuit after the application of a test vector V_k can be expressed by:

$$P_{inst}(V_k) = E_{Vk}/t_{small} \tag{7.6}$$

The *peak power* corresponds to the highest value of instantaneous power measured during test. It can be expressed in terms of the highest energy consumed during a small instant of time during the test session:

$$P_{peak} = Max_k\, P_{inst}(V_k) = Max_k(E_{Vk}/t_{small}) \tag{7.7}$$

Finally, the *average power* consumed during the test session can be calculated from the total energy and the test time. Considering that the test time is given by the product $L.T$, where T corresponds to the nominal clock period of the circuit, the average power can be expressed as follows:

$$P_{average} = E_{total}/(L.T) \tag{7.8}$$

The above expressions of power and energy, although based on a simplified model, are accurate enough for the intended purpose of power analysis during test. According to these expressions, and assuming a given technology and a supply voltage for the considered circuit, it appears that the switching activity factor S_i is the only parameter that has impact on the energy, peak power, and average power. This explains why most of the methods proposed so far for reducing power and/or energy during test are based on a reduction of the switching activity factor.

Concerning static power dissipation during test, there is no clear evidence that it can be higher than static power in functional mode, except for I_{DDQ} test (sensitivity is reduced in this case) or burn-in test (the exponential dependence of sub-threshold leakage on temperature leads to higher static power dissipation that can result in thermal runaway condition and hence yield loss). Though depending on the logic values of test patterns (but not on input transition or load capacitance), static power dissipation does not necessarily increase during test. Modeling of static power during test is similar to modeling of static power during functional mode.

7.2.3 Test Power Estimation

During conventional design, power consumption in functional mode is estimated by using (i) architectural-level power estimation, (ii) RT-level power estimation, and/or (iii) gate-level power estimation (Najm 1994). Each one of these estimation strategies represents different tradeoffs between accuracy and estimation time (see Fig. 7.3).

Estimation of test power consumption is not only required for sign-off (and avoid destructive testing) but also to facilitate power-aware test space exploration (during DFT or ATPG) early in the design cycle (Ravi et al. 2008). However, as scan insertion and test generation are commonly done at the gate level in today's design flows, only gate-level estimators for test power are used in practice. Though accurate, a limitation of gate-level estimation is that it prevents better decisions regarding test power early in the design cycle. Moreover, these industrial estimators are often simulation-based. Though manageable for small size circuits, this approach may be impractical for multi-million gate SoCs as a complete simulation of ATPG test patterns is too much time and memory consuming.

Quick and approximate models of test power have also been suggested in the literature. The weighted transition metric proposed in Sankaralingam et al. (2000) is a simple and widely used model for scan testing, wherein transitions at flip-flops

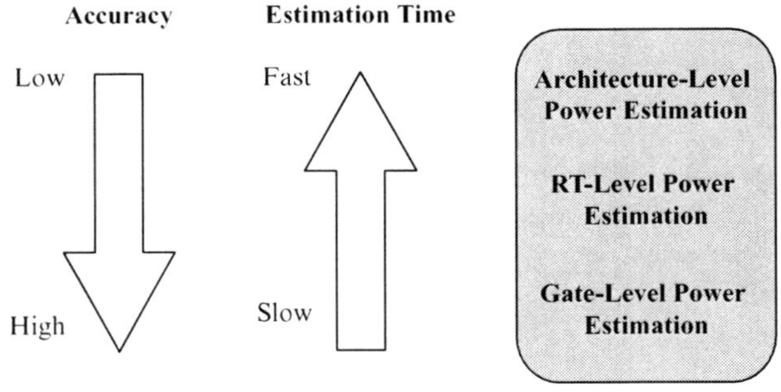

Fig. 7.3 Accuracy versus time in power estimation

weighted by their position in the scan chain are counted to provide a rough estimate of test power. Though the correlation with the overall circuit test power is quite good, a drawback of this metric however is that it does not provide an accurate value of test power dissipation as it neglects combinational logic toggling. Nevertheless, this metric remains an efficient mean to compare different solutions (DFT/ATPG) in terms of test power dissipation.

In order to quickly power-analyze ATPG patterns and better define the final DFT architecture, stochastic power estimation techniques based on using transition probabilities at flip-flop outputs can be adopted (Ravi et al. 2008). Alternatively, RT-level test power estimators can be used but only if DFT insertion and test generation can be done at the RT-level (Midulla and Aktouf 2008).

7.3 Overview of Power Issues During Test

Power issues during test application may occur when the circuit switching activity is higher than the switching activity during functional mode of operation. In this case, the circuit may be unable to behave properly as power constraints considered during the design process have been violated. These power issues are mainly due to two reasons: excessive average power consumption and excessive peak power consumption during test. In this section, we explain the origins and the consequences of these power issues during scan testing.

7.3.1 Issues due to Elevated Average Power

As explained in the previous section, the switching operations of a circuit always lead to heat dissipation. The heat is produced by the collision of carriers with the conductor molecular structure (Joule effect) and is responsible for die temperature increase observed during operation. There is a well-known relationship between die temperature and power dissipation that can be formulated as follows (Weste and Eshraghian 1993):

$$T_{die} = T_{air} + \theta \times P_{Average} \tag{7.9}$$

Where T_{die} is the die temperature, T_{air} is the temperature of surrounding air, θ is the package thermal impedance expressed in °C/Watt, and $P_{Average}$ is the average power dissipated by the circuit.

From the above expression, it is clear that an increase of average power dissipation will increase the circuit temperature. If the temperature is too high, even during the short duration of a manufacturing or on-line test session, it may have the following impacts on the circuit (see Fig. 7.4):

- *Chip damage*. The excessive heat related to high temperature may lead to hot spots, which appear during test data application and may result in premature and permanent damage (referred to as infant mortality) of the circuit (Pouya and Crouch 2000).

- *Reduced reliability*. Another type of structural degradations, which are accelerated gradually over time (ageing), may affect circuit performance or cause functional failures after a given lifetime (Hertwig and Wunderlich 1998; Shi and Kapur 2004). In this case, the main mechanisms leading to these structural degradations are corrosion, hot-carrier-induced defects, electro-migration or dielectric breakdown (Altet and Rubio 2002). The occurrence of these degradation mechanisms will therefore affect long-term circuit reliability.
- *Yield loss*. High switching activity during test may lead not only to elevated average power and hence temperature increase, but also to temperature variations that may differ from those in functional mode. These temperature variations may induce timing variations and in some cases may lead to yield loss (also called overkill).

Besides the above circuit related issues, excessive average power during test may also have the following impact on the test process, and hence on test cost (see Fig. 7.5):

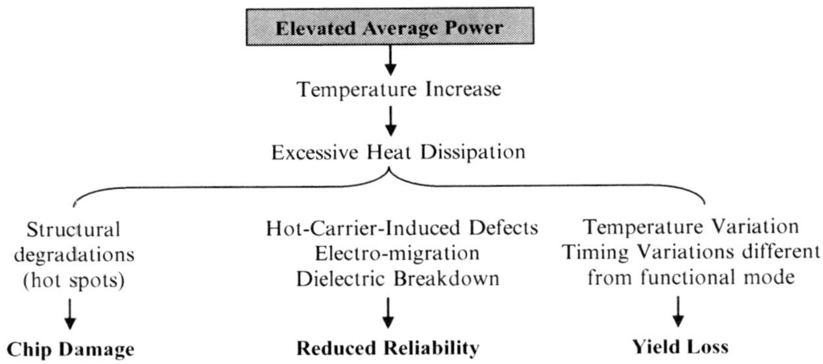

Fig. 7.4 Impact of elevated average test power on the circuit

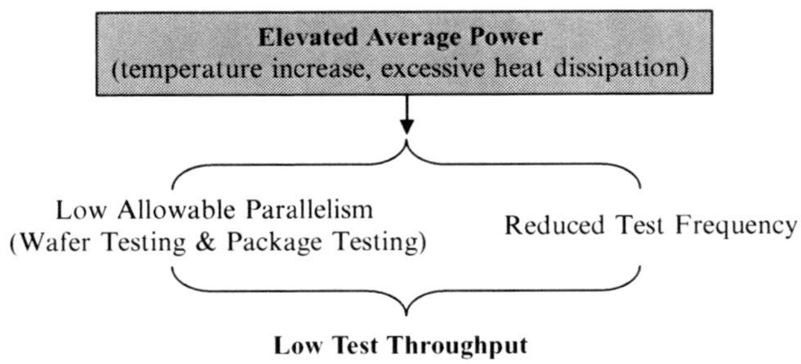

Fig. 7.5 Impact of elevated average test power on the test process

- *Low test throughput.* In order to avoid the above degradation mechanisms, heat dissipation has to be kept under a given safety limit. This limit is determined from the knowledge of the thermal capacity of the package, the use of cooling systems, etc. Restricting the amount of power that can be dissipated will have a negative impact on the level of parallelism that can be used during wafer or chip testing, or on the test frequency that can be used. A reduced test concurrency or test frequency will therefore lead to a lower test throughput.

7.3.2 Issues due to Elevated Peak Power

As for average power, excessive peak power consumption may occur during test and lead to undesirable and abnormal behavior of the circuit. These issues may occur when testing the circuit at the wafer level or at the chip level. Excessive peak power consumption comes with a high instantaneous current demand due to high switching activity during test, and may lead to considerable drops in voltage levels at power grid nodes.

Voltage drop in the power grid, also referred to as Power Supply Noise (PSN), is mainly due to two components: IR-drop and L(di/dt) (Arabi et al. 2007). IR-drop refers to the amount of decrease in the power rail voltage and is linked to the existence of a non negligible resistance between the rail and each node in the circuit under test. R represents the resistances of the power mesh network, power pads and device package. L(di/dt) is an inductive noise and refers to current variations occurring during switching through inductive connections. In this expression, *L* represents the inductances of the power mesh network, power pads and device package, and *di/dt* represents the magnitude of the variation of the current flowing through this connection. These two types of noise are illustrated in Fig. 7.6. Note that L(di/dt) also refers to voltage glitches or surge/droop phenomena known as Ground Bounce or Simultaneous Switching Noise (SSN) (Chang et al. 1997). Crosstalk is another noise phenomenon that refers to capacitive coupling between neighboring lines within a circuit. Crosstalk is known to be less significant during test compared to IR-drop and L(di/dt) (Saxena et al. 2003).

With high peak current demands during test, PSN may become much higher than during functional mode and then is no longer negligible. With increased PSN (see Fig. 7.7), the voltages at some gates in the circuit are reduced and these gates exhibit higher delays (performance degradation), possibly leading to test fails (good dies

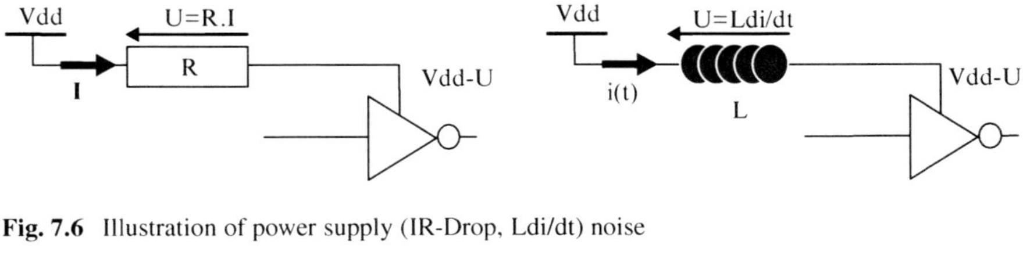

Fig. 7.6 Illustration of power supply (IR-Drop, Ldi/dt) noise

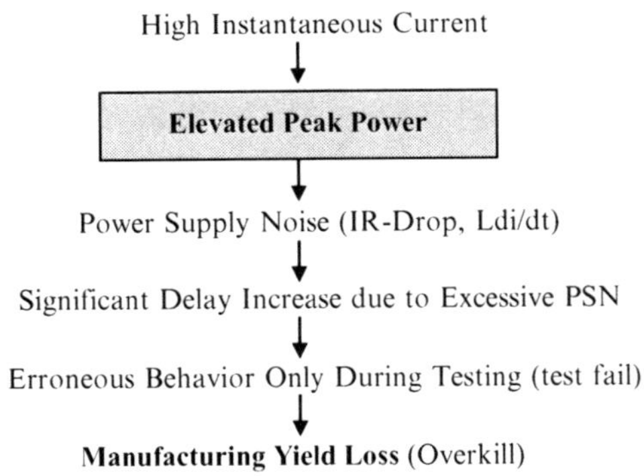

Fig. 7.7 Impact of elevated peak power during test

are declared faulty) and hence manufacturing yield loss (Butler et al. 2004). These phenomena have been widely reported in the literature, in particular when at-speed scan testing is performed (Saxena et al. 2003).

In order to avoid excessive peak power consumption and its related issues, it is important to reduce the level of switching activity during test. This can be done by dedicated and efficient solutions as those described in the next sections.

7.4 Power-Aware Test Generation

Power-aware test generation can be used to create patterns that are inherently optimized to achieve minimum switching activity. In this section, we first provide an overview of power-aware test pattern based approaches. The description will be limited to approaches where test patterns are generated by an ATPG tool and target scan-based designs. Next, we describe a solution for low energy Built-In Self-Test (BIST) where the parameters of the test generator are tuned to provide low switching on-chip test patterns.

7.4.1 Overview of Power-Aware Test Generation Solutions

Power-aware test pattern generation for scan testing can be used for either shift power reduction or capture power reduction. Solutions proposed so far can be categorized as described below.

7.4.1.1 Low-Power ATPG

Low-power ATPG consists in modifying an existing ATPG algorithm or developing a new one for generating low-power test patterns that still meet the original ATPG objectives (maximum fault coverage and minimum pattern length with reasonable run time). An example is given in Wang and Gupta (1994) where the path-oriented decision-making (PODEM) algorithm is modified so that don't care bits are maximized and are then assigned in a clever manner to minimize the number of flip-flop transitions between two consecutive test patterns. This solution reduces both average and peak power dissipation during scan shift operations. Another example is given in Wen et al. (2006a) where the PODEM algorithm is modified for efficient capture power reduction during scan testing. The primary objective of the modified algorithm is the detection of targeted faults and the secondary one is the minimization of the difference between before-capture and after-capture output values of scan cells. This is achieved by introducing the concept of a capture conflict (C-conflict) in addition to the conventional detection conflict (D-conflict). A C-conflict occurs when a difference between the before-capture and after-capture output values of a scan cell is created by logic value assignment during ATPG. A C-conflict, in the same manner as a D-conflict, may be avoided through the backtrack operation. However, backtracking for a C-conflict may make fault detection impossible. In this case, the backtracking for the C-conflict is reversed, and the transition at the scan cell is tolerated since the primary goal is fault detection.

7.4.1.2 Power-Constrained ATPG

When the ATPG algorithm cannot be modified but is capable of accepting constraints (this feature is offered in many commercial ATPG tools), the problem of generating power-aware test patterns can be viewed as a constrained ATPG problem. In this context, a typical constraint is a user-specified toggling activity limit that needs to be satisfied for each generated pattern. During the test generation process, the ATPG tool evaluates the toggling activity generated by each test pattern, and only replaces don't care bits needed to keep the toggling activity under the specified limit (Ravi et al. 2007).

7.4.1.3 Power-Aware X-filling

The number of don't care (X) bits in test cubes generated by an ATPG tool is usually a very high fraction of the total number of bits. Even after static or dynamic test compaction and test relaxation, this number can still be high enough to be exploited by power-aware X-fill techniques (Wohl et al. 2003). These techniques propose a judicious filling of don't care (X) bits to achieve significant reduction in test power consumption.

Various X-fill techniques have been proposed so far, including 0-fill (fill all X bits with 0), 1-fill (fill all X bits with 1), and Minimum Transition fill (MT-fill) also called adjacent fill (fill any X bit with the nearest care bit from the input side). Though only addressing shift-in power reduction (shift-out power is not considered), they can significantly reduce average power consumption during scan shifting (Butler et al. 2004). Some of these fill options are available today in commercial ATPG tools.

In the context of at-speed scan testing, X-filling solutions have also been proposed to reduce capture power (and not necessarily shift power) and thus avoid IR-drop-induced yield loss (Wen et al. 2005a). These solutions were developed to provide power-aware launch-off-capture delay tests, and are based on estimation-based X-filling (Wen et al. 2006b), probability-based X-filling (Remersaro et al. 2006), justification-based X-filling (Wen et al. 2005b) or a mix of them (Wen et al. 2007).

X-filling techniques have no impact on design flow and do not incur any area overhead. On the other hand, they reduce test power sometimes at the expense of an increase in pattern count (to achieve a given fault coverage) and they offer limited gain for compressed patterns (Ravi et al. 2008).

7.4.2 Test Generation for Low Energy BIST

A Linear Feedback Shift Register (LFSR) is a low-overhead structure commonly used during BIST to generate and apply pseudo-random test patterns to the circuit under test (CUT). Various low power LFSR-based test generators have been proposed so far, such as (among others) the Dual-Speed LFSR-based generator (Wang and Gupta 1997) or the Low Transition Random Test Pattern Generator (Wang and Gupta 1999). The main objective of these solutions is to reduce heat dissipation during testing, which is achieved by reducing switching activity without increasing test time.

Energy is another important dissipation parameter and represents the total switching activity generated during application of the complete test sequence. Beside issues as those described in Section 7.3, an energy increased during BIST has impact on the battery lifetime of battery operated devices, particularly those equipped with on-line test facilities or those requiring self-test procedures during power-up for system integrity checking (such as cellular phones).

In this section, we describe a study initially proposed in Girard et al. (1999) to analyze the impact of tuning LFSR parameters on the energy consumed during test. A solution for low energy BIST generation is then proposed.

7.4.2.1 Impact of LFSR Parameters on Energy

The aim of LFSR tuning is to find a way of decreasing the energy consumed during BIST by appropriately selecting the parameters of the LFSR, i.e., the seed and the characteristic polynomial. However, the minimization problem, constrained by two functions, Energy and Fault Coverage, may show different sensitivity to the variables under analysis (polynomial and seed). For this reason, two sub-problems were considered: (i) impact of LFSR polynomial selection on energy, and (ii) impact of LFSR seed selection on energy.

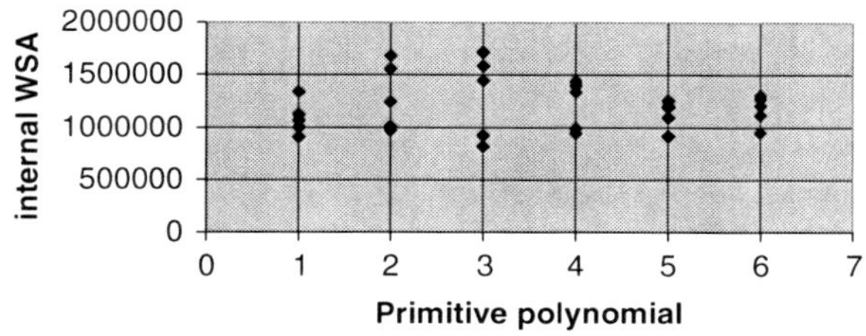

Fig. 7.8 Impact of LFSR polynomial on energy

To analyze the impact of these parameters on the switching activity generated in the CUT, a number of experiments were conducted on various types of circuit. For each circuit, several characteristic polynomials were used for the LFSR, and for each of these polynomials, several seeds were tried. Polynomials were taken from the list of primitive polynomials of an n-stage LFSR (n being the number of primary inputs of the CUT), and seeds were randomly chosen for each selected polynomial. In each experiment, the length of the test sequence required to reach the target fault coverage was determined through fault simulation. Results of these experiments are reported in Fig. 7.8 for circuit c1908 of the ISCAS'89 benchmark suite. The stuck-at fault coverage is equal to 99%. Each value on the X axis corresponds to a particular primitive polynomial of the LFSR, and each dot corresponds to the internal WSA resulting from a randomly selected seed for the particular polynomial. Note that the internal WSA refers to the Weighted Switching Activity of the internal nodes of the CUT.

As can be seen, the WSA obtained for a given primitive polynomial of the LFSR strongly depends on the seed selected. Indeed, the deviation between best seeds and worst seeds is very significant in terms of WSA. On the other hand, sensitivity of the WSA to a given primitive polynomial is much lower; the value of the minimum WSA is almost the same regardless of which primitive polynomial is used. Therefore, selecting a primitive polynomial to minimize energy dissipation during BIST is not as crucial as selecting a good seed for the LFSR.

7.4.2.2 LFSR Seed Selection for Energy Minimization During BIST

Finding the best seed of a given primitive polynomial LFSR to achieve the lowest energy and a given fault coverage is a rather complex problem. Consequently, a nearly optimal solution proposed in Girard et al. (1999) is based on a simulated annealing algorithm that follows the basic strategy of Johnson et al. (Johnson et al. 1989). This method is built on an analogy with thermodynamics, and allows to converging to a final solution by only considering a limited number of partial solutions that are selected partly based on randomness.

The algorithm is reported in Fig. 7.9. The inputs to this algorithm are the number of stages of the LFSR, which is given by the number of primary inputs of the circuit,

```
Algorithm Search-A-Seed (#prim_inputs, target_FC)
BestCost = +∞;
P_find_better_solution = 1;
p = 0;
temperature = 30 000;
begin
        while (P_find_better_solution > p)
        {
                seed = new_seed(#prim_inputs) ;
                Cost = Return_#patterns(seed, target_FC) ;
                Δ = Cost - BestCost ;
        if ( Δ ≤ 0 ) {
                        BestCost = Cost ;
                        BestSeed = seed ;
        } else {
                 p = Return_random_number() ;
                P_find_better_solution = Exp(- Δ / temperature) ;
        }
        decrease temperature ; ** temperature = temperature × 0.85
}
Output(BestCost, BestSeed) ;
end
```

Fig. 7.9 The Search-A-Seed algorithm

```
Function Return_#patterns (seed, target_FC)
begin
FC = 0 ; #patterns = 0 ;
            while (FC < target_FC)
            {
                    FC = generate_one_pattern (seed) ;
                    #patterns ++ ;
            }
return #patterns;
end
```

Fig. 7.10 The *Return_#patterns* function

and the targeted fault coverage. The output is a seed of the LFSR that leads to a pseudo-random test sequence with the shortest length required to achieve this fault coverage.

We denote *BestCost* the global cost function to be minimized, which represents the smallest number of patterns in a test sequence required to achieve the targeted fault coverage, and which corresponds to the minimum energy solution. The function *new_seed()* is a trivial function that determine a new seed randomly. The function *Return_#patterns()* (Fig. 7.10) generates a pseudo-random sequence of input test patterns from a given seed, and returns the number of patterns in the test sequence that is required to achieve the targeted fault coverage (this is done by fault simulation). We denote Δ the variation of the objective function. This variation is used to determine whether a current solution is advantageous or not. If advantageous, the solution is definitively accepted as local minimum. Otherwise, it may only be accepted if the acceptance probability is greater than a given threshold.

In order to determine whether a disadvantageous transformation will be actually accepted or not, we use a number denoted p which is randomly chosen between 0 and 1 with a uniform distribution. This number is then compared with $P_{find\_better\_solution}$ that represents the probability to find a better solution and which is calculated from the Gibbs-Boltzmann distribution. The temperature in the Gibbs-Boltzmann distribution is a control parameter that fixes the number of disadvantageous solutions that can be accepted. The temperature takes a high value at the beginning of the annealing process, and decreases after each acceptance of a local solution. When the temperature is low enough, only advantageous solutions are accepted.

The above simulated annealing algorithm proposed to find a "good" LFSR seed has been validated experimentally and results can be found in Girard et al. (1999). These results show that energy during BIST can be significantly reduced with no loss of fault coverage. On the other hand, average power consumption can not be minimized with such approach. Only issues related to battery lifetime can hence be addressed in this case.

7.5 Design for Power-Aware Testing

Dedicated Design-For-Test (DFT) solutions offer a structured and configurable means to reduce test power consumption. In this section, we first give an overview of the various low-power DFT solutions proposed so far. Next, we present one of these solutions based on staggered clocking.

7.5.1 Overview of Power-Aware DFT Solutions

Low power DFT has been an active research area for more than one decade. Numerous solutions have been proposed so far and some of them are surveyed below. A more complete presentation of low power DFT solutions can be found in Girard et al. (2007).

7.5.1.1 Gated Scan Cells

The combinational logic toggling that happens during scan shifting can be eliminated (or reduced) by incorporating a blocking circuitry at all (or some) outputs of the scan flip-flops. Although there exists various ways of implementing the blocking logic (muxes, transmission gates, etc.), a typical solution consists in adding a NOR gate as a blocking element, with the scan enable of the flip-flop also serving as blocking enable signal (Gerstendörfer and Wunderlich 1999). During scan shifting, the NOR gate can hence prevent data in the scan cells from propagating to the combinational logic. Obviously, this technique is very effective in reducing shift power. However, it induces design overheads (area and functional timing) and timing closure issues related to the scan enable signal.

Alternative techniques such as first-level power supply gating have emerged that do not require modifications of the scan cells. Power switches are added between the power supply and the first level of combinational logic. Such a technique also helps in reducing leakage power consumption.

7.5.1.2 Scan Cell Reordering

Shift power can also be reduced by changing the order of the scan cells in each scan chain of a design. Of course, changing the order of the scan cells in the scan chains implies a change of the bit order in each test vector to preserve the initial fault coverage. Finding the best (less power consuming) order of scan cells in the scan chains is a complex problem and polynomial-time approximation algorithms have therefore to be used for large designs.

Scan cell ordering has many advantages as it does not require additional hardware, the fault coverage and test time are left unchanged, the impact on the design flow is very low, and significant reduction in test power can be obtained. However, power-driven stitching of the scan cells may result in longer interconnections between the scan cells and congestion issues during scan routing. To solve these problems, physical design constraints can be included in the reordering algorithm (Bonhomme et al. 2003).

7.5.1.3 Logic Insertion in Scan Chain

This technique consists in inserting logic elements (XOR gates) between the scan cells so as to minimize the occurrence of transitions in the scan chains (and hence in the combinational logic) during shift operations. Adding logic elements in the scan chains transforms the logic values that need to be shifted in. By doing this intelligently, it is possible to transform the scan vectors so that they contain fewer transitions. Although efficient to reduce shift power (in both combinational logic and scan path), the computational demand of this solution increases with the number of scan cells and the pattern count. Moreover, the scan chains are built based on the available test set, so that adding top-off patterns can be a problem.

7.5.1.4 Scan Chain Segmentation

The basic concept in scan segmentation is to divide a given chain into two (or N) scan chain segments, and then activating one segment at a time when loading and unloading test data (Whetsel 2000; Saxena et al. 2001). During the shift in/out of one scan chain segment, all other segments can have their clocks disabled. When one scan chain segment has been completely loaded/unloaded, then the next scan chain segment is activated.

This technique requires clock gating and the use of bypass multiplexers for segment-wise access. It drastically reduces shift power (both average and peak) dissipated in the combinational logic. It can be applied to circuits with multiple scan chains (e.g. STUMPS architectures), even when test compression is used. It has no impact on the test application time and the fault coverage, and requires minimal modifications to the ATPG flow.

The main drawback of scan segmentation is that capture power remains a concern that needs to be addressed. This problem can be partially solved by creating a data dependency graph based on the circuit structure and identifying the strongly connected components (SCC). Flip-flops in an SCC must load responses at the same time to avoid capture violations. This way, capture power can be minimized (Rosinger et al. 2004).

Low power scan partitioning has been shown to be feasible on commercial designs such as the CELL processor (Zoellin et al. 2006).

7.5.2 Staggered Clocking

Various staggered clock schemes can be used to reduce test power consumption (Sankaralingam and Touba 2003; Lee et al. 2000; Huang and Lee 2001). Staggering the clock during shift or capture achieves power savings without significantly affecting test application time. Staggering can be achieved by ensuring that the clocks to different scan flip-flops (or chains) have different duty cycles or different phases, thereby reducing the number of simultaneous transitions. The biggest challenge to these techniques is its implications on the clock generation, which is a sensitive aspect of chip design. In this section, we describe a staggering clocking scheme proposed in Bonhomme et al. (2001) that can achieve significant power reduction with a very low impact and cost on the clock generation.

7.5.2.1 Basic Principle

The technique proposed in Bonhomme et al. (2001) is based on reducing the operating frequency of the scan cells during scan shifting without modifying the total test time. For this purpose, a clock whose speed is half of the normal (functional) clock speed is used to activate one half of the scan cells (referred to as "Scan Cells A" in Fig. 7.11) during one clock cycle of the scan operation. During the next clock cycle, the second half of the scan cells (referred to as "Scan Cells B") is activated by another clock whose speed is also half of the normal speed. The two clocks are synchronous with the system clock and have the same period during shift operation except that they are shifted in time. During capture operation, the two clocks operate

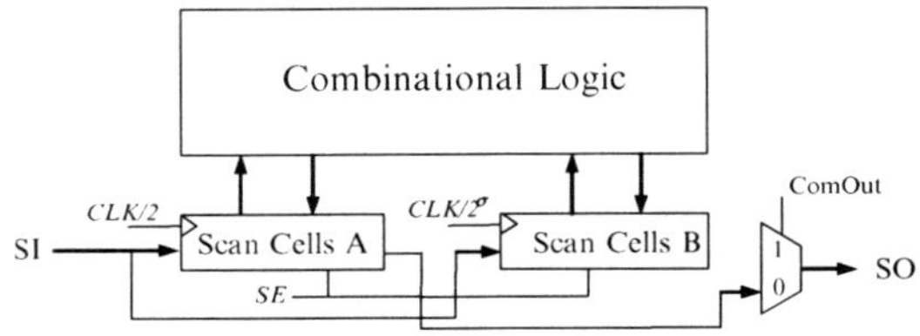

Fig. 7.11 Staggered clocking scheme

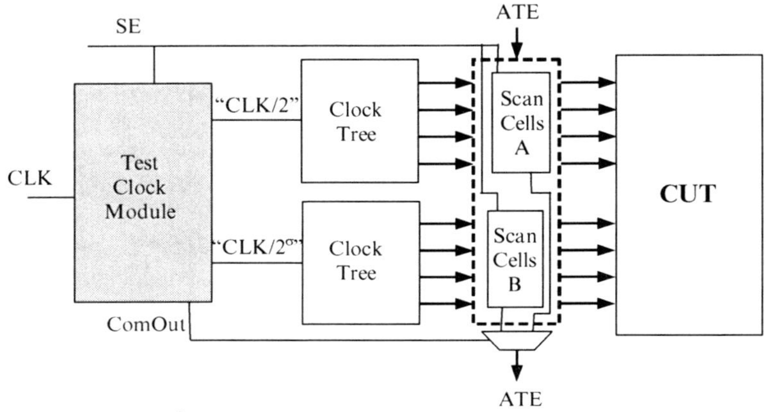

Fig. 7.12 The complete structure

as the system clock. The serial outputs of the two groups of scan cells are connected to a multiplexer that drives either the content of Scan Cells A or the content of Scan Cells B to the ATE during scan operations. As values coming from the two groups of scan cells must be scanned out alternatively, the multiplexer has to switch at each clock cycle of the scan operations.

With such a clock scheme, only half of the scan cells may toggle at each clock cycle (despite the fact that a shift operation is performed at each clock cycle of the whole scan process). Therefore, the use of this scheme lowers the transition density in the combinational logic (logic power), the scan chain (scan power) and the clock tree feeding the scan chain (clock power) during shift operation. Both average power consumption and peak power consumption are significantly minimized in all of these structures. Moreover, the total energy consumption is also reduced as the test length with the staggering clocking scheme is exactly the same as the test length with a conventional scan design to reach the same stuck-at fault coverage.

7.5.2.2 Design of the Staggered Clock Scheme

The complete low power scan structure is depicted in Fig. 7.12. This structure is first composed by a test clock module which provides test clock signals CLK/2 and CLK/2σ from the system clock CLK used in the normal mode. Signal SE allows to switching from the scan mode to the normal or capture mode. Signal ComOut controls the MUX allowing to alternatively outputting test responses from Scan Cells A and Scan Cells B during scan operations. As two different clock signals are needed for the two groups of scan cells, two clock trees are used. These clock trees are carefully designed so as to correctly balance the clock signals feeding each group of scan cells.

The test clock module which provides the control signal ComOut and the test clock signals CLK/2 and CLK/2σ from the system clock CLK is given in Fig. 7.13. This module is formed by a single D-type flip-flop and six logic gates, and allows to generating non-overlapping test clock signals. This structure is very simple and requires a small area overhead. Moreover, it is designed with minimum impact on performance and timing. In fact, some of the already existing driving buffers of the clock tree have to be transformed into AND gates as seen in Fig. 7.13. These gates mask each second phase of the fast system clock during shift operations.

As two different clock signals are used by the two groups of scan cells, the clock tree feeding these scan cells has to be modified. For this purpose, two clock trees are implemented, each with a clock speed which is half of the normal speed. Let us assume a scan chain composed of six scan cells. The corresponding clock trees in the test mode are depicted in Fig. 7.14. Each of them has a fanout of 3 and is composed of a single buffer. During the normal mode of operation, the clock tree feeding the input register at the normal speed can therefore be easily reconstructed as shown in

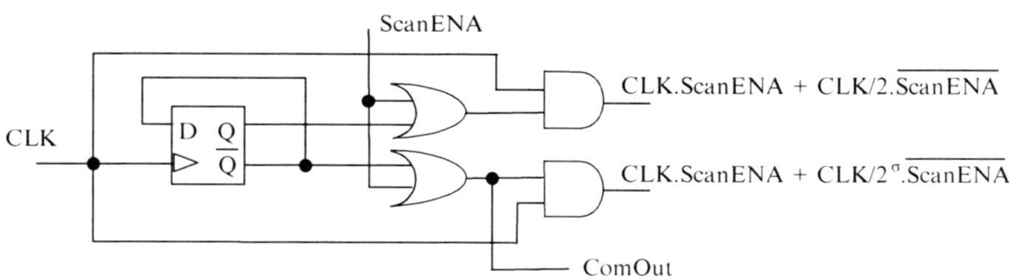

Fig. 7.13 Test clock module

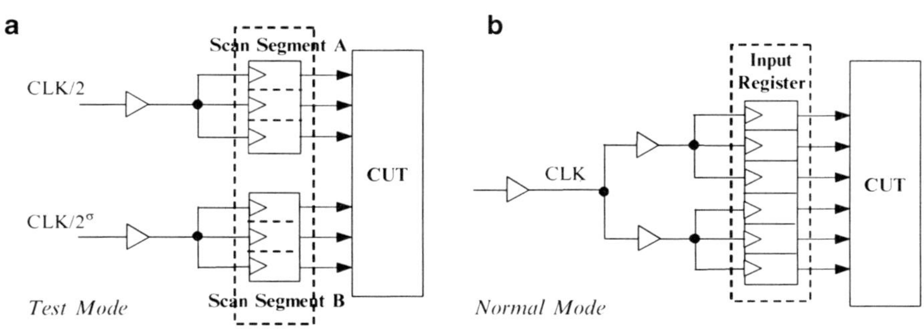

Fig. 7.14 The clock tree in test mode (**a**) and normal mode (**b**)

Fig. 7.14. Note that using two clock trees driven by a slower clock (rather than a single one) allows to further drastically reduce the clock power during scan testing.

The area overhead, which is due to the test clock module and the additional routing, is negligible. The proposed scheme does not require any further circuit design modification and is very easy to implement. Therefore, it has a low impact on the system design time and has nearly no penalty on the circuit performance. Further details about this staggered clock scheme can be found in Bonhomme et al. 2001; Girard et al. 2001).

7.6 Power-Aware Test Data Compression

Test Data Compression (TDC) is an efficient solution to reduce test data volume. It involves encoding a test set so as to reduce its size. By using this reduced set of test data, the ATE limitations, i.e., tester storage memory and bandwidth gap between the ATE and the CUT, may be overcome. During test application, a small on-chip decoder is used to decompress test data received from the ATE as it is fed into the scan chains.

Although reducing test data volume and test application time, TDC increases test power during scan testing. To address this issue, several techniques have been proposed so far to simultaneously reduce test data volume and test power during scan testing. In this section, we first give an overview of power-aware TDC solutions proposed so far. Next, we present one of these solutions based on selective encoding of scan slices.

7.6.1 Overview of Power-Aware TDC Solutions

As proposed in Wang et al. (2006), power-aware TDC techniques can be classified into the three following categories: code-based schemes, linear-decompression-based schemes, and broadcast-scan-based schemes.

7.6.1.1 Code-Based Schemes

The goal of power-aware code-based TDC is to use data compression codes to encode the test cubes of a test set so that both switching activity generated in the scan chains after on-chip decompression and test data volume can be minimized. In the approach presented in Chandra and Chakrabarty (2001), test cubes generated by an ATPG are encoded using Golomb codes. All don't care bits of the test cubes are filled with 0 and Golomb coding is used to encode runs of 0's. For example, to encode the test cube "X0X10XX0XX1", the Xs are filled with 0 and the Golomb coding provides the compressed data (codeword) "0111010". More details about

Golomb codes can be found in Wang et al. (2006). Golomb coding efficiently compresses test data, and the filling of all don't cares with 0 reduces the number of transitions during scan-in, thus significantly reducing shift power. One limitation is that it is very inefficient for runs of 1's. In fact, the test storage can even increase for test cubes that have many runs of 1's. Moreover, implementing this test compression scheme requires a synchronization signal between the ATE and the CUT as the size of the codeword is of variable length.

To address the above limitations, an alternating run-length coding scheme was proposed in Chandra and Chakrabarty (2002). While a Golomb coding only encodes runs of 0's, an alternating run-length code can encode both runs of 0's and runs of 1's. The remaining issue in this case is that the coding becomes inefficient when a pattern with short runs of 0's or 1's has to be encoded. Another technique based on Golomb coding is proposed in Rosinger et al. (2001) but uses a MT filling of all don't care bits rather than a 0-filling at the beginning of the process. The Golomb coding is then used to encode runs of 0's, and a modified encoding is further used to reduce the size of the codeword.

7.6.1.2 Linear-Decompression-Based Schemes

Linear decompressors are made of XOR gates and flip-flops (see Wang et al. (2006) for a comprehensive description) and can be used to expand data coming from the tester to fed the scan chains during test application.

When combined with LFSR reseeding, linear decompression can be view as an efficient solution to reduce data volume and bandwidth. The basic idea in LFSR reseeding is to generate deterministic test cubes by expanding seeds. Given a deterministic test cube, a corresponding seed can be computed by solving a set of linear equations – one for each specified bit – based on the feedback polynomial of the LFSR. Since typically 1% to 5% of the bits in a test cube are care bits, the size of the corresponding seed (stored in the tester memory) will be very low (much smaller than the size of the test cube). Consequently, reseeding can significantly reduce test data volume and bandwidth. Unfortunately, it is not as good for power consumption because the don't care bits in each expanded test cube are filled with pseudo-random values thereby resulting in excessive switching activity during scan shifting. To solve this problem, Lee and Touba (2004) takes advantage of the fact that the number of transitions in a test cube is always less than its number of specified bits. A transition in a test cube is defined as a specified 0 (1) followed by a specified 1 (0) with possible X's between them, e.g., X10XXX or XX0X1X. Thus, rather than using reseeding to directly encode the specified bits as in conventional LFSR reseeding, the proposed encoding scheme divides each test cube into blocks and only uses reseeding to encode blocks that contain transitions. Other blocks are replaced by a constant value which is fed directly into scan chains at the expense of extra hardware.

Unlike reseeding-based compression schemes, the solution proposed in Czysz et al. (2007) uses the Embedded Deterministic Test (EDT) environment (Rajski et al. 2004) to decompress the deterministic test cubes. However, rather than doing random fill of each expanded test cube, the proposed scheme pushes the decompressor into the self-loop state during encoding for low power fill.

7.6.1.3 Broadcast-Scan-Based Schemes

These power-aware TDC schemes are based on broadcasting the same value to multiple scan chains. Using the same value reduces the number of bits to be stored in the tester memory and the number of transitions generated during scan shifting. The main challenge is to achieve this goal without sacrificing the fault coverage and the test time.

The segmented addressable scan architecture presented in Fig. 7.15 is an efficient power-aware broadcast-scan-based TDC solution (Al-Yamani et al. 2005). Each scan chain in this architecture is split into multiple scan segments thus allowing the same data to be loaded simultaneously into multiple segments when compatibility exists. The compatible segments are loaded in parallel using a multiple-hot decoder. Test power is reduced as segments which are incompatible within a given round, i.e., during the time needed to upload a given test pattern, are not clocked.

Power-aware broadcast-scan-based TDC can also be achieved by using the progressive random access scan (PRAS) architecture proposed in Baik and Saluja (2005) that allows individual accessibility to each scan cell. In this architecture, scan cells are configured as an SRAM-like grid structure using specific PRAS scan cells and some additional peripheral and test control logic. Providing such accessibility to every scan cell eliminates unnecessary switching activity during scan, while reducing test time and data volume by updating only a small fraction of scan-cells throughout the test application.

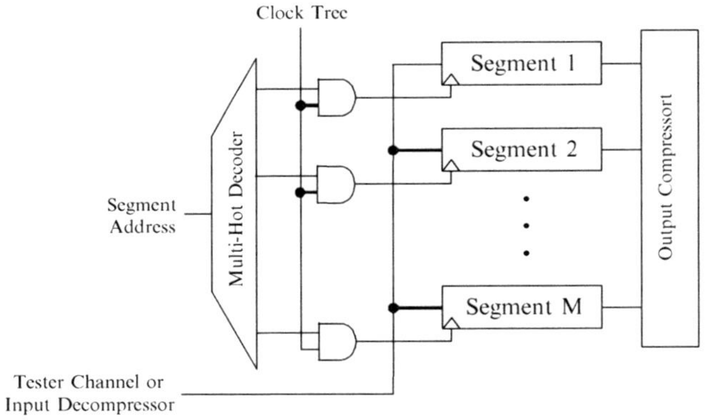

Fig. 7.15 The segmented addressable scan architecture

7.6.2 Power-Aware TDC Using Selective Encoding of Scan Slices

The section describes an efficient code-based TDC solution initially proposed in Badereddine et al. (2008) to simultaneously address test data volume and test power reduction during scan testing of embedded Intellectual Property (IP) cores.

7.6.2.1 TDC Using Selective Encoding of Scan Slices

The method starts by generating a test sequence with a conventional ATPG using the non-random-fill option for don't-care bits. Then, each test pattern of the test sequence is formatted into scan slices. Each scan slice that is fed to the internal scan chains is encoded as a series of c-bit slice-codes, where $c = K + 2$, $K = [\log 2\,(N + 1)]$ with N being the number of internal scan chains of the IP core. As shown in Fig. 7.16, the first two bits of a slice-code form the control-code that determines how the following K bits, referred to as the data-code, have to be interpreted.

This approach only encodes a subset of the specified bits in a slice. First, the encoding procedure examines the slice and determines the number of 0- and 1-valued bits. If there are more 1s (0s) than 0s (1s), then all don't-care bits in this slice are mapped to 1 (0), and only 0s (1s) are encoded. The 0s (1s) are referred to as *target-symbols* and are encoded into data-codes in two modes: *single-bit-mode* and *group-copy-mode*.

In the *single-bit-mode*, each bit in a slice is indexed from 0 to $N-1$. A target-symbol is represented by a data-code that takes the value of its index. For example, to encode the slice "XXX10000", the Xs are mapped to 0 and the only target-symbol 1 at bit position three is encoded as "0011". In this mode, each target-symbol in a slice is encoded as a single slice-code. Obviously, if there are many target-symbols that are adjacent or near to each other, it is inefficient to encode each of them using separate slice-codes. Hence the group-copy-mode has been designed to increase the compression efficiency.

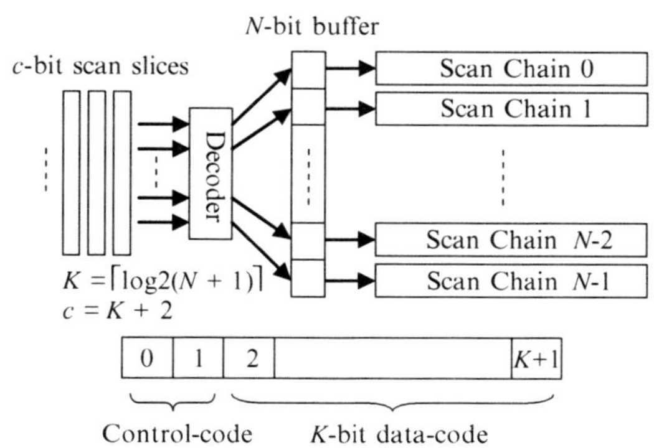

Fig. 7.16 Principle of scan slice encoding

In the *group-copy-mode*, an N-bit slice is divided into $M = N/K$ groups, and each group is K-bits wide with the possible exception for the last group. If a group contains more than two target-symbols, the group-copy-mode is used and the entire group is copied to a data-code. Two data-codes are needed to encode a group. The first data-code specifies the index of the first bit of the group, and the second data-code contains the actual data. In the group-copy-mode, don't-care bits can be randomly filled instead of being mapped to 0 or 1 by the compression scheme. For example, let $N = 8$ and $K = 4$, *i.e.* each slice is 8-bits wide and consists of two 4-bit groups. To encode the slice "X1110000", the three 1s in group 0 are encoded. The resulting data-codes are "0000" and "X111", which refer to bit 0 (first bit of group 0) and the content of the group, respectively.

Since data-codes are used in both modes, control-codes are needed to avoid ambiguity. Control-codes "00", "01" and "10" are used in the single-bit-mode and "11" is used in the group copy-mode. Control-codes "00" and "01" are referred to as *initial control-codes* and they indicate the start of a new slice. Table 7.1 shows a complete example to illustrate the encoding procedure. The first column shows the scan slices. The second and third ones show the resulting slice-codes (control- and data-codes) and the last column describes the compression procedure.

A property of this compression method is that consecutive c-bit compressed slices fed by the ATE are often identical or compatible. Therefore, ATE pattern-repeat can be used to further reduce test data volume after selective encoding of scan slices. More details about ATE pattern-repeat can be found in Wang and Chakrabarty (2005).

7.6.2.2 Test Power Considerations

The above technique drastically reduces test data volume (up to 28x for a set of experimented industrial circuits) and test time (up to 20x). However, power consumption is not carefully considered, especially during the filling of don't-care bits in the scan slices. To illustrate this problem, let us consider the 4 slice-code example given in Table 7.2 with $N = 8$ and $K = 2$.

Table 7.1 A slice encoding – example 1

Slices	Slice Codes: *Data Code*	Slice Codes: *Control Code*	Descriptions
XX00 010X	00	0101	Start a new slice, map Xs to 0, set bit 5 to 1
1110 0001	00	0111	Start a new slice, map Xs to 0, set bit 7 to 1
	11	0000	Enter group-copy-mode starting from bit 0
	11	1110	The data is 1110
XXXX XX11	01	1000	Start a new slice, map Xs to 1, no bits are set to 0

Table 7.2 A slice encoding – example 2

Slices	Slice Codes: *Control Code*	Slice Codes: *Data Code*	Descriptions
XX00 010X	00	0101	Start a new slice, map Xs to 0, set bit 5 to 1
XXXX XX11	01	1000	Start a new slice, map Xs to 1, no bits are set to 0
X00X XXXX	00	1000	Start a new slice, map Xs to 0, no bits are set to 1
11XX 0XXX	01	0100	Start a new slice, map Xs to 1, set bit 4 to 0

Table 7.3 Scan-slices obtained after decompression

Slices after performing decompression								
SC1	SC2	SC3	SC4	SC5	SC6	SC7	SC8	Descriptions
0	0	0	0	0	1	0	0	Xs are set to 0
1	1	1	1	1	1	1	1	Xs are set to 1
0	0	0	0	0	0	0	0	Xs are set to 0
1	1	1	1	0	1	1	1	Xs are set to 1
6	6	6	6	5	3	6	6	WT
44								Total WT

Table 7.4 Slice encoding with the 0-filling option

Slices								Slice Codes	
0	0	0	0	0	1	0	0	00	0101
0	0	0	0	0	0	1	1	00	1000
								11	0100
								11	0011
0	0	0	0	0	0	0	0	00	1000
1	1	0	0	0	0	0	0	00	1000
								11	0000
								11	1100
				15				Total WT	

The scan slices obtained after decompression and applied to the internal scan chains are given in Table 7.3. The two last lines give the number of weighted transitions (WT) in each internal scan chain (SC) and the total number of weighted transitions generated at the circuit inputs after application of all test patterns. As can be seen, the toggle activity in each scan chain is very high, mainly because Xs in the scan slices are set alternatively to 0 and 1 before performing the compression procedure.

By modifying the assignment of don't-care bits in our example, and filling all don't care with 0 (0-filling) or 1 (1-filling) for the entire test sequence, the total number of WT is greatly reduced (15 with the 0-filling option and 19 with the 1-filling option). Results are shown in Tables 7.4 and 7.5 respectively.

Table 7.5 Slice encoding with the 1-filling option

Slices								Slice Codes	
1	1	0	0	0	1	0	1	01	1000
								11	0000
								11	1100
								11	0101
1	1	1	1	1	1	1	1	01	1000
1	0	0	1	1	1	1	1	01	1000
								11	0000
								11	1001
1	1	1	1	0	1	1	1	01	0100
				19				Total WT	

Consequently, test power considerations in this technique will consist in modifying the initial selective encoding procedure by using one of the following X-filling heuristics to fill don't-care bits:

- 0-filling: all Xs in the test sequence are set to 0s
- 1-filling: all Xs in the test sequence are set to 1s
- MT-filling (Minimum Transition filling): all Xs are set to the value of the last encountered care bit (working from the top to the bottom of column)

A counterpart of this positive impact on test power is a possible negative impact on the test data compression rate. By looking at the results in Tables 7.4 and 7.5, we can notice that the number of slice-codes obtained after compression is 8 and 9 respectively, which is much higher than 4 obtained with the original procedure (shown in Table 7.2). In fact, the loss in compression rate is much lower than it appears in this example. Experiments performed on industrial circuits and reported in Badereddine et al. (2008) have shown that test data volume reduction factors (12x on average) are in the same order of magnitude than those obtained with the initial compression procedure (16x on average). On the other hand, test power reduction with respect to the initial procedure is always higher than 95%. Moreover, this method does not require detailed structural information about the IP core under test, and utilizes a generic on-chip decoder which is independent of the IP core and the test set.

7.7 Summary

Reliability, yield, test time and test costs in general are affected by test power consumption. Carefully modeling the different types and sources of test power is a prerequisite of power aware testing. Test pattern generation, design for test, and test data compression have to be implemented with respect to their impacts on power. The techniques presented in this chapter allow power restricted testing with minimized hardware cost and test application time.

References

Altet J, Rubio A (2002) Thermal testing of integrated circuits. Springer Science, New York

Al-Yamani A, Chmelar E, Grinchuck M (May 2005) Segmented addressable scan architecture. In Proceedings of VLSI test symposium, pp 405–411

Arabi K, Saleh R, Meng X (May–Jun 2007) Power supply noise in SoCs: metrics, management, and measurement. IEEE Des Test Comput 24(3)

Athas WC, Svensson LJ, Koller JG, Tzartzanis N, Chin Chou EG (Dec 1994) Low-power digital systems based on adiabatic-switching principles. IEEE Trans VLSI Sys 2(4):398–416

Badereddine N, Wang Z, Girard P, Chakrabarty K, Virazel A, Pravossoudovitch S, Landrault C (Aug 2008) A selective scan slice encoding technique for test data volume and test power reduction. JETTA J Electron Test – Theory Appl 24(4):353–364

Baik DH, Saluja KK (Oct 2005) Progressive random access scan: a simultaneous solution to test power, test data volume and test time. In Proceedings of international test conference. Paper 15.2

Bonhomme Y, Girard P, Guiller L, Landrault C, Pravossoudovitch S (Nov 2001) A gated clock scheme for low power scan testing of logic ics or embedded cores. In Proceedings of Asian Test Symposium, pp 253–258

Bonhomme Y, Girard P, Guiller L, Landrault C, Pravossoudovitch S (Oct 2003) Efficient scan chain design for power minimization during scan testing under routing constraint. In Proceedings of international test conference, pp 488–493

Borkar SY, Dubey P, Kahn KC, Kuck DJ, Mulder H, Pawlowski SP, Rattner JR (2005) Platform 2015: Intel processor and platform evolution for the next decade. In Intel White Paper Platform 2015

Butler KM, Saxena J, Fryars T, Hetherington G, Jain A, Lewis J (Oct 2004) Minimizing power consumption in scan testing: pattern generation and DFT techniques. In Proceedings of international test conference, pp 355–364

Chandra A, Chakrabarty K (Jun 2001) Combining low-power scan testing and test data compression for system-on-a-chip. In Proceedings of design automation conference, pp 166–169

Chandra A, Chakrabarty K (Jun 2002) Reduction of SOC test data volume, scan power and testing time using alternating run-length codes. In Proceedings of design automation conference, pp 673–678

Chang YS, Gupta SK, Breuer MA (Apr 1997) Analysis of ground bounce in deep sub-micron circuits. In Proceedings of VLSI test symposium, pp 110–116

Cirit MA (Nov 1987) Estimating dynamic power consumption of CMOS circuits. In Proceedings of international conference on computer-aided design, pp 534–537

Czysz D, Tyszer J, Mrugalski G, Rajski J (May 2007) Low power embedded deterministic test. In Proceedings of VLSI test symposium, pp 75–83

Gerstendörfer S, Wunderlich HJ (Sep 1999) Minimized power consumption for scan-based BIST. In Proceedings of international test conference, pp 77–84

Girard P, Guiller L, Landrault C, Pravossoudovitch S, Figueras J, Manich S, Teixeira P, Santos M (1999) Low energy BIST design: impact of the LFSR TPG parameters on the weighted switching activity. In Proceedings of international symposium on circuits and systems, CD-ROM

Girard P, Guiller L, Landrault C, Pravossoudovitch S, Wunderlich HJ (May 2001) A modified clock scheme for a low power BIST test pattern generator. In Proceedings of VLSI test symposium, pp 306–311

Girard P (May–Jun 2002) Survey of low-power testing of VLSI circuits. IEEE Des Test Comput 19(3):82–92

Girard P, Wen X, Touba NA (2007) Low-power testing. In: Wang LT, Stroud CE, Touba NA (eds) System-on-chip test architectures: nanometer design for testability. Morgan Kaufmann Publishers, pp 307–350

Hertwig A, Wunderlich HJ (May 1998) Low power serial built-in self-test. In Proceedings of European test workshop, pp 49–53

Huang T-C, Lee K-J (1989) A token scan architecture for low power testing. In Proceedings of international test conference, pp 660–669

Johnson DS, Aragon C, McGeoch L, Schevon C (1989) Optimisation by simulated annealing : an experimental evaluation; part I, graph partitioning. Oper Res 37(865–892)

Lee K-J, Huang T-C, Chen J-J (Dec 2000) Peak-power reduction for multiple-scan circuits during test application. In Proceedings of Asian test symposium, pp 453–458

Lee J, Touba NA (Oct 2004) Low power test data compression based n LFSR reseeding. In Proceedings of international conference on computer design, pp 180–185

Midulla I, Aktouf C (Dec 2008) Test power analysis at register transfert level. ASP J Low Pow Electron 4(3):402–409

Najm F (Dec 1994) A survey of power estimation techniques in VLSI circuits. IEEE Trans VLSI Sys 2(4):446–455

Nicolici N, Al-Hashimi B (2003) Power-constrained testing of VLSI circuits. Springer Science, New York, NY

Pedram M, Rabaey J (eds) (2002) Power aware design methodologies. Kluwer Academic Publishers

Pouya B, Crouch A (Oct 2000) Optimization trade-offs for vector volume and test power. In Proceedings of international test conference, pp 873–881

Rajski J, Tyszer J, Kassab M, Mukherjee N (May 2004) Embedded deterministic test. IEEE Trans Computer-Aided Des 23:776–792

Ravi S, Devanathan VR, Parekhji R (Nov 2007) Methodology for low power test pattern generation using activity threshold control logic. In Proceedings of international conference on computer-aided-design, pp 526–529

Ravi S, Parekhji R, Saxena J (Apr 2008) Low power test for nanometer system-on-chips (SoCs). ASP J Low Power Electron 4(1):81–100

Remersaro S, Lin X, Zhang Z, Reddy SM, Pomeranz I, Rajski J (Oct 2006) Preferred fill: a scalable method to reduce capture power for scan based designs. In Proceedings of international test conference, paper 32.2

Rosinger P, Gonciari T, Al-Hashimi B, Nicolici N (2001) Simultaneous reduction in volume of test data and power dissipation for systems-on-a-chip. IEE Electron Lett 37(24):1434–1436

Rosinger P, Al-Hashimi B, Nicolici N (Jul 2004) Scan architecture with mutually exclusive scan segment activation for shift- and capture-power reduction. IEEE Trans Computer-Aided Des 23(7):1142–1153

Roy K, Mukhopadhaya S, Mahmoodi-Meimand H (2003) Leakage current mechanisms and leakage reduction techniques in deep-submicrometer CMOS circuits. In Proceedings of IEEE, pp 305–327

Sankaralingam R, Oruganti R, Touba NA (May 2000) Static compaction techniques to control scan vector power dissipation. In Proceedings of VLSI test symposium, pp 35–42

Sankaralingam R, Touba NA (Feb 2003) Multi-phase shifting to reducing instantaneous peak power during scan. In Proceedings of Latin American Test Workshop, pp 78–83

Saxena J, Butler KM, Whetsel L (Oct 2001) A scheme to reduce power consumption during scan testing. In Proceedings of internatinal test conference, pp 670–677

Saxena J, Butler KM, Jayaram VB, Kundu S, Arvind NV, Sreeprakash P, Hachinger M (Oct 2003) A case study of ir-drop in structured at-speed testing. In Proceedings of international test conference, pp 1098–1104

Sde-Paz S, Salomon E (Oct 2008) Frequency and power Correlation between At-Speed Scan and Functional Tests. In Proceedings 39th IEEE international test conference (ITC) 2008, pp 13.3

Shi C, Kapur R (2004) How power aware test improves reliability and yield. IEEDesign.com, Sep. 15

Wang Z, Chakrabarty K (Oct 2005) Test data compression for IP embedded cores using selective encoding of scan slices. In Proceedings of international test conference, paper 24.3

Wang S, Gupta SK (Oct 1994) ATPG for heat dissipation minimization during test application. In Proceedings of international test conference, pp 250–258

Wang S, Gupta SK (Oct 1997) DS-LFSR: a new BIST TPG for low heat dissipation. In Proceedings of international test conference, pp 848–857

Wang S, Gupta SK (Oct 1999) LT-RTPG: a new test-per-Scan BIST TPG for low heat dissipation. In Proceedings of international test conference, pp 85–94

Wang CY, Roy K (Jan 1995) Maximum power estimation for CMOS circuits using deterministic and statistical approaches. In Proceedings of VLSI conference, pp 364–369

Wang L-T, Wu C-W, Wen X (2006) Vlsi test principles and architectures: design for testability. Morgan Kaufmann, San Francisco

Wen X, Suzuki T, Kajihara S, Miyase K, Minamoto Y, Wang L-T, Saluja KK (Dec 2005a) Efficient test set modification for capture power reduction. ASP J Low Pow Electron 1(3):319–330

Wen X, Yamashita Y, Morishima S, Kajiihara S, Wang L-T, Saluja KK, Kinoshita K (May 2005b) On low-capture-power test generation for scan testing. In Proceedings of VLSI test symposium, pp 265–270

Wen X, Kajihara S, Miyase K, Suzuki T, Saluja KK, Wang L-T, Abdel-Hafez KS, Kinoshita K (May 2006) A new ATPG method for efficient capture power reduction during scan testing. In Proceedings of VLSI test symposium, pp 58–63

Wen X, Miyase K, Suzuki T, Yamato Y, Kajihara S, Wang L-T, Saluja KK (Oct 2006) A highly-guided x-filling method for effective low-capture-power scan test generation. In: Wen X et al. (eds) Proceedings of international conference on computer design, pp 251–258

Wen X, Miyase K, Kajihara S, Suzuki T, Yamato Y, Girard P, Oosumi Y, Wang LT (Oct 2007) A novel scheme to reduce power supply noise for high-quality at-speed scan testing. In Proceedings of international test conference, paper 25.1

Weste NHE, Eshraghian K (1993) Principles of CMOS VLSI design: a systems perspective, 2nd edn. Addison-Wesley

Whetsel L (Oct 2000) Adapting scan architectures for low power operation. In Proceedings of international test conference, pp 863–872

Wohl P, Waicukauski JA, Patel S, Amin MB (Jun 2003) Efficient compression and application of deterministic patterns in a logic BIST architecture. In Proceedings of design automation conference, pp 566–569

Zoellin C, Wunderlich HJ, Maeding N, Leenstraa J (Oct 2006) BIST power reduction using scan-chain disable in the CELL processor. n Proceedings of international test conference, Paper 32.3

Zorian Y (Apr 1993) A distributed BIST control scheme for complex VLSI devices. Proceedings of 11th IEEE VLSI test symposium, pp 4–9

Chapter 8
Physical Fault Models and Fault Tolerance

Jean Arlat and Yves Crouzet

Abstract Dependable systems are obtained by means of extensive testing procedures and the incorporation of fault tolerance mechanisms encompassing error detection (on-line testing) and system recovery. In that context, the characterization of fault models that are both tractable and representative of actual faults constitute an essential basis upon which one can efficiently verify, design or assess dependable systems. On one hand, models should refer to erroneous behaviors that are as abstract and as broad as possible to allow for the definition and development of both generic fault tolerance mechanisms and cost-effective injection techniques. On the other hand, the models should definitely aim at matching the erroneous behaviors induced by real faults.

In this chapter, we focus on the representativeness of fault models with respect to physical faults for deriving relevant testing procedures as well as detection mechanisms and experimental assessment techniques. We first discuss the accuracy of logic fault models with respect to physical defects in the implementation of off-line/on-line testing mechanisms. Then, we show how the fault models are linked to the identification and implementation of relevant fault injection-based dependability assessment techniques.

Keywords Defect characterization · Fault models · Testability improvement · Testing procedures · Test sequences generation · Layout rules · Coding · Error detection · Self-checking · Fault-injection-based testing · Dependability assessment

8.1 Introduction

The proper characterization of component defects and related fault models during the development phase and during normal operation is a main concern. In order to be appropriate and efficient, methodologies and procedures have to rely on models

J. Arlat (✉) and Y. Crouzet
LAAS-CNRS; Université de Toulouse; 7, avenue du Colonel Roche, F-31077 Toulouse, France
e-mail: jean.arlat@laas.fr

H.-J. Wunderlich (ed.), *Models in Hardware Testing: Lecture Notes of the Forum in Honor of Christian Landrault*, Frontiers in Electronic Testing 43,
DOI 10.1007/978-90-481-3282-9_8,

reflecting as much as possible the real defects and faults that are likely to affect both the production and the operational phases. Hardware testing was initially based on the assumption that defects could be adequately modeled by stuck-at-0 and stuck-at-1 logical faults associated with the logic diagram of the circuit under test. Nevertheless, with the increasing integration density, this hypothesis has become less and less sound. Similar concerns about fault representativeness apply to the definition of suitable fault tolerance mechanisms (error detection and recovery) meant to cope with faults occurring during normal operation (on-line testing). Fault representativeness issues also impact the specific testing methods (classically, fault injection techniques), that are specifically intended to assess the fault tolerance mechanisms against the typical sets of inputs they are a meant to cope with: the faults and errors induced. Such techniques are to be related to the simulation techniques described in Chapter 4 for estimating the quality of test sets, with respect to manufacturing defects.

This chapter addresses fault representativeness issues at large, i.e., encompassing the definition and application of various forms of testing: *off-line testing* with respect to manufacturing defects and *on-line testing* mechanisms to cope with faults occurring during normal operation (Section 8.2), and a recursive form of testing designed to assess the coverage of the fault tolerance mechanisms (Section 8.3). Finally, Section 8.4 concludes the chapter.

It is worth noting that the results reported in Section 8.2 are based on seminal research work carried out at LAAS-CNRS during years 1975–1980 and directed by Christian Landrault (first work by Christian devoted to hardware testing). These studies were dedicated to the design of easily testable and self-checking LSI circuits. We voluntarily maintained the historical and pioneering perspective of that work in keeping the original figures, among which some are from Christian's hand.

Before moving to the next section of this chapter, we will provide here some basic definitions and terminology about hardware dependability issues that will be used throughout the paper, and that are compliant with the currently widely accepted taxonomy in the domain (Avižienis et al. 2004). In this process, we assume the recursive nature attached to the notions of *failure*, *fault*, *error*, *failure*, *fault*, etc.:

a. **Defect**: a physical defect is a *failure* occurring in the manufacturing process or in operation (e.g., short, open, threshold voltage drift, etc.).
b. **Fault**: a fault is the direct consequence of a defect. At the logical level, the most popular fault model has been for long time the stuck-at-X fault model – $X \in \{0, 1\}$. A defect is equivalent to a stuck-at X of a line l (l/X) if the behavior of the defective circuit is identical to the behavior of a perfect circuit with the line maintained at logical value X.
c. **Error**: an error corresponds to the activation of a fault that induces an incorrect operation of the target system (IC or system including the IC). A line presents an error at a value X if, during normal operation, it is at the logical value X instead of the value $\overline{X}$. The error observed at a given point of a target IC, depends not only on the type of fault, but also on the structure of the circuit (logical function), as well as the logical inputs and outputs of the circuit. A defect may induce:

- A single error, if it only influences one output
- A unidirectional error, if it impacts several outputs in the same manner
- A multiple error, if it influences several outputs in different ways

d. **Failure**: A failure occurs when the service delivered by the target system is perceived by its users as deviating from the correct one. This has to be related to the definition that illustrates the recursion attached to the concepts governing the fault-error-failure pathology.

8.2 Fault Models and Off-Line/On-Line Testing

Off-line and on-line testing techniques have been based for long time on the assumption that defects may be modeled by stuck-at-0 and stuck-at-1 logical faults associated with the logic diagram of the circuit to be tested. This hypothesis is becoming less and less sound with the advance of integration technology. This section is based on a pioneering study aimed at addressing this problem (Galiay 1978; Galiay et al. 1980; Crouzet 1978; Crouzet and Landrault 1980).

Section 8.2.1 derives a set of fault assumptions motivated by the physical origin of the defects observed by direct inspection of 4-bit microprocessor chips. A great majority of the defects affecting complex gates are shorts and opens that cannot be accounted for by the commonly used logic level models. Section 8.2.2 deals with the generation of **(off-line) test sequences** for such defects. Section 8.2.3 proposes layout rules aimed at facilitating testing procedures. These rules aim at decreasing the variety of possible defects and at avoiding those that are not easily testable. By adhering to these rules, then logic level models are again able to accurately represent the effects of actually observed physical defects. Sections 8.2.4 and 8.2.5 address the problem of designing fault-tolerant systems able to cope with defect manifestations during operation. Proposals helping the design of circuits better adapted to the realization of fault-tolerant systems (Sedmak and Liebergot 1978; Rennels et al. 1978) are provided. These sections focus on **on-line testing** issues for detecting errors induced by the physical defects in operation. Suitable error models and related implementation rules aimed at facilitating the efficiency of the detection are briefly presented. Finally, Section 8.2.6 provides concluding remarks.

8.2.1 Defects Analysis for MOS LSI

The problem of test generation can be formulated as follows: given a description of the target circuit and a list of faults, derive the shortest sequence of input vectors enabling the detection of the faults in the list. This detection must be ensured by observing the primary outputs of the circuit, only. The nature of the considered list of faults strongly influences the test sequence generation. The more these faults are related to the physical nature of the circuit, the higher the quality of the test, but as a general consequence, more effort will be required for the generation of the test sequence. The list of faults must therefore be carefully selected to satisfy these conflicting requirements: sufficient fault coverage and easy test generation.

Rather than considering each kind of physical defect of the circuit individually, it is custom practice to deal with a more general fault model able to represent all of them. At the time when the considered study was carried out, all testing approaches were based on stuck-at-0 or stuck-at-1 of any connection of the logic diagram of the circuit. Even if this model was relatively satisfactory for small-scale integration, it was clearly no longer valid for large-scale integrated circuits. To tackle this problem, we have first tried to carry out a characterization of LSI failure modes by analyzing a set of failed circuits (Crouzet et al. 1978).

The considered application circuit is a 4-bit microprocessor designed by EFCIS[1] and realized with PMOS technology. It is able to manage four processes with different priorities, and it includes all the basic functions of an LSI IC: (1) scratchpad memory, (2) arithmetic and logic unit, (3) program counters, and (4) control unit realized with PLA. The internal architecture is based on a classical bus structure represented by Fig. 8.1. Two blocks are specific to the application circuit: the allocation unit, and the timing unit. The allocation unit enables management of interruptions

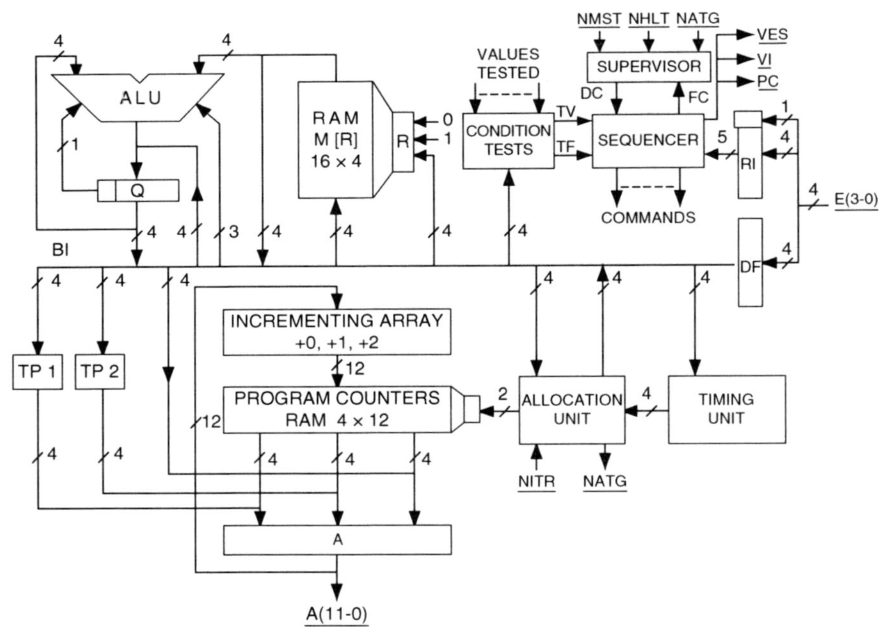

Fig. 8.1 Functional internal architecture of the application circuit

[1] EFCIS: *Société pour l'Etude et la Fabrication de Circuits Intégrés Spéciaux*, that has evolved to form ST Microelectronics, in the late 1990s.

from the four processes. The timing unit furnishes real time clocks acting as internal interrupts for the allocation unit. The remaining part of the circuit is composed of three main blocks:

1. The addressing system, composed of the incrementing array, the program counters, the output register A, and the buffers TP1 and TP2
2. The processing unit, including the ALU, the accumulator Q, the input buffer DF, the RAM M[R], and the condition test block
3. The control block, including the sequencer and the supervisor

Pinpointing defects simply by direct observation of the chip is a very complex task. Thus, to reduce the region of investigation, an initial step aimed at a prelocalization of the failures was introduced. This specific test sequence is hierarchically organized using a "start small" approach:

- The total sequence is divided into subsequences each dedicated to the test of a specific microprocessor block whose size is as small as possible.
- The ordering of the subsequences is such that a fault detected by any of them cannot be induced by one of the blocks tested by the previous subsequences.

The second step of the analysis consists of a direct observation of the chip in the region determined by the prelocalization sequence. Different techniques were applied:

1. Parametric measurements, giving information about process quality
2. Research of the shmoo plot domain (i.e., the domain of correct operation) for different parameters, e.g., temperature, frequency, and supply voltage
3. Visual inspection with an optical microscope
4. Potential cartography with a scanning electron microscope
5. Electrical analysis of the circuits nodes by placing probes onto the chip

This method has been applied to a set of 43 defective chips. The two main results obtained from this study are as follows: (1) defects are randomly distributed and no block is more vulnerable than any other, and (2) insights about the typical physical defect modes were derived. Table 8.1 depicts the observed defect modes. They consist mainly of shorts and opens concerning either the metallizations or the diffusions. It should be noted that no short was observed between metallization and diffusion. For 10% of the cases, a logical error was clearly observed, but no defect could be identified. For another 15%, the chips presented a very large imperfection (e.g., a scratch from one side to the other of the chip) which can be considered as

Table 8.1 Observed failure modes

Short between metallizations	39%
Open of a metallization	14%
Short between diffusions	14%
Open of a diffusion	6%
Short between metallization and substrate	2%
Non identified	10%
Non significant	15%

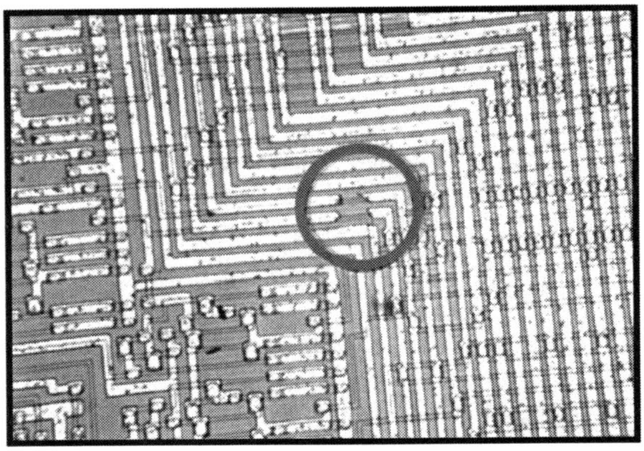

Fig. 8.2 Example of open defects in the application circuit

"non significant" for test purposes because such faults can be easily revealed by any test sequence. Figure 8.2 illustrates opens affecting two metallization lines in the Timing Unit.

Two alternative approaches have been followed to cope with defects that cannot be handled by logical fault models:

1. Try to *generate test sequences* accounting directly for the defects (shorts and opens) at the electrical level.
2. Propose *restrictive layout rules*, so that defects results essentially to stuck-at faults at the logic level.

8.2.2 Generation of Test Sequences for Shorts and Opens

8.2.2.1 Basic Consequences from the Failure Mode Analysis

Concerning test sequence generation and fault simulation, the results of the failure mode analysis have two very important consequences.

Not All Defects Can Be Modeled by Stuck-at Faults This can be clearly illustrated by the following example. Figure 8.3a represents the electrical diagram of a MOS gate on which two possible shorts (#1 and #2) and two possible opens (#3 and #4) are indicated. Short #1 and open #3 can be modeled by a stuck-at-1 at input e and by a stuck-at-0 at input e (or input f or both), respectively. On the other hand, short #2 and open #4 cannot be modeled by any stuck-at-fault because they lead to a modification of the function realized by the gate. For the same reason, a short between the outputs of two gates (Fig. 8.3b) cannot be modeled by any stuck-at faults.

Representing the Circuit as a Logic Diagram Is Not Adequate Taking into account physical defects such as shorts and opens implies the consideration of the actual topology of the circuit. This advocates for the consideration of an electrical

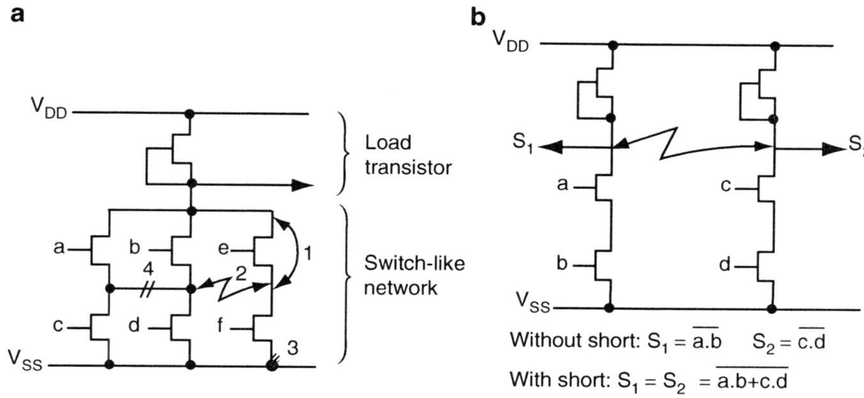

Fig. 8.3 (**a**) Failure examples in a MOS gate. (**b**) Short between the outputs of two gates

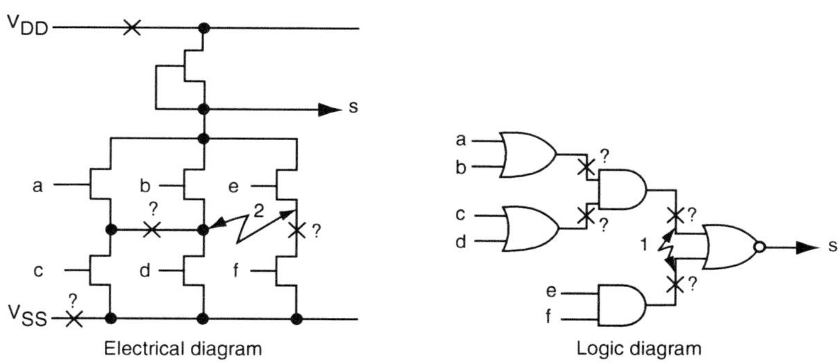

Fig. 8.4 Relations between electrical and logic diagrams

diagram rather than a logic diagram, since the latter does not constitute a real model of the physical circuit. Some connections of the real circuit are not represented in the logic diagram, whereas some connections appearing on the logic diagram may be missing in the physical circuit.

As an example, Fig. 8.4 shows the logic and electrical diagrams of the same gate. The faults considered in each diagram are those that cannot be represented on the other or even cannot occur.

For instance, short #2, which is physically possible, cannot be represented on the logic diagram, and short #1 in the logic diagram has no physical meaning.

Consequently, all methods for test sequence generation and fault simulation based on a stuck-at fault model at the logic diagram level are not well adapted. A possible approach for fault simulation may be to introduce short defects or, better, to work directly with the transistor diagram. For test sequence generation, it is necessary to use a new method accounting directly for the faults at the gate and blocks levels.

8.2.2.2 Test Sequence Generation at the Gate Level

For testing at the gate level, we first need to define the notion of a conduction path that will be used later. MOS technology enables the realization of complex gates including several cascaded AND/OR basic functions (Fig. 8.3a). Schematically, such a gate can be divided into two parts: a load transistor and a set of "control" transistors that can be considered as a switch-like network.

The switch-like network constitutes the active part of the gate. It allows, by applying convenient input patterns, the realization of a set of conduction paths between the output node and the V_{SS} power supply node. A conduction path is activated when all of its control transistors are on. Conversely, a conduction path is blocked when at least one of its control transistors is off. For the whole gate, when one (or more) paths between the output node and the V_{SS} node is (are) activated, the output of the gate is at V_{SS}, i.e., logical state 0, while, when all conduction paths are blocked, the output of the gate is at V_{DD}, i.e., logical state 1.

Opens Depending on its location, an open in the switch-like network corresponds to the removal of one or more conduction paths. In order to detect such an open, two conditions are required: (1) activate at least one of the conduction paths between the output node and the V_{SS} node which connects to the open line, and (2) block all the conduction paths between the output node and the V_{SS} node not connected to the open line.

When these conditions are fulfilled, if the considered open is not present, at least one conduction path is really activated and the output of the gate is at logical state 0. Otherwise, if this open is present, no conduction path is activated and the output of the gate remains at logical state 1. The potential presence of the open is indeed observed at the output of the gate.

In order to derive a test sequence able to detect all possible opens in the switch-like network, a systematic procedure deduced from the general graph theory is given in Galiay (1978). It consists of first listing all the conduction paths between the output node and the V_{SS} node and then successively activating one of these conduction paths and simultaneously blocking all the others. For example, Table 8.2 gives a set of five test vectors obtained for the gate of Fig. 8.3a.

The test sequence obtained with such a systematic procedure is generally redundant, but minimizing its length is only feasible if we have information about the actual layout of the control transistors. For instance, if the layout is exactly the one of Fig. 8.3a, only the three tests T2, T3, and T5 are required to detect all opens in the switch-like network.

Table 8.2 Test sequence for opens in gate of Fig. 8.3a

	a b c d e f	Actived path
T1	1 0 1 0 0 –	a c
T2	1 0 0 1 0 –	a d
T3	0 1 1 0 0 –	b c
T4	0 1 0 1 0 –	b d
T5	0 0 – – 1 1	e f

Table 8.3 Set of test vectors detecting short #2 of Fig. 8.3a

a	b	c	d	e	f
1	0	0	0	0	1
0	1	0	0	0	1
1	1	0	0	0	1
0	0	1	0	1	0
0	0	0	1	1	0
0	0	1	1	1	0

Shorts A short between any couple of nodes in the switch-like network corresponds to the creation of one or more conduction paths. Two conditions are required to detect a short between nodes i and j: (1) activate at least one conduction path between the output node and the i (respectively, j) node and at least one conduction path between the j (respectively, i) node and the V_{SS} node, and (2) block all other conduction paths of the network.

When these two conditions are realized, if nodes i and j are shorted, the output node is at logical state 0. If there is no short, the output node is at logical state 1: the potential presence of the short is observed at the output of the gate. For a given short, usually, several test vectors enabling its detection exist. For instance, Table 8.3 gives the set of six tests enabling detection of short #2 of Fig. 8.3a. To obtain a complete test sequence for all shorts of the switch-like network, it is first necessary to determine in this way the set of test vectors for each of them, and then to search for a minimal coverage enabling detection of all these shorts. If n is the number of nodes in the network, this minimal cover presents a maximum number of $\binom{n}{2}$ vectors.

8.2.2.3 Test Sequence Generation at the Block Level

Two specific problems can be identified for testing a block interconnecting several gates.

Controllability and Observability Assuming that a complete test sequence has been determined for each gate of the block, we must now apply to each gate its test sequence even if not all inputs are easily controllable and/or observable. This can be done by using a path sensitization method based on *propagation* to primary outputs of the block and *consistency* according to primary inputs (Roth et al. 1978).

Additional Failure Modes Specific to the Block Structure We address in sequence the problems related to opens and shorts:

a. In addition to the opens in the gates, the block structure introduces opens in interconnections between gates. Such an interconnection always connects the output of a gate to the gates of one or more control transistors of other gates. The open of such a connection thus leads to a floating gate potential of the transistors that are located after the open. In static operation, the leakage current is sufficient to

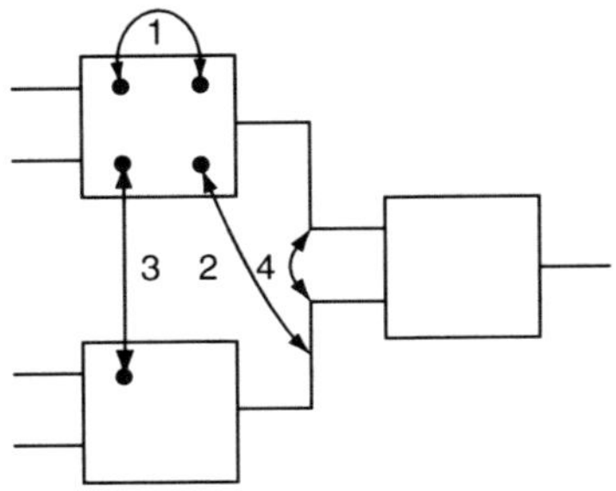

Fig. 8.5 Shorts in a block structure

set that potential to V_{SS}, and the transistors concerned are therefore always off. The open will thus be detected while testing the gates including these transistors if a low testing rate is assumed.

b. The problem concerning shorts is more difficult. Figure 8.5 represents all types of shorts occurring in the block structure. In addition to the shorts inside a gate (1), there are shorts between a connection inside a gate and the output of this or another gate (2), shorts between two internal connections belonging to two different gates (3), and shorts between the outputs of two different gates (4).

Each of these shorts involves a specific erroneous behavior (introduction of asynchronous sequential loop, modification of the function, introduction of analog behavior, etc.). To determine suitable test sequences, these errors must be individually analyzed and, as the number of possible shorts is very high in LSI circuits, this requires a very large amount of work.

Accordingly, it is more realistic to simplify the problem by reducing the number of potential shorts and preferably removing those that are the most difficult to test, even if this leads to a chip area increase. Such an approach (see Section 8.2.3) is based on restricting the circuit layout by a set of rules concerning arrangement of the gates at the block level and of nodes at the gate level.

8.2.3 Improvement of Circuit Testability

The layout rules aimed at improving circuit testability are based on a set of failure hypotheses that we will first justify by an analysis of the manufacturing process of the technology used for the application circuit.

8.2.3.1 Manufacturing Process Analysis

Basically, a MOS chip consists of three interconnection levels (assuming an aluminum gate): (1) a lower level made by diffusions in an insulating substrate, (2) an upper level made by metallizations, and (3) a medium level of oxide insulating the two previous levels and presenting two kinds of discontinuities: holes enabling contact between a metallization of the upper level and a diffusion of the lower level and thindowns corresponding to transistor gates.

The realization of diffusions and metallizations requires selective masking of very precise regions on the surface of the chip. The inherent failure mode of such a process consists of diffusing (or etching) regions that do not have to be diffused (or etched) or vice-versa. On a manufactured chip, such defects involve shorts between diffusions or metallizations and opens of diffusions or metallizations, only.

The growth of thick and thin oxide levels uniformly over the whole surface of the chip implies homogeneous levels with relatively few defects. The thick oxide constitutes a very good insulator between the diffusion level and the metallization level and, as a consequence, shorts between a metallization and a diffusion are very unlikely. Concerning the thin oxide, two kinds of defects can occur: local thindowns enabling breakdown by electrostatic discharge, and local contamination involving threshold voltage drift for the corresponding transistor. Electrostatic breakdown affects mainly input and output buffers. It can be detected by parametric testing, and more rarely affects transistors in the middle of the chip. Threshold voltage drift is a gradual aging phenomenon and implies, at the logical level, that the concerned transistor is either off or on. This has the same effect as an open of the drain or source diffusion or a short between these two diffusions.

Finally, for pin holes in the oxide, the only possible failure consists of a bad contact between the diffusion and the metallization and acts as an open of one of these connections.

Accordingly, an incomplete, but satisfactory fault coverage will be ensured for monochannel MOS integrated circuits when considering the following two failure assumptions: (A1) all possible defects consist of opens of diffusions or metallizations and shorts between two adjacent diffusions or metallizations, and (A2) No short can involve a metallization and a diffusion.

It is worth noting that this evaluation qualitatively agrees with the experimental results described in Section 8.2.1.

8.2.3.2 Rules for Improving Testability

Block level. Using the failure assumptions defined above, we will first define five layout rules governing the relative arrangement of gates. Rule 1 is based on failure assumption A2. It aims at avoiding shorts between an internal connection between the gates (short # 2 in Fig. 8.5).

R1: Make all internal gate connections entirely with diffusions and all interconnections between gates entirely with metallization.

Rules 2 and 3 are intended to control the short possibilities between two internal connections of two different gates (short #3 in Fig. 8.5).

R2: Arrange all internal gate connections (which are made by diffusions according to rule 1) inside a domain whose external limits are either the output diffusion or a V_{SS} diffusion of this gate (the latter diffusion can be commonly used by two adjacent gates).

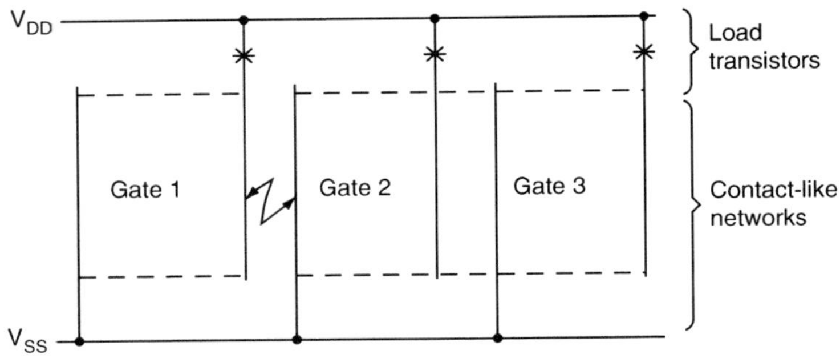

Fig. 8.6 Rules at block level

R3: Arrange any two adjacent gates to ensure that, if the output diffusion of the one constitutes one of its external limits, then it adjoins a V_{SS} diffusion of the other gate.

Figure 8.6 illustrates rules 2 and 3.

All connections of each gate are enclosed in a domain bounded by the vertical lines representing diffusions and the horizontal dotted lines. With such a layout, the only possible short between two gates, for instance, gates numbered 1 and 2 on the figure always involve the output diffusion of the first and a V_{SS} diffusion of the second. This short can be modeled by a stuck-at-0 fault of the output of the former.

Rules 4 and 5 are intended to control the short possibilities between the outputs of two different gates (short #4 in Fig. 8.5).

R4: Arrange the gates of the block along a given direction in increasing level order according to their logic level in the block.

R5: Arrange the interconnections between the gates (which are realized by metallizations according to **R1** to avoid shorts that can introduce asynchronous sequential loops. For instance, given three interconnections A, B, and C, if a short between A and B leads to a loop and shorts between A and C on one hand and B and C on the other hand do not lead to a loop, then A and B can be isolated by placing C between them. As a practical matter, this rule is not very systematic, and therefore is not as easy to apply as **R4**.

Gate Level We now describe rules for arrangements of equipotential lines within a gate. First, remember that, according to **R1** all connections in a gate are made by diffusions.

For **opens**, as reported in Section 8.2.1, most of the opens in a complex gate can be modeled by stuck-at faults of either the output or one or more inputs of that gate. For instance, for the gate of Fig. 8.3a, only open #4 cannot be modeled by any stuck-at fault because it leads to a modification of the logical function of the gate. Several solutions are applicable for the layout. Figure 8.7 shows two of these possibilities for the part of the gate related to inputs a, b c and d. With the first solution, open

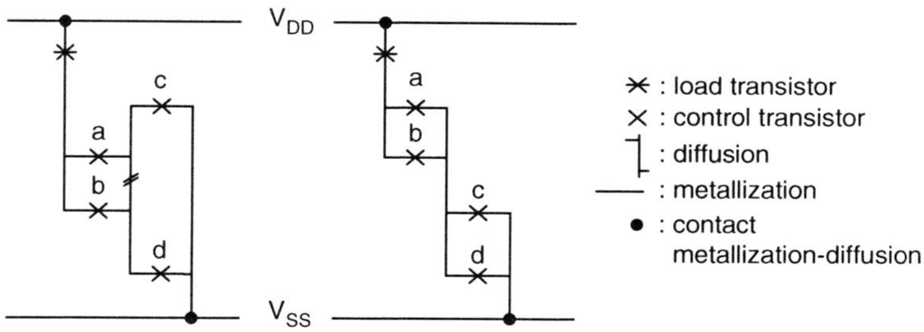

Fig. 8.7 Rules for opens at gate level

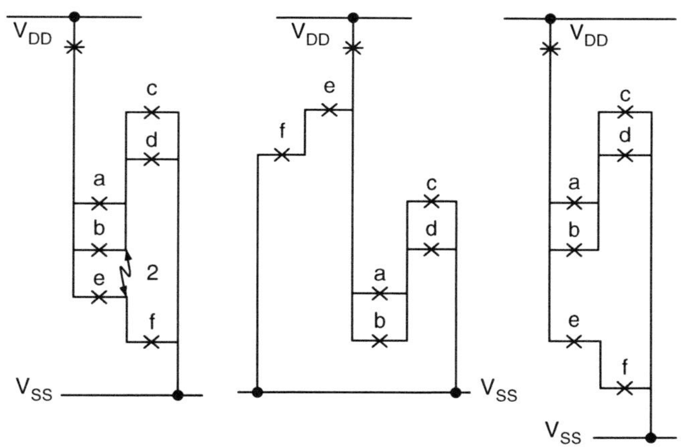

Fig. 8.8 Rules for shorts at gate level

number 4 is physically possible but with the second this open is no longer feasible and, in this latter case, all opens can be modeled by stuck faults of either the output, or one or more inputs of the gate.

A similar reasoning applies to the only type of **short** that cannot be modeled by a stuck-at fault (i.e., short #2, for the gate of Fig. 8.3a). Among the three layout options shown in Fig. 8.8, the first enables such a short to occur, while the other two prevent its occurrence either because the diffusions concerned are separated by the output diffusion or simply because they are more separated physically. It is also worth pointing out that for the two latter layouts, any short can also be modeled by a stuck-at fault of either the output, or one or more inputs.

According to the whole set of rules concerning the connections within the gates therefore consist of arranging these connections so that any defect can be modeled by a stuck-at fault.

8.2.3.3 Test Strategies

According to whole set of rules that has been defined, two different test strategies can be considered:

S1 Apply **only** the layout rules related to the arrangement of the gates.

In this case, the possible defects of a block are: (1) shorts and opens inside the gates that cannot all be modeled by stuck-at faults, (2) opens of gate interconnections that can all be modeled as stuck-at faults, and (3) shorts between two gate interconnections that cannot induce asynchronous sequential loops and can be modeled by OR wiring of the two shorted connections.

The determination of a test sequence for the whole block can thus be divided into three steps:

1. Determine a complete test sequence for each gate of the block with the method described in Section 8.2.2.2 using the transistor diagram of this gate.
2. Determine a test sequence at the level of the whole block, applying to each gate its own test sequence. This can be done by using a path sensitization method on the logic diagram of the block. Notice that this sequence also detects all interconnection opens.
3. The test sequence obtained after step 2 will generally also detect some shorts between interconnections. For each undetected short, a specific test vector must be added to that sequence. Using the logic diagram of the block, this test vector can be derived by first setting anyone of the shorted interconnections at logical state 0, the other at logical state 1, and then propagating the logical state of the latter interconnection to an observable output of the block.

S2 Apply **all** rules concerning both the arrangement of the gates and the arrangement of the connections inside the gates.

The main difference with respect to **S1** is that all shorts and opens inside the gates can be modeled by stuck-at faults. This offers the advantage of saving step 1 in the previous testing procedure, thus reducing computation time and avoiding the need for the transistor diagram. Generating a test sequence for **S2** can therefore be achieved simply in two steps:

1. Determination on the logic diagram of a test sequence for stuck-at faults of the connections of that diagram.
2. Same as step 3 for **S1**.

8.2.3.4 Application

The application of these layout rules involves, for a given function, an increase of the chip area. To evaluate this approach, these layout rules were applied to a set of basic building blocks of the application circuit. There is a significant difference between blocks with a relatively regular layout, such as programmable logic arrays (PLA), and blocks with a random layout.

For PLAs, due to the array nature, the gate arrangement is fixed and concurs with all the layout rules that we have defined. For such blocks, it is therefore easy to determine test sequences with satisfactory fault coverage.

The situation is more difficult for random logic. Although it is always feasible to apply all rules, the relative freedom offered by some of them means that, for a given function, several possible layouts are possible, each with different area. Determining the minimal one is therefore a complex task. In the years when the work was conducted, no design tools able to provide an automatic solution to such an optimization problem were available, and the search for a solution was essentially based on designer's skills. Thus, to reduce the design periods required for a circuit, two methods can be retained, depending on the nature of the block. For the specific blocks appearing only on one particular chip (custom design), one can systematically apply all layout rules without optimization. Conversely, for the basic blocks appearing on many chips (flip-flops, registers, counters, RAM, etc.), an optimization can lead to a block library that will then be used by the designer for future circuits.

These general remarks can be illustrated by a particular block in the application circuit: a master-slave flip-flop used in the realization of registers or counters. Because at that time, conventional layout rules did not consider testing requirements, no preestablished test sequence can be formed for 30% of the shorts between two metallizations and for 28% of the shorts between two diffusions. This situation arises from the fact that these defects introduce analog behaviors. Systematic application of all layout rules leads to a design for which all shorts and opens can be detected by a very simple test sequence (Set-Reset-Set), but with an area increase of 30% after optimization.

This area increase does not, however, apply to the whole circuit because at a chip level we have to account for the following factors: (1) the I/O buffers and the interconnections take up a non negligible area which is not influenced by the application of these rules, and (2) certain blocks (e.g., PLA) have an arrangement such that the respect of the layout rules leads to a very slight or even negligible increase.

8.2.4 Modeling of Errors Induced in Operation

We now address defects acting during the operation of the chip. In this case, the single defect assumption is a realistic one. Thus, we focus on studying, in isolation, the error induced by each type of defect, such as: shorts, opens, threshold voltage drift and degradation of propagation time.

As previously depicted by Fig. 8.3a, a MOS single channel gate can be represented by a load transistor and a switch-like network arranged between the two power supply lines: V_{DD} and V_{SS}.

8.2.4.1 Shorts

Each short that can affect the switch-like network (including the output connection) increases the number of conduction paths and thus creates an error at 0 of the output of the gate. This error is transient or permanent depending whether the short is resistive or not. The error is of the single type if the short does not affect more than two gates. It is unidirectional if it affects more than two gates. A short of the load transistor decreases the load resistance and degrades the logical level 0 which in turn may produce a transient or permanent error at 1 whether the short is clean or not.

8.2.4.2 Opens

An open provokes a dual effect when compared to a short. Thus, an open in the switch-like network leads to a permanent error at 1 and an open of the load transistor leads to a permanent error at 0.

An open of a connection between gates leads to a floating potential on the gates of the control transistors located below; the induced transient error is unidirectional if the cut connection is a divergent one.

Opens of the supply lines V_{DD} or V_{SS} lead respectively to a permanent stuck-at-0 (s-a-0) or stuck-at-1 (s-a-1) of the outputs of the gates and give a permanent unidirectional error at 0 or 1.

8.2.4.3 Threshold Voltage Drift

A threshold voltage drift renders any action on the transistor gate ineffective. The error induced by this fault is a 0 or 1 error whether the threshold voltage is decreased or increased. Such a kind of fault can simultaneously affect several transistors in the same way and thus can create a single (or unidirectional) error(s) depending whether the transistors belong to the same gate or not.

Such faults can be due to external disturbances such as temperature or power supply variations. In this case, the resulting error is transient and its duration is equal to the duration of the external disturbance.

8.2.4.4 Degradation of Propagation Time

As for the threshold voltage drift, the degradation of the propagation time can affect several transistors and can be induced by external phenomena. The main difference is that the error in this case depends on the variable generated by the faulted gate. This error multiple if the gates impacted by the failure generate complementary logic variables.

Table 8.4 Characterization of defects/faults according error classes and error types

Error	Permanent	Transient
Single	- Short or open in switch-like network - Open of load transistor - Clean short between two gates	- Open of connection between gates - Threshold voltage drift of one or more transistors inside one gate
Unidirectional	- Open of supply line - Short between more than two gates	- Threshold voltage drift of one or more transistors inside one gate
Multiple	- Resistive short between two gates leading to an undefined logic level - Degradation of propagation time	- Degradation of propagation time

8.2.4.5 Synthesis

Summing up, the faults can be divided into three classes depending whether they induce single, unidirectional or multiple errors. Each of these classes can also be further divided into two groups: permanent and transient errors. Table 8.4 characterizes the defects/faults according to these error classes and types.

8.2.5 Constraints for Designing Self-Checking Circuits

Among the different properties that a self-checking (SC) IC can possess, we study here the constraints related to the *fail-safe* property. This property, like the others, is based on the use of detecting codes.

We say that a circuit using a detecting code C is fail-safe if, for all the faults belonging to the class considered and whatever the input pattern applied, the output is correct or the output does not belong to the code C.

The failure of the circuit is signaled by the means of a checker verifying that the output belongs to the code C. This checker must also be fail-safe itself with a coded output; in general the used code is a two-rail 1-out-of-2 code (two variables taking complementary values during normal operation). The use of an error detection code enables an immediate on-line testing and signaling of any incorrect operation of a circuit independently of the permanent or transient nature of the error. Thus, in the following, we no longer make any distinction between the temporal characteristics (permanent or transient) of the errors induced. We describe in turn hereafter the implementation rules related to the detection techniques of the three classes of errors: defects inducing single, unidirectional and multiple errors.

8.2.5.1 Implementation Rules for Detecting Single Errors

For detection techniques targeting single errors, the main functional constraint is that the various outputs of the circuit should be produced by independent circuits (slices), i.e., circuits that have no common link except possibly input connections. Such a constraint enables the detection of all the faults that induce single errors, only.

A set of implementation rules enables the detection of opens of interconnections or of supply lines that can produce unidirectional errors when they are shared by more than one output. These rules concern the delivery of a common signal to several slices (common variables, power supplies).

They can be summarized as follows:

R1′: Check the signal;

R2′: Distribute the signal in such a way that an open only affects one slice or if it affects more than one slice it affects also the checker (no supply to the checker means that the two outputs are at the same value which corresponds to the detection of an error).

In Fig 8.9, we illustrate the two main alternatives. Figure 8.9a depicts the use of a splitting node and Fig. 8.9b describes the use of a main line with the checker located at the physical end of this line. In the latter case, the divergences are only allowed if they supply several gates inside the same slice.

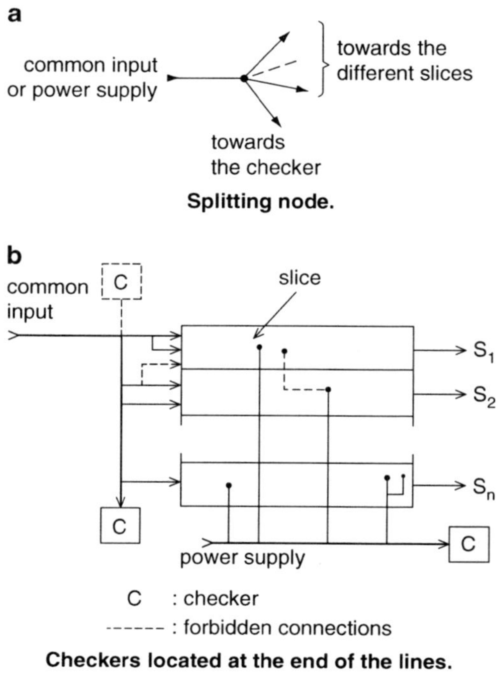

Fig. 8.9 Main alternatives for single errors (**a**) splitting node. (**b**) checkers located at the end of the lines.

8.2.5.2 Implementation Rules for Detecting Unidirectional Errors

To make the detection of all the unidirectional errors feasible, the implementation of the circuit should be inverter-free. This is impossible with MOS technology, because all basic gates are inverting ones. Thus, unidirectional errors internal to the circuit can induce multiple errors at the output.

As for single errors, the detection efficiency can be improved by means of implementation rules, mainly targeting the supply lines. Using of the same principle as the one proposed for single errors, it is possible to guarantee the detection of all unidirectional errors induced by an open of a supply line. Conversely, as there exists no means of telling which gates can be affected by a threshold voltage drift, it is impossible to detect all the unidirectional internal errors induced by such a fault as they can finally produce a multiple error at the outputs of the circuit.

8.2.5.3 Implementation Rules for Detecting Multiple Errors

The detection of multiple errors is based on the use of the duplex paradigm, i.e., a structure made of two identical units performing the same task. With such a structure the detection of multiple errors affecting one of the two units is only ensured if the two units are fault independent. For preventing a design fault (over-loaded gate inducing a bad noise immunity) or a manufacturing defect to simultaneously affect both units, it is desirable for the two units to be *diversified* (distinct implementations, one unit realized with normal logic and the other with complementary logic (Crouzet et al. 1978; Crouzet and Landrault 1980).

When the two units are rigorously similar it is necessary to separate as much as possible during the implementation those elements that have the same function in the two units: so that a local degradation will not affect these elements.

As for the two previous cases, it is necessary that all opens of a supply line do not affect both units without impacting the checker.

8.2.6 Concluding Remarks

It is recognized that the results presented are specific of the proposed example and IC technology. However, regardless of this particular technology, one can retain the proposed procedure and reproduce it for any circuit realized with any technology. In that respect, note that Wadsack (1978) deals with fault modeling for the CMOS technology.

To test a circuit, the first step must include an analysis of the failure mechanisms of this circuit to obtain information about their nature and their probability. Then, to facilitate test sequence generation, it is essential to derive a general model rather than to individually consider all types of defects. However, as manufacturing processes become more and more sophisticated, it appears that the stuck-at model, very often used because of its practical interest, will cover a more and more reduced part of the defect modes. One can thus adopt two different approaches: the first consists of defining a specific test generation method taking directly into account the defects of the circuit, and the second consists of submitting the layout of the circuit to a set of rules in order to cover all the defects by the stuck-at fault model. As the first solution generally leads to very great complexity, the second one appeared more realistic for most cases, although it implies layout constraints and an increase in chip area. The conducted study showed that this second approach appears to be quite efficient.

As for improving the efficiency of testing procedures based on the stuck-at model, several implementation rules have been derived at the level of fail-safe circuits, which can greatly improve the efficiency of the on-line testing techniques and thus increase the percentage of detected faults.

These rules naturally lead to an increase of the surface area occupied by the circuit that is not possible to precisely evaluate in advance. However, due to the fast evolution at the integration level, we had anticipated that this increase should not be a great handicap as it could be easily envisaged for many current and future circuits.

8.3 Fault Models and Fault Tolerance Testing

For almost 40 years, many successful efforts were reported on the use of fault injection for contributing to the assessment of fault-tolerant systems, sometimes in cooperation with other dependability validation techniques (e.g., formal verification or analytical modeling). Building on these advances, fault injection made progressively its way to industry, where it is actually part of the development process of many manufacturers, integrators or stakeholders of dependable computer systems (Benso and Prinetto 2003). This confirms the pertinence of the approach.

Nevertheless, one key concern that is often related to fault injection-based experiments is usually termed as *fault representativeness*, i.e., the plausibility of the supported fault model with respect to real faults (Gil et al. 2002). The investigations carried out concerning the comparison of the impact of (1) specific injection techniques with respect to real faults, e.g., see Daran and Thévenod-Fosse (1996); Durães and Madeira (2006), and (2) several injection techniques, e.g., see Stott et al. (1998), Folkesson et al. (1998), Moraes et al. (2006), have shown mixed results. Some techniques demonstrated to be quite equivalent, while others were rather complementary. The fault representativeness issue remains therefore a concern and is still a matter of research.

In this context, the goal of this section is fourfold: (1) introducing a conceptual frame characterizing the notion of fault injection (Section 8.3.1), (2) briefly describing the main fault injection techniques, with an emphasis on techniques suitable to target physical faults (Section 8.3.2), (3) discussing the pertinent criteria to assess the extent to which injection techniques are suitable to induce erroneous behaviors that are representative of the consequences of the activation or occurrence of real physical faults (Section 8.3.3), (4) summarizing the results of a comprehensive study, aimed at comparing four injection techniques (Section 8.3.4). Finally, Section 8.3.5 concludes this part by providing some additional insights derived from the study.

8.3.1 Some Rationale About Fault Injection

The successful deployment of a dependable computing system heavily relies on various forms of hardware and/or software redundancies that are aimed at handling faults/errors, i.e., which embody the fault tolerance features of the system. A large number of studies (both theoretical and experimental) have shown that the adequacy and the efficiency, i.e., the *coverage* (Bouricius et al. 1969), of the fault tolerance mechanisms (FTMs) have a paramount influence on the dependability and in particular on the measures (reliability, availability, etc.) usually considered for assessing the level of dependability actually obtained.

For a pragmatic and objective assessment of the coverage of the FTMs, it is essential to be able to test them against the typical sets of "inputs" they are a meant to cope with: the faults and resulting errors; hence, the rationale for applying test sequences consisting in fault injection experiments. Moreover, the difficulty in accurately modeling/simulating the erroneous behaviors of a complex computing system sustain the need of relying on experimental techniques in complement to more formal approaches. Moreover, the scarcity of the fault events prevents from relying on the natural occurrence of faulty conditions: controlled experiments that speed-up the occurrence of errors are needed.

Fault injection, i.e., the deliberate introduction of faults into a system (the target system) is applicable every time fault and/or error notions are concerned in the development process. Classically, fault injection testing is based on the design and realization of a *test sequence.* More precisely, a fault injection test sequence is characterized by an *input* domain and an *output* domain (Arlat et al. 1990).

8.3.1.1 The FARM Attributes

The input domain ***I*** corresponds to a set of injected *faults* ***F*** and a set ***A*** that specifies the data used for the *activation* of the target system and thus, of the injected faults. Both ***F*** and ***A*** are the lever to provoke errors suitable to exercise the FTMs.[2] The output domain ***O*** corresponds to a set of *readouts* ***R*** that are collected to characterize the target system behavior in presence of faults and a set of *measures* ***M*** that are derived from the analysis and processing of the ***FAR*** sets. Together, the ***FARM*** sets

[2] Recent work oriented towards the development of (fault injection-based) dependability benchmarks (e.g., see Kanoun and Spainhower 2008) has adapted the notions attached to the ***A*** and ***F*** domains to the ones of *Workload* and *Faultload*, respectively.

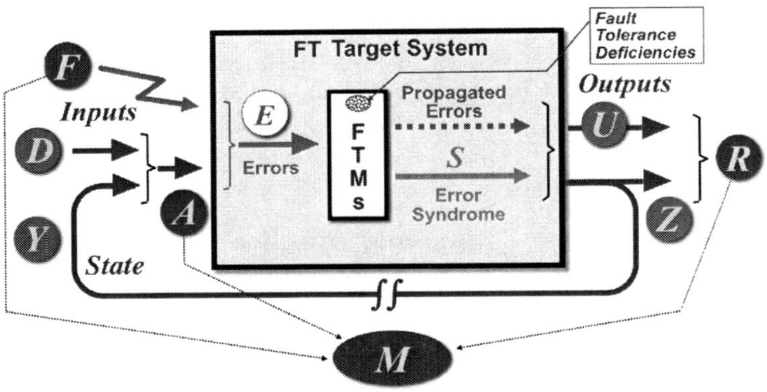

Fig. 8.10 The fault injection attributes and the fault-tolerant target system

constitute the major attributes that fully characterize a fault injection test sequence. In practice, the fault injection test sequence is made up of a series of *experiments*; each experiment specifies a point of the ***FxAxR*** space.

Figure 8.10 exemplifies these notions and further details them, in particular, to illustrate how the attributes relate to the state space of the target system (Mealy-style state machine). Indeed, the ***A*** set encompasses the primary ***D*** and secondary (current state) ***Y*** inputs. The **A** and ***F*** sets fully characterize the input domain ***I*** and combine together to induce errors that are the patterns meant to test the FTMs. An additional insight shown relates to the fact that the output domain ***O*** extends to the primary ***U*** (delivery of functional service to the users) and secondary ***Z*** (next state) outputs. Note also the explicit observation, as part of ***R***, of the error signaling (syndrome) provided by the FTMs when subjected to the error patterns. The figure also identifies deficiencies in the FTMs: incapacities in handling some error situations. Such "fault-tolerance deficiencies" are the target of the fault injection testing experiments.

8.3.1.2 Modeling the Fault Pathology

The behavior of the target system can be described by a sequence of states characterized by a function linking these extended attributes as $\phi\,(\boldsymbol{I}) = \boldsymbol{O}$, *with* $\boldsymbol{I} = \{\boldsymbol{F}, \boldsymbol{D}, \boldsymbol{Y}\}$ and $\boldsymbol{O} = \{Z, U\}$ (Arlat et al. 1990). To account for discrepancies in *value* and *time*, we also consider the time dimension t. For the sake of brevity, the system function $\phi\,(d, y, f; t)$ can be decomposed according to the output domain sets as $\phi_z\,(d, y, f; t) = z\,(t+1)$ and $\phi_u\,(d, y, f; t) = u\,(t+1)$.

For example, the impact of a fault vector at time t (denoted $f\,(t)$) can be perceived when the fault is activated:

$$\forall t, \exists d(t) \text{ and/or } y(t) \; such\;that\; \phi\,(d, y, f; t) \neq \phi\,(d, y, f_0; t) \qquad (8.1)$$

where f_0 (t) designates the vector "absence of fault".

This activation corresponds to the deviation from the nominal trace:

– either as an internal *error* when only the state vector $\boldsymbol{Z}$ is altered

$$\phi\,(d,\ y,\ f\ ;t) = \left(z',u;t+1\right) \neq (z,u;t+1) \tag{8.2}$$

where $z'\,(\bullet)$ denotes an internal state distinct from the nominal one;

– or as an error impacting the service delivered when the vector from U is also altered (which thus corresponds to the *failure* of the target system):

$$\phi\,(d,\ y,\ f\ ;t) = \begin{cases} (z,u';t+1) \neq (z,u;t+1) \\ (z',u';t) \neq (z,u;t+1) \end{cases} \tag{8.3}$$

where $u'\,(\bullet)$ denotes an output distinct from the nominal one $u\,(\bullet)$.

This modeling frame is also useful to describe the *equivalence* of the impact on the behavior caused by a fault and by an erroneous state, as follows:

$$\phi\,(d,y,f;t) = \phi\left(d,y',f_0;t\right) \tag{8.4}$$

Another useful refinement is related to the fact that the evolution of a system does not depend at any time on all its internal states. This leads to make a partition of the state sets $\boldsymbol{Y}$ and $\boldsymbol{Z}$ that distinguishes:

– $\boldsymbol{Y_d}$ and $\boldsymbol{Z_d}$ the dynamic part, characterizing the state variables that actually impact the evolution of the behavior of the system at time t;
– $\boldsymbol{Y_s}$ and $\boldsymbol{Z_s}$ the static part, including the variables that are not sensitized at time t.

Such a distinction is useful in practice to account for dormant faults and latent errors. In particular, it essential to describe the evolution of the erroneous behavior caused by a transient fault after it has disappeared:

$$\phi(d,y_d,y_s,f;t) = \left(z_d,z_s',u;t+1\right) \Rightarrow \phi\left(d,y_d,y_s',f_0;t\right) = (z,u;t+1) \tag{8.5}$$

Clearly, dormant faults may not create erroneous behaviors and all erroneous states do not necessarily cause a failure. This has a direct impact on the *controllability* for the definition of the fault/error injection method to produce an *error* set suitable to sensitize the FTMs and on the *observability*, in particular with respect to the control of the activation of the injected fault as an error and of the subsequent errors induced by its propagation. Moreover, it is helpful for the design and implementation of the fault-tolerant system since in practice it is not necessary neither to observe nor to recover all system's states, which is especially important for the observation of the reaction of the target system in presence of injected faults.

As another example, let us consider the case of an error detection mechanism (EDM). The detection is only possible when an error is activated. It is based either on the direct observation of an alteration of the dynamic state:

$$\forall t, \exists (d, y; t) : \phi_z (d, y, f; t) = \left(z'_d, z_s, u; t + 1\right) \tag{8.6}$$

or via the explicit sensitization (e.g., via a specific test program) of a an erroneous static state and on the observation of the resulting modification of the dynamic state:

$$\forall t, \exists (d, y; t) : \phi_z (d, y, f; t) = \left(z'_d, z'_s, u; t + 1\right) \tag{8.7}$$

8.3.2 *The Fault Injection Techniques*

Numerous injection techniques have been proposed (Benso and Prinetto, 2003), ranging classically from (1) simulation-based techniques at various levels of representation of the target system (physical, logical, RTL, PMS, etc.), (2) hardware-implemented techniques (*HWIFI*, for short), e.g., pin-level injection, heavy-ion radiation, laser injection, EMI, power supply alteration, etc., and (3) software-implemented fault injection (also known as *SWIFI*) techniques that are meant to corrupt the execution of a software program either at compile time (code mutation) or at run time. In particular, the latter supports the bit-flip model in register/memory elements. Many tools were developed to facilitate experiments based on these techniques.

Most of the work on fault injection focused on the injection of faults/errors intended to "mimic" the consequences of hardware faults (stuck-at, opens, bridging, logical inversion, bit-flips, voltage spikes, etc.). Only during the past decade, several efforts were devoted to the analysis of software faults. Indeed, besides the SWIFI technique was primarily targeting hardware faults, the erroneous behaviors that can be provoked by applying this technique can also simulate (to some extent) the consequences of software faults (Durães and Madeira 2006; Crouzet et al. 2006). A typical branch of work on this area concerns the investigation of dependability benchmarks aimed at characterizing the robustness of software executives, e.g., microkernels, OSs, middleware (Kanoun and Spainhower 2008). More recently, some studies addressed the analysis of cryptographic circuits with respect to malicious attacks targeting potential vulnerabilities including also side channels procured by scan chain test devices (Hély et al. 2005), as well as via fault injection applied to VHDL models (Leveugle 2007).

Due to the context of this book, we focus on typical techniques targeting hardware faults. Hereafter, we emphasize the four injection techniques – heavy-ion radiation, pin-level injection, electromagnetic interferences, as well as a compile-time SWIFI – that were applied in the multi-site cooperative work carried out in the late 1990s in the framework of the ESPRIT PDCS project. The objective was

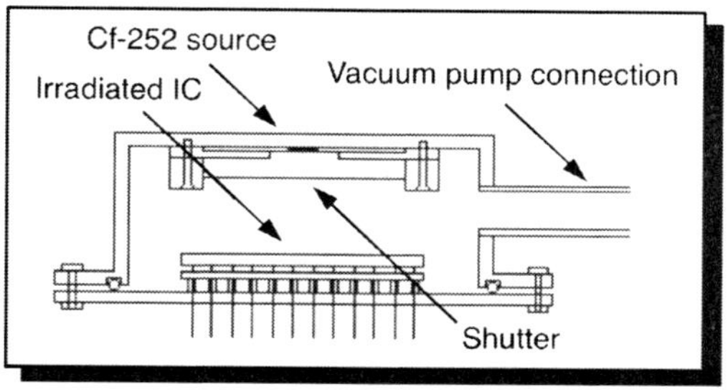

Fig. 8.11 Cross-sectional view of the miniature vacuum chamber

to compare these techniques by running experiments on the same testbed architecture and a common test scenario. The results of these experiments are presented in Section 8.3.4.

8.3.2.1 Heavy-Ion Radiation

The fault injection experiments with heavy-ion radiation (HI, for short) were carried out at Chalmers University of Technology in Göteborg, Sweden. A Californium-252 source can be used to inject single event upsets, i.e., bit-flips at internal locations of a target IC using a miniature vacuum chamber. Figure 8.11 depicts the cross-sectional view of the miniature vacuum chamber. The pins of the target IC are extended through the bottom plate of the vacuum chamber, so that the chamber with the circuit can be directly plugged into the socket of the circuit under test. The vacuum chamber contains an electrically controlled shutter, which is used to shield the circuit under test from radiation during bootstrapping.

A major feature of the HI injection technique is that faults can be injected into VLSI circuits at locations that are difficult (and mostly impossible) to reach by other techniques. The transient faults produced are also reasonably well spread at random locations within an IC, as there are many sensitive memory elements in most VLSI circuits. As device feature size of integrated circuits is shrinking, radiation induced bit-flips, also known as *soft errors*, constitute an increasingly important source of failures in computer systems (Baumann 2005). For the target IC (the 68070 CPU, see Section 8.3.4.1), the heavy-ions from Cf-252 mainly provoke single bit upsets. The percentage of multiple bit errors induced in the main registers was found to be less than 1% in the experiments reported in Johansson (1994).

8.3.2.2 Pin-Level Fault Injection

The experiments with the pin-level fault injection technique were conducted at LAAS-CNRS, in Toulouse, France using the MESSALINE tool. Figure 8.12 depicts the principle of the pin-forcing technique (PF). In this case, the fault is directly applied on the pin(s) of the target IC.

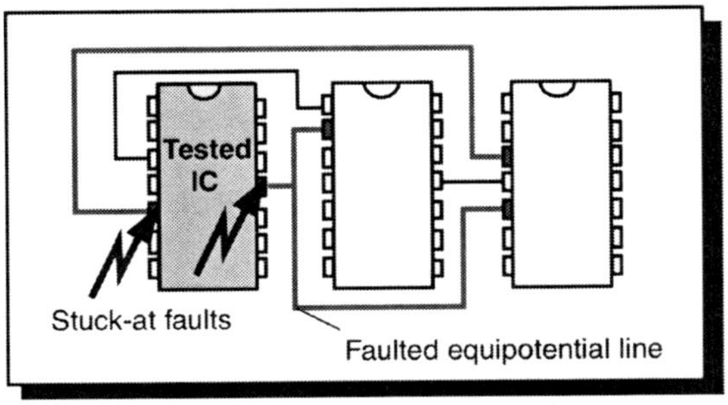

Fig. 8.12 Principle of pin-forcing fault injection

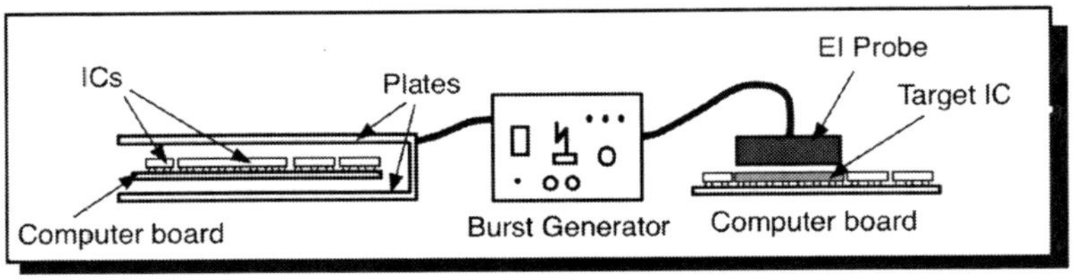

Fig. 8.13 Application of electromagnetic interferences

It is noteworthy that the pins of the ICs connected, by means of an equipotential line, to an injected pin are faulted as well. Accordingly, to simplify the accessibility to the pins of the microprocessor, the target ICs were mainly the buffer ICs directly connected to it. The supported fault models include temporary stuck-at faults affecting single or multiple pins. Indeed, temporary faults injected on the pins of the ICs can simulate the consequences of internal faults on the pins of the faulted IC(s).

8.3.2.3 Electromagnetic Interferences

Electromagnetic interferences (EI) are common disturbances in automotive vehicles, trains, airplanes, or industrial plants. Such a technique is widely used to stress digital equipment.

These experiments were carried out at the Vienna University of Technology, Austria. Thanks to the use of a commercial burst generator this technique is easy to implement.

Two different forms of application of this technique were considered (Fig. 8.13). In the first form, the single computer board of the target MARS node (see Section 8.3.4.1) was mounted between two metal plates connected to the burst generator. In this way, the entire node was affected by the generated bursts.

Because the Ethernet transceivers turned out to be more sensitive to the bursts than the node under test itself, a second configuration was set up, which used a special probe that was directly placed on top of the target circuit. In this way the generated bursts affected only the target circuit (and some other circuits located near the probe).

8.3.2.4 Software-Implemented Fault Injection

For these experiments, the compile-time version of SWIFI was selected: faults were injected at the machine code level and the mutilated application (code segment or data segment) was loaded to the target system afterwards. Two main reasons led us to select such an approach (Fuchs 1996):

1. The intrusiveness is reduced to a minimum, since faults are injected only into the application software (no additional code, which could probably interfere with the behavior of the application software, is needed).
2. Fault injection at the machine code level is capable of injecting faults that cannot be injected at higher levels by using source code mutations.

The SWIFI experiments started at the Vienna University of Technology, Austria and continued at the Research and Technology Institute of Daimler Benz AG (then DaimlerChrysler) in Berlin, Germany.

Both the code and data segments of the application software used as the workload for the experiments were targeted by the SWIFI technique. Within each segment, the bit to be faulted was selected randomly to achieve a uniform distribution over the whole segment. To facilitate the comparison with the HWIFI techniques, we only consider here the single bit-flip experiments, because they constitute a reasonable fault scenario for the comparison with these techniques (e.g., heavy-ion radiation generates, to a large extent, single bit-flips).

8.3.3 Representativeness with Respect to the F Set

In this section, we describe a general framework (Arlat and Crouzet 2002) that is meant to help address comprehensively the representativeness issue.

From a pragmatic viewpoint, the main objective is to identify the technology that is both *necessary* and *sufficient* to generate the F set to conduct a fault injection test sequence. Several important issues have to be accounted for in this effort.

8.3.3.1 System Levels and Fault Pathology

As shown in Fig. 8.14, several relevant levels of a computer system can be identified where faults can occur and errors can be identified (e.g., physical-device, logic, RTL, algorithmic, kernel, middleware, application, operation). Concerning faults, these levels may correspond to levels where real faults are considered and (artificial) faults can be injected. Concerning errors, the FTMs (especially, the error detection mechanisms, EDMs) provide convenient built-in monitors.

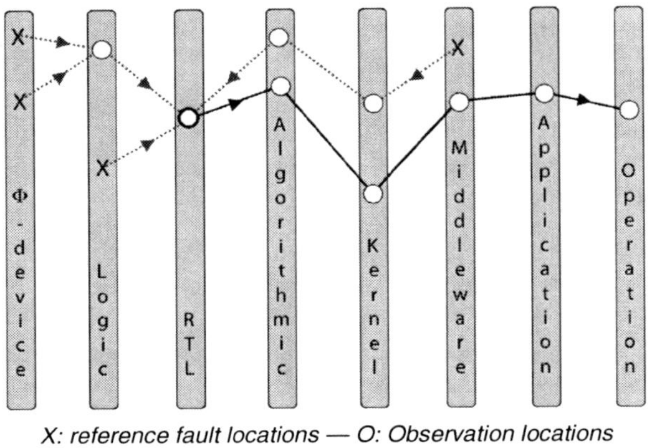

Fig. 8.14 Target system levels and fault pathology

8.3.3.2 Error Equivalences

For characterizing the behavior of a computer system in presence of faults, it is not necessary a priori that the injected fault be "close" to the target (reference) fault. It is sufficient that it induces similar behaviors. Similar errors can actually be induced by different types of faults (e.g., a bit-flip in a register or memory cell can be provoked by an heavy-ion or as the result of a glitch provoked by a software fault). What is important is not to establish an equivalence in the fault domain, but rather in the error domain (see expression 8.4 in Section 8.3.1.2).

8.3.3.3 Distances

What matters is that the respective error propagation paths *converge* before the level where the behaviors are observed. Two important parameters can be defined on these various levels (Fig. 8.15):

- the *distance* ***dr***, separating the level where faults are injected from the ***reference*** fault level(s);
- the *distance* ***do***, separating the level where the faults are injected from the levels their effects are ***observed***.

The shorter ***dr*** and the longer ***do*** are, the more likely the injected faults will exhibit behaviors similar to those provoked by the targeted reference faults.

8.3.3.4 Constraints on Error Propagation

In practice, the presence of a specific FTM may alter the error propagation paths. This has a significant impact on the scope of (real) faults actually covered by the

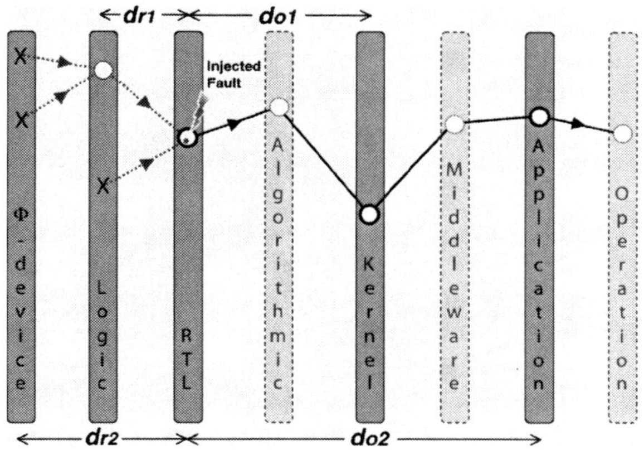

Fig. 8.15 Reference fault and observation distances

injected faults, whenever the FTM is implemented at a level located between the level of the targeted faults and the level where the faults are injected and thus intercepts the error propagation paths. Indeed, assuming a perfect (100%) coverage for the FTM, then representativeness (with respect to the targeted faults) would then be null. This could be simply accounted for by introducing another distance parameter: the distance ***dm*** separating the level where faults are injected from the level where the FTM is acting.

Another typical concern is whether the error propagation process is monotonously (usually upstream) directed or not; the first alternative prevents FTMs characterized by negative values of ***dm*** from being exercised. For example, the SWIFI technique can seldom be used to test hardware-implemented ECCs.

8.3.4 Comparative Study of Fault Injection Techniques

As already pointed out in Section 8.3.3, we advocate that the study of the impact and consequences of a fault (i.e., the error propagated) offers a pragmatic and sensible means to address the representativeness issue. Two categories of approaches can be distinguished, depending whether the analysis concerns the erroneous behaviors provoked by (1) some specific fault injection technique with respect to a set of real faults or (2) the concurrent application of several fault injection techniques.

The study considered hereafter – see Arlat et al. (2003) for additional details – refers to the second approach. While the comparison of several fault injection techniques provides only an indirect means for assessing their representativeness, such an approach is well suited to obtain extensive error data sets from which useful insights can be derived. Should the experiments using different injection techniques lead to similar behaviors, then the techniques can be considered as "equivalent", and thus the one that exhibits the most suitable practical properties (e.g., reachability, controllability, reproducibility, intrusiveness, etc.) should be preferred. However, if different behaviors are observed, then the techniques are rather complementary. Such an insight is very helpful in the light of the recent work devoted to developing dependability benchmarks (Kanoun and Spainhower 2008),[3] in particular to substantiate which kind of relevant "faultload" should be considered for such benchmarks.

The four techniques – heavy-ion radiation, pin-level injection, electromagnetic interferences, as well as a compile-time SWIFI – described in Section 8.3.2 were jointly applied and analyzed. It is worth noting that in order to carry out all the fault injection experiments on a consistent basis, we used the same distributed testbed architecture featuring five MARS[4] nodes and a common test scenario. The assessment of the fault injection techniques is supported by using the EDMs built-in into a MARS node as "observers" to characterize the erroneous behaviors induced by the faults injected by the techniques considered. The EDMs are meant to procure the MARS node with self-checking properties suitable to confer it a *fail-silent* behavior[5]. The experiments conducted were also aimed at assessing the extent to which this property was ensured.

In the sequel, we briefly present the relevant features of the target system and of the testbed. Finally, some typical results are presented and the main insights gained are summarized.

8.3.4.1 The Target System and the Testbed

We focus here on the fault tolerance feature of a special-purpose processing node designed to support the *fail-silent* property. Special attention is paid to the identification and characterization of the error detection mechanisms (EDMs) built-in into a MARS node.

The MARS Processing Node This study uses a single-board implementation of the MARS node. More details on MARS features and on the architecture of the processing nodes can be found in Reisinger et al. (1995). Each node consists of two independent processing units: the application unit and the communication unit. Each unit is based on a 68070 CPU, featuring a memory management unit (MMU). The application unit also contains a dynamic RAM, and two bidirectional FIFOs, one of which serves as an interface to external add-on hardware, the other one connecting the application unit to the communication unit. Additional hardware for the communication unit comprises a static RAM, two Ethernet controllers (LANCEs), each

[3] These efforts include also the IFIP WG 10.4 SIG on Dependability Benchmarking (http://www.dependability.org/wg10.4/SIGDeB) and the European Project on Dependability Benchmarking – DBench - Project IST 2000–25425 (http://www.laas.fr/dbench).

[4] MARS (MAintainable Real-time System) is the distributed system developed at Vienna Univ. of Technology, that has evolved to the TTA and TTP concepts (Kopetz and Bauer 2003).

[5] *Fail silence* is intended to describe the behavior of a computer that fails "cleanly" by just stopping to send messages in case a failure occurs (Powell, 1994).

coupled to a Clock Synchronization Unit (CSU) for maintaining a global time base, and a Time Slice Controller (TSC) for controlling the access to the system bus.

Whenever an error is detected, the subsequent error processing activity of the node is to save the error information into non-volatile memory and then turn itself off. Upon restart, the node writes its previously saved error information to two serial ports (one for each unit), from where it can be read for diagnosis purpose. This feature was exploited in the context of this study to precisely monitor and characterize the consequence of the injected faults.

Three levels of error detection mechanisms (EDMs) are implemented in the MARS nodes: (1) the hardware EDMs, (2) the system software EDMs implemented in the operating system and support software (i.e., the Modula/R compiler), and (3) the application-level (end-to-end) EDMs at the highest level. They are briefly described in the following paragraphs.

Hardware EDMs Whenever an error is detected by one of the hardware EDMs, an *exception* is usually raised and the two CPUs then wait for a reset issued by a watchdog timer. This timer is the only device that may cause a reset of all devices including the CPUs. Two categories of hardware EDMs can be distinguished: the CPU built-in mechanisms and those provided by special hardware on the processing board. In addition, faults can also trigger "unexpected" exceptions (i.e., neither the EDMs built into the CPUs nor the mechanisms provided by special hardware are mapped to these exceptions).

The EDMs built into the CPUs are: bus error, address error, illegal op-code, privilege violation, zero-divide, stack format error, non-initialized vector interrupt and spurious interrupt. These errors cause the processor to jump to the appropriate exception handling routines, which save the error state to the non-volatile memory and then reset the node.

The following errors are detected by mechanisms implemented by special hardware on the node: silent shutdown of the CPU of the communication unit, power failure, parity error, FIFO over/underflow, access to physically non-existing memory, write access to the real-time network at an illegal point in time (monitored by the TSC), error of an external device and error of the other unit. We globally call these "NMI mechanisms", as they raise a Non-Maskable Interrupt (a specific exception number) when an error is detected. An NMI leads to the same error handling as EDMs built into the CPUs and can only be cleared by resetting the node, which is carried out by the watchdog timer.

System Software EDMs These mechanisms consist of mechanisms implemented by the operating system or special system tasks. They include:

- Assertions built into the operating system (OS), such as integrity checks on data or processing time overflow
- Mechanisms inserted by the compiler (i.e., Compiler Generated Run-Time Assertions – CGRTA) to implement concurrent checks, such as value range overflow of a variable and loop iteration bound overflow

When an error is detected by any of these mechanisms, a "trap" instruction is executed leading to a node reset.

End-to-End EDMs These mechanisms include end-to-end checksums for message data and multiple (basically, double) executions of tasks.

The end-to-end checksums are used to detect the mutilation of message data exchanged between two nodes of an FTU and are therefore used by the receiving task, for extending the fail silence property of the MARS nodes.

Double execution of tasks in time redundancy can detect errors caused by transient faults that cause different output data of the two instances of the task. Combined with the concept of message checksums, task execution in time redundancy forms the highest level in the hierarchy of the error detection mechanisms. These mechanisms also trigger the execution of a trap instruction, which causes a reset of the node.

8.3.4.2 The Experimental Framework

The testbed that has supported the fault injection experiments at each site features five MARS nodes (Fig. 8.16). The node under test (NUT, for short) is the node subject to the injection of a fault during each experiment run.

Another node (*golden* node) serves as a reference and a third node (*comparator* node) is used to compare the messages sent by the two previous nodes. When a discrepancy is observed by the comparator node (fail silence violation) or the NUT detects an error, the NUT is declared to be failed and then shut down by the comparator node to clear all error conditions for the subsequent experiment run. After some time, power is reinstalled and the NUT is reloaded for the next run. The *data generation* node simulates the data corresponding to the real-time application that is being used to activate the NUT and the golden node during each fault injection experiment.

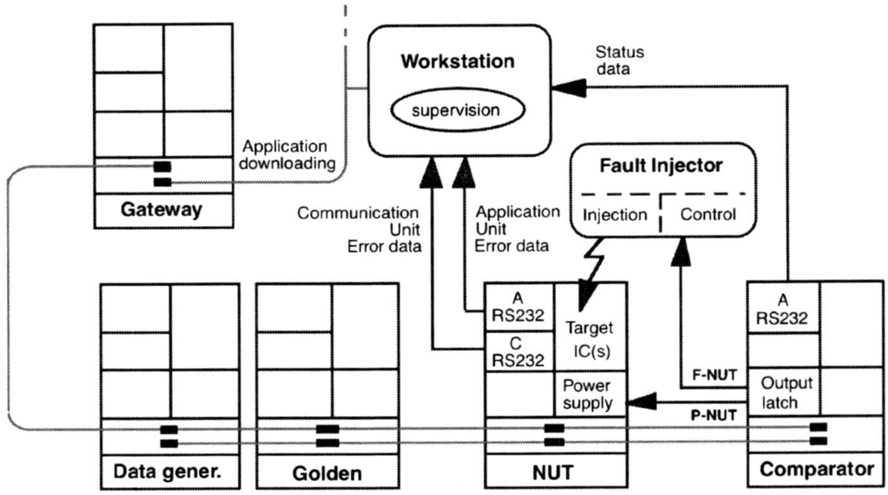

Fig. 8.16 The testbed architecture featuring five MARS nodes

The application is taken from the rolling ball demonstration (Kopetz et al. 1991): a ball is kept rolling along a circular path on a tiltable plane by controlling the two horizontal axes of the plane by servo motors and observing the position of the ball with a video camera. However, the tiltable plane and the camera are not present in the set-up used in the fault injection experiments; instead, the data from the camera is simulated by a data generation task running on the *data generation* node. The task provides the nominal and actual values of the position, speed and acceleration of the ball.

A fifth node is included that serves as a *gateway* between a local area network (LAN) and the MARS network. It is required for loading the entire application and for reloading the NUT. A host computer (Unix workstation) connected to the LAN is used for supervising the experiments, i.e., reloading failed nodes and collecting data from each experiment for further analysis.

Figure 8.16 also depicts the specific interactions with the HWIFI devices. The experiments are managed by the workstation and controlled by the comparator node. When the comparator node detects an error, it reports the error type to the workstation and turns off the power to the NUT with the signal P-NUT. Signal F-NUT is used to discontinue fault injection.[6] Then, the NUT is powered-up again and restarted. Upon restart, the memorized error data is sent to the workstation via two serial lines (one for each processing unit).[7] Once the NUT has been restarted, the workstation immediately initiates the downloading of the application via the gateway node. When the application has been restarted, the comparator node enables fault injection (signal F-NUT) and a new experiment run begins.

8.3.4.3 Examples of Results

Several combinations of the end-to-end EDMs were used for the four fault injection techniques considered:

- NOAM: no application level mechanisms, i.e., single execution and no checksums
- SEMC: single execution, message checksums
- DEMC: double executions, message checksums
- TEMC: triple executions, message checksums; only for HI experiments

We focus only on the NOAM case for which all techniques provide a large ratio of hardware error detection (more than 70%). Indeed, when the application-level EDMs are enabled, an important percentage of hardware detections is maintained for the HWIFI techniques (and to some extent for SWIFI on the code segment – more than 60%), but that percentage is significantly reduced (less than 5%) when injecting on the data segment. This suggests that the faults injected in the code

---

[6] Such a direct control on the injected fault is not possible in the case of the software-implemented fault injection technique used (see Section 8.3.2.4).

[7] If the error was not detected by the NUT itself, then the node has no error information available and sends only a status message.

Table 8.5 Detailed analysis of the hardware EDMs activated by the injection techniques

Hardware EDMs	HI	PF	EI	SWIFI
# of CPU EDMs activated	7	4	5	5
# of NMI types (and combinations)	13 (26)	17 (34)	10 (16)	0
# of exceptions activated	73	65	14	Data not available

segment provoke rather control flow errors that, to a large extent, better simulate the consequences of hardware faults. SWIFI experiments targeting the data segment generate mainly data flow errors that are different from the ones created by the other techniques.

Nevertheless, it is worth noting that these experiments provide also a rather high level of activation of the hardware EDMs in the NOAM configuration, i.e., when the end-to-end mechanisms are inhibited. A closer examination of the results is reported in Table 8.5 that summarizes some of the main differences observed.

Concerning CPU EDMs, although out of the eight mechanisms supported, a different number of mechanisms were activated, the same three mechanisms dominated (bus error, address error and illegal opcode) for all the fault injection techniques.

The number of NMI types and their combinations (i.e., the simultaneous occurrences of several triggering events) vary significantly for the HWIFI techniques.[8] The results indicate that PF may be more effective than the other techniques in exercising hardware EDMs located outside of the CPU chip. Moreover, the most frequent NMIs observed differ: while "unavailable memory" significantly dominates for HI (more than 60%), "memory parity" dominates for PF and EI (more than 50%). Both PF and EI also exhibited a significant proportion of NMIs triggered by the TSC (more than 15%).

The differences observed are further exemplified by the variations in the number of different types of exceptions (including CPU-related and NMI) activated during the various experiments, out of the 255 possible exceptions. This is illustrated by Fig. 8.17 that shows the distribution of the exceptions observed for the three HWIFI techniques considered.

However, it is worth noting that for EI, most of the experiments exercised CPU EDMs (especially when using the probe without antennas), which reveals the very restricted spectrum of the type of errors generated by this technique. However, the variation in the error set was somewhat enhanced when the antennas were used.

8.3.5 Some Lessons Learned

Fault injection is a powerful and suitable technique for objectively assessing the efficiency of fault tolerance mechanisms. Based on the analysis reported here and other related studies conducted at different levels, e.g., see Moraes et al. (2006), the quest for the "unique best technique" looks somewhat unproductive.

[8] Due to the small number of NMIs observed, such an analysis was not carried out for SWIFI.

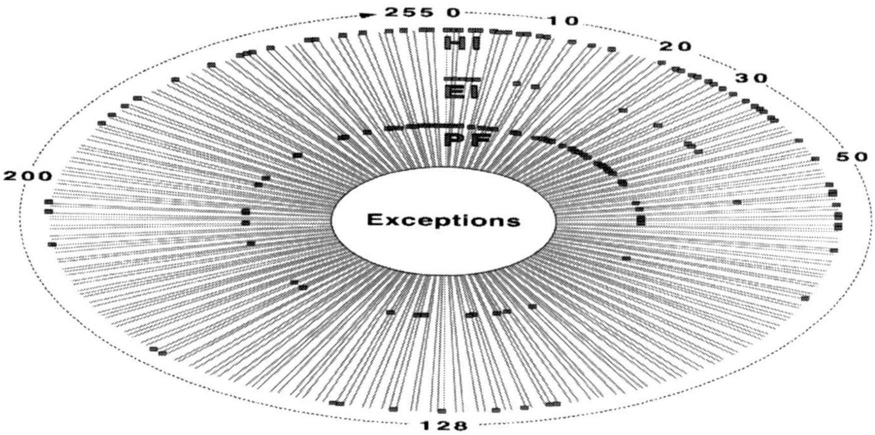

Fig. 8.17 Distribution of exceptions provoked by the HI, PF and EI techniques

The results of the analysis conducted on the MARS self-checking node showed that the four injection techniques are rather complementary, i.e., they exercise, to a large extent, different types of error detection mechanisms. This advocates for the application of various techniques to improve the confidence in the assessment of the properties of a target fault-tolerant system.

Indeed, the extensive sets of test sequences significantly contributed to get conviction in the ability of the MARS nodes to sustain the "fail silence" property. It was also shown that, beyond the hardware and system software error detection mechanisms (EDMs), the application-level detection mechanisms are necessary for achieving a very high coverage on the fail silence assumption. Indeed, although the time-slice controller effectively prevents fail silence violations in the time domain, fail silence violations in the value domain were observed for all four injection techniques when double execution of tasks was not used.

We conclude by addressing some practical issues that have also to be taken into account when selecting a fault injection technique. In addition to *fault representativeness* (i.e., the plausibility of the supported fault model with respect to actual faults) that is one concern that is often raised in conjunction with fault injection experiments, and for the study of which we provided a conceptual frame and objective insights, a wide range of criteria can be considered to assess the merits of the fault injection techniques. Without any claim for an exhaustive analysis, we have considered the following eight basic properties: *reachability*, *controllability*, with respect to *space* and *time*, *repeatability* (with respect to experiments), *reproducibility* (with respect to results), *non-intrusiveness*, the possibility for *time measurement* (e.g., error detection latency) and the *efficacy* to generate significant experiments.

A characterization of the considered fault injection techniques with respect to these eight basic properties is shown in Table 8.6. Further insights can be found in Arlat et al. (2003). For each property, the techniques are graded according to the scale *none*, *low*, *medium* and *high*. It is worth noting that, although it is quite generic

Table 8.6 Properties of the fault injection techniques used

Properties	Heavy-ion	Pin-level	EI (with probe)	SWIFI Compile time
Reachability	High	Medium	Medium	Low to medium
Controllability wrt space	Low	High	Low	High
Controllability wrt time	None	Low to medium	Low	Medium to high
Repeatability	None to low	Medium to high	None to low	High
Reproducibility	Medium to high	High	Low	High
Non-intrusiveness	Low	Medium	High	High
Time measurement	Low to medium	High	Low	Medium to high
Efficacy	High	High	High	Low

in scope, this analysis builds up on insights gained during the experiments carried out on the MARS system.

The table shows that *reachability* and *controllability* properties exhibit rather distinct ratings for each technique. Moreover, the rating of pin-level injection as *medium* and *high* with respect to reachability and space controllability is very much dependent upon the integration level of the technologies of the ICs implementing the target system. Indeed, recent highly integrated ICs would pose more problems from these respects.

Recently, novel techniques have emerged that allow improving both reachability while featuring a high level of controllability, including with respect to time. They correspond to: (1) the scan chain-implemented fault injection technique, e.g., see Aidemark et al. (2001), that builds upon the testability-support capabilities featured by most moderns VLSI devices and (2) FPGA-implemented fault injection technique (de Andrés et al. 2008) that rely on the flexibility offered by FPGA devices to fairly emulate a wide range of real hardware faults, including delay faults.

8.4 Summary

The representativeness of fault models with respect to real physical defects affecting the manufacturing process or faults occurring in operation, is a major challenge for the developers of dependable computer systems. In particular, as was clearly evidenced for the first time by the study summarized in Section 8.2.1, while real physical defects could well be identified at the electrical diagram level, however they could not be adequately accounted for by relying only on logical fault models established at the gate level.

Among the seminal approaches proposed concerning off-line testing (either account directly of the defects at electrical level for test generation or impose layout rules so that the test generation can still be based on the stuck-at fault model), the major developments made in the subsequent years have rather followed the first approach. This is evidenced by the large body of work reported in other chapters.

The same concerns about fault representativeness apply in turn to the testing techniques aimed at assessing the adequacy and efficiency of the fault tolerance mechanisms meant to ensure the dependability of a computerized-system. Fault injection testing – i.e., the explicit exposure of a fault-tolerant system to artificially induced faulty situations – offers a pragmatic and now well-recognized approach to test the fault tolerance mechanisms with respect to the specific inputs they are intended to cope with: the faults. The extent to which the errors provoked by injected faults match those induced by real faults is an essential dimension to ensure the soundness of the inferences derived from a fault injection experiment. To illustrate this issue we have described the main results of a series of experiments meant to compare the errors induced by four injection techniques – namely, heavy-ion radiation, pin-forcing, electromagnetic interference and SWIFI. Applying these techniques to a common target system (a real time distributed fault-tolerant architecture) using a common testbed, has allowed for a comprehensive and fair assessment. The results show that, to a large extent, these techniques are complementary: they provoke rather distinct distributions in the activation of the error detection mechanisms included into the target system. Accordingly, their joint application is to be recommended to increase the confidence in the evaluation of (i) the coverage provided by these mechanisms and (ii) the dependability properties of the target system. In spite of such results and other related efforts, further research is still needed for an improved characterization of the representativeness of the injected fault models.

Acknowledgment The pioneering research reported in Section 8.2 was led by Christian Landrault at LAAS-CNRS. Incidentally, it constitutes his first work on hardware testing, topic on which he has eagerly contributed since then at LIRMM. We are really pleased that we have been given the opportunity to participate in this way to this special book!

The authors would like to thank several colleagues and friends from EFCIS (now ST Microelectronics), ESPRIT project PDCS, IST project DBench and from IFIP WG 10.4 on Dependable Computing and Fault Tolerance, for the fruitful exchanges along the years on the various topics addressed in the Chapter. In particular, we are grateful to Alain Costes and Michel Diaz (LAAS-CNRS), and also X. Messonnier, P. Rousseau, and Michel Vergniault (EFCIS) for their helpful comments, suggestions, and assistance for the study reported in Section 8.2. For what concerns Section 8.3, thanks go to Jean-Claude Laprie and Karama Kanoun (LAAS-CNRS), Johan Kalrsson and Peter Folkesson (Chalmers U.), Hermann Kopetz, Günther Leber and Emmerich Fuchs (Vienna UT), for their contributions to the reflections carried out and to the comprehensive comparative study reported.

This work was supported in part by DRET, EFCIS, ESPRIT project PDCS, IST project DBench, and IST network of excellence ReSIST.

In Memoriam. Jacques Galiay, whose contribution to the work on offline testing was essential, sadly deceased in the early 1980s, during a hike in the Alps mountains.

References

Aidemark JL, Vinter JP, Folkesson P, Karlsson J (2001) GOOFI: A generic fault injection tool. Proceedings of IEEE/IFIP DSN. Göteborg, Sweden, pp 83–88

Arlat, J (1990) Dependability validation by fault injection: method, implementation, application, Doctorat d'Etat Dissertation, INP, Toulouse, France (In French, available from LAAS)

Arlat J, Aguera M, Amat L, Crouzet Y, Fabre J-C, Laprie J-C, Martins E, Powell D (1990) Fault injection for dependability validation – a methodology and some applications. IEEE TSE 16(2):166–182

Arlat J, Crouzet Y (2002) Faultload representativeness for dependability benchmarking. Suppl. Volume IEEE/IFIP DSN. Washington, DC, USA, pp F.29–F.30

Arlat J, Crouzet Y, Karlsson J, Folkesson P, Fuchs E, Leber GH (2003) Comparison of physical and software-implemented fault injection techniques. IEEE TC 52(9):1115–1133

Avižienis A, Laprie J-C, Randell B, Landwehr C (2004) Basic concepts and taxonomy of dependable and secure computing. IEEE TDSC 1(1):11–33

Baumann R (2005) Soft errors in advanced computer systems. IEEE Des Test Comput 22(3): 258–266

Benso A, Prinetto P (eds) (2003) Fault injection techniques and tools for embedded systems reliability evaluation. Frontiers in electronic testing, 23. Kluwer Academic, London, UK, 245p.

Bouricius WG, Carter WC, Schneider PR (1969) Reliability modeling techniques for self-repairing computer systems. Proceedings of the 24th ACM National Conference, pp 295–309

Crouzet Y, Landrault C (1980) Design of self-checking LSI circuits – application to a 4-bit microprocessor. IEEE TC C-29(6):532–537

Crouzet Y (1978) Design of self-checking large scale integration circuits Ph.D Dissertation, INP, Toulouse, France (in French, available from LAAS)

Crouzet Y, Galiay J, Landrault C, Rousseau P Vergniault M (1978) Definition and design of easily testable or self-testing LSI circuits Contract Rep. DRET 77/008, LAAS Report 1787, 334 p (in French)

Crouzet Y, Waeselynck H, Lussier B, Powell D (2006) The SESAME experience: from assembly languages to declarative models. Proceedings of Mutation 2006, Raleigh, NC, USA

Daran M, Thévenod-Fosse P (1996) Software error analysis: a real case study involving real faults and mutations. Proceedings of ISSTA'96, San Diego, CA, USA, 1996, pp 158–171

de Andrés D, Ruiz JC, Gil D, Gil P (2008) Fault emulation for dependability evaluation of VLSI systems. IEEE TVLSIS 16(4):422–431

Durães J, Madeira H (2006) Emulation of software faults: a field data study and a practical approach. IEEE TSE 32(11):849–867

Folkesson PS, Svensson S, Karlsson J (1998) A comparison of simulation based and scan chain implemented fault injection. Proceedings of FTCS. Munich, Germany, pp 284–293

Fuchs E (1996) An evaluation of the error detection mechanisms in MARS using software-implemented fault injection. Proceedings of EDCC. Taormina, Italy, pp 73–90

Galiay J (1978) Design of easily testable LSI circuits PhD Dissertation, UPS, Toulouse, France (in French, available from LAAS)

Galiay J, Crouzet Y, Vergniault M (1980) Physical versus logical faults models in MOS-LSI circuits – impact on their testability. IEEE TC C-29(6):527–531

Gil P, Arlat J, Madeira H, Crouzet Y, Jarboui T, Kanoun K, Marteau T, Durães J, Vieira M, Gil D, Baraza JC, Gracia J (2002) Fault representativeness. IST Project DBench, Deliverable ETIE2, 101p.. Available at www.laas.fr/dbench/deliverables.html

Hély D, Bancel F, Flottes M-L, Rouzeyre B (2005) Scan design and security: can they work together? Sophia Antipolis MicroElectronics, Sophia Antipolis, France

Johansson R (1994) On single event upset error manifestation. Proceedings of EDCC. Berlin, Germany, pp 217–231

Kanoun K, Spainhower L (Eds) (2008) Dependability benchmarking for computer systems. IEEE CS Press and Wiley, 362 p.,

Kopetz H, Bauer G (2003) The time-triggered architecture. Proceedings of IEEE 91(1):112–126

Kopetz H, Holzer P, Leber G, Schindler M (1991) The rolling ball on MARS. Vienna University of Technology, Research Report, No 13/91

Leveugle R (2007) Early analysis of fault-based attack effects in secure circuits. IEEE TC 56(10):1431–1434

Moraes R, Barbosa R, Durães J, Mendes N, Martins E, Madeira H (2006) Injection of faults at component interfaces and inside the component code: are they equivalent? Proceedings of EDCC. Coimbra, Portugal, pp 53–64

Powell D (1994) Distributed fault-tolerance – lessons from Delta-4. IEEE Micro 14(1):36–47

Reisinger J, Steininger A, Leber G (1995) The PDCS implementation of MARS hardware and software. In Predictably dependable computing systems, pp 209–224

Rennels DA, Avižienis A, Ercegovac M (1978) A study of standard building blocks for the design of fault-tolerant distributed computer systems Proceedings of IEEE FTCS Toulouse, France, pp 144–149

Roth JP, Bouricius WG Schneider PR (1978) Programmed algorithms to compute tests to detect and distinguish between failures in logic circuits IEEE TC C-16(10):567–580

Sedmak RS Liebergot HL (1978) Fault-tolerance of a general purpose computer implemented by very large scale integration Proceedings of FTCS Toulouse, France, pp 137–143

Stott DT, Ries G, Hsueh M-C, Iyer RK (1998) Dependability analysis of a high-speed network using software-implemented fault injection and simulated fault injection. IEEE TC 47(1): 108–119

Wadsack RL (1978) Fault modelling and logic simulation of CMOS and MOS integrated circuits Bell Syst Tech J 57:1449–1473

Index

Lightning Source UK Ltd.
Milton Keynes UK
14 March 2010

151358UK00007B/84/P

9 789048 132812